Our vanishi relative

Our vanishing relative

The status of wild orang-utans at the close of the twentieth century

H.D. Rijksen
E. Meijaard

A Tropenbos Publication

The Tropenbos Foundation
Lawiekse Allee 11
P.O. Box 232
6700 AE Wageningen
THE NETHERLANDS
Phone: +31 (317) 426262
Fax: +31 (317) 423024
E-mail: tropenbos@iac.agro.nl
Internet URL:
http://www.tropenbos.nl

Kluwer Academic Publishers
Dordrecht / Boston / London

Sponsored by:

ibn-dlo

ISBN 978-90-481-5239-1

©1999 Stichting Tropenbos/H.D. Rijksen
Softcover reprint of the hardcover 1st edition 1999
Our vanishing relative
The status of wild orang-utans at the close of the twentieth century
© 1999 H.D. Rijksen and E. Meijaard
A Tropenbos Publication

Layout and cover design: Studio Imago, Amersfoort
Cover photo (inset): A male Sumatran orang-utan (photo by H.D. Rijksen)

The opinions expressed in this publication are those of the authors and do not necessarily reflect the views of the Tropenbos Foundation.

No part of this publication, apart from bibliographic data and brief quotations in critical reviews, may be reproduced, re-recorded or published in any form including print photocopy, microform, electronic or electromagnetic record without written permission.

Every form of life is unique, warranting respect regardless of worth to man, and to accord other organisms such recognition man must be guided by a moral code of action.
The World Charter for Nature –
General Assembly of the United Nations, October 1982

Wild fauna and flora in their many beautiful and varied forms are an irreplaceable part of the natural systems of the earth which must be protected for this and the generations to come.
Preamble of the Convention on International Trade in
Endangered Species of Wild Fauna and Flora (CITES), 1975

The genetic viability of the earth shall not be compromised; the population levels of all life forms, wild and domesticated, must be at least sufficient for their survival, and to this end the necessary habitats shall be safeguarded.
World Charter for Nature, 2nd principle -
General Assembly of the United Nations, October 1982

If the orang-utan were to become extinct due to human expansion and persecution, it would be grossly unfair to our ape relative; it is an insult to human reason, and a blasphemous curse on our shared ancestry.

As a nation that attaches high importance to the preservation of biological resources and the environment, it is our duty to preserve them well. I would like to make an appeal to the international community to preserve the great apes of the world (...) by providing direct assistance or by developing the science and technology related to the preservation of the species and their habitats...
Suharto, President of the Republic of Indonesia
(Address to the Great Apes Conference, December 18, 1991)

Our nation will continue to insist on the safety of the great ape who lives in our forests ...
H. Harahap, Minister of Forestry
(Address to the Great Apes Conference, December 15, 1991)

Dedicated to:

Ms. Barbara Harrisson, who was most prominent in activating the conservation of the orang-utan in the 1960s.

Sri Suci Utami, the first Indonesian woman field-scientist to study the socio-biology of wild orang-utans; may her *semangat* (i.e. soul- spirit or dedication) inspire many people in Southeast Asia to help save our wild Hominoid relative from extinction.

Emily Mae, born 1998; so that in her future she can still wonder at the beauty and mysterious forces of the natural world.

About the authors

Herman D. Rijksen is head of the section International Nature Conservation of the Department of International Affairs at the Institute for Forestry and Nature Research, IBN-DLO in The Netherlands, and visiting lecturer for international nature conservation at the University of Amsterdam. He established the Ketambe field research station in the Aceh Tenggara regency (the Leuser Ecosystem) in northern Sumatra, where he conducted a field study on orang-utans from 1971-1974. The results of this field study were laid down in a PhD dissertation entitled *A field study of Sumatran Orang-utans, ecology, behaviour and conservation* (1978). Between 1979 and 1989 he was involved as a lecturer in Nature Conservation at the central training institute (BLK) of the Ministry of Forestry in Indonesia (School of Environmental Conservation Management), and maintained his interest in the conservation and social evolution of orang-utans while expanding his interest to include human forest dwellers both in Sumatra and Borneo, spending extended periods of field study in many different rainforest locations. He participated in the design and production of innovative master plans and management programmes for major conservation areas inhabited by orang-utans (e.g. Lanjak Entimau wildlife reserve, the designated Kutai national park and the Leuser Ecosystem) and guided the preparation of a new policy for orang-utan conservation (i.e. Orang-utan Survival Strategy), including a new design for rehabilitation.

Erik Meijaard is a biologist, now acting as freelance consultant in tropical ecology and nature conservation. Under assignment to IBN-DLO and the Golden Ark Foundation, and under the *aegis* of the Ministry of Forestry International Tropenbos Programme for the present study, he has travelled throughout Java, Borneo and northern Sumatra, penetrating into remote regions in order to verify reports of orang-utan sightings. Since 1992 he has specialised in large mammal conservation and wildlife surveying methodology. He is co-director of Ecosense Consultants, and has acted until February 1999 as coordinator and mediator between WWF Netherlands and WWF Indonesia. Parallel to the conducting of surveys for this report, he addressed the conservation status of the Sumatran rhino, Malayan sunbear and proboscis monkey in Kalimantan.

Contents

Acknowledgements XIII
    Apeology XVII
Summary XIX

Section I: The orang-utan 23

Introduction
    Identity of an ape in danger of extinction 25
    The protected status of the orang-utan 27
  The orang-utan's place in nature
    The Pongids 30
    Two subspecies? 32
A history of the orang-utan's distribution
    Prehistoric expansion 37
    Geomorphological barriers and bottlenecks for expansion 42
  Historic distribution
    Kalimantan 50
    Sarawak, Sabah and Brunei 52
    Sumatra 55
    The orang-pendek enigma 60
Ecology and natural history
    Feeding ecology 65
  Patchy orang-utan distribution
    Altitude 68
    Habitat distribution 68
    Temporal variation in food productivity 73
    Representing the structure of biodiversity 78
  Variable orang-utan ranging patterns
    Empirical facts 80
    Socio-ecology 82
    The arena; reproductive market of the orang-utan 88
    Density estimates 91
    Contrasts between the islands 93
  The impact of logging
    Introduction 96
    The impact of logging on the ecosystem 98
    The possible secondary effects of logging 102
    The impact of logging on the socio-ecology of the orang-utan 106
A history of hunting and poaching
    Prehistoric hunting 109
    Traditional hunting 109

Scientific collecting   116
Commercial collecting   117
Poaching   122
The impact of persecution   124

History of orang-utan conservation
The administrative structure of protection   129
Reserves   132
Towards the brink of extinction?   137
Conservation awareness?   142
The impact of numbers?   147

Rehabilitation of orang-utans
Introduction   151
The value and pitfalls of orang-utan rehabilitation   152
The historical development of the concept   154
Rehabilitation and the King-Kong archetype   157
Rehabilitation and the commercial trap   163
Modern rehabilitation: reintroduction   171

Section II: Orang-utan distribution   177

Survey methods
Introduction   179
Inquiries   180
Literature, documents and thematic map research   181
Checking presence   184
Criteria for estimating orang-utan numbers   186

The present distribution
Introduction   188
West Kalimantan
General description   192
Orang-utan distribution in West Kalimantan   194
Central Kalimantan
General description   204
Orang-utan distribution in Central Kalimantan   209
East Kalimantan
General description   223
Orang-utan distribution in East Kalimantan   224
Sarawak
General description   234
Orang-utan distribution in Sarawak   234
Sabah
General description   237
Orang-utan distribution in Sabah   238

Brunei
  General description   246
  No orang-utan distribution   246
Sumatra
  General description   247
  Orang-utan distribution in Sumatra   248
Evaluation of survey data
  Processing   267
  The facts   268

Section III: The decline   281

Summary of survey results
  The decline   283
Discussion
  Direct (proximate) causes of the decline   287
    Habitat loss   287
    Habitat degradation; the crucial role of timber concessions   293
    Habitat fragmentation   296
    Amplifier effect   298
    The value of the conservation area network?   299
    Evidence of poaching   309
  Diagnosis of the underlying (ultimate) causes of decline
    General   315
    Land-use conflict   318
    The legal framework: asset or liability?   321
    Land: between subsistence and speculation   326
    Law-enforcement?   329
    International support? 332
    The value of law enforcement   340
    Institutional impediments   343
    Behind the proximate impediments to conservation 347

Section IV: The future of the wild orang-utan   351

Prospects of survival
  The possibilities   353
    The protected area system   353
    The permanent wildland forest: integrating conservation of biological diversity and timber exploitation   358
    Regeneration   365
  Conditions for survival
    Recognition of global responsibility   367
    The role of research   371

    Solving problems
        Redressing misconceptions   375
        The permanent State forest estate: How to integrate ape conservation and sustainable forestry?   381
        Upgrading the organisation for conservation   387
        Funding   391
The orang-utan survival programme
    Introduction   395
    Objective and major goals   395
    Overview framework overview of the programme (Table XXVI)   398
    Structure of programme implementation
        General   400
        Reorganisation of rehabilitation; reintroduction   401
        Establishment of new conservation areas   404
        Conservation of habitat outside conservation areas; integration of conservation and sustainable forest management   408
        Awareness: an appeal for public support   411
        Development of eco-tourism with respect to orang-utans   411

Section V: Appendices and references   417

Appendix I
    Vernacular names   419
Appendix II
    What's in a name?   421
    Addendum 1: Transcript Bontius   428
    Addendum 2: Transcript Tulp   429
Appendix III
    IUCN criteria   433
Appendix IV
    Estimates of orang-utan numbers per fragment (Table XXVII)   437
Appendix V
    Research at Ketambe (Aceh, Sumatra)(Table XXVIII)   441

References   443

Index   478

Acknowledgements

We express our sincere gratitude to the Minister of Forestry of the Republic of Indonesia, Jamaludin Suryohadikusomo, for his stimulating interest and the constructive support given to initiate an Orang-utan Survival Programme and to facilitate development of a revolutionary new approach to conservation of the Sumatran orang-utan in the greater Leuser Ecosystem. We also gratefully acknowledge the Directorate-General Forest Protection and Nature Conservation (PHPA), notably its former Director General, Sutisna Wartaputra, and his subdirector, Flora and Fauna Conservation Officer Widodo S. Ramono, who in the early 1990s transformed their concern about the orang-utan's plight into active commitment. They provided the framework for action even when resources were scarce. We are also grateful for the support of the current Director General, R. Soemarsono and his staff, notably the Director of Fauna and Flora Conservation Office Dwiatmo Siswomartono. Special thanks in this regard are owed to W. Daniel Sinaga as well, and we also wish to acknowledge the guidance of Tony Suhartono in certain administrative matters.

For their invaluable support, we are indebted to several staff members at the provincial offices of the Forestry Service (*Dinas Kehutanan*), the Agency for Forestry Research and Development (*Puslitbang Hut*) and the Directorate-General Forest Protection and Nature Conservation (BKSDA and UPT-TN). Our thanks are due in particular to Komar Sumarna, head of Kanwil Sumatera Utara, who smoothed the bureaucratic pathway for the surveys in northern Sumatra. We are also especially grateful for the commitment of the staff members Abdul Muin and Bahtiar, who skilfully arranged, and assisted in, the field surveys in Kalimantan.

We are particularly indebted to Carel van Schaik, professor of biological anthropology, zoology and conservation at Duke University (Durham, NC, USA) and supervisor of the unique Suaq-Balimbing orang-utan field-research station in the Leuser Ecosystem, for his contributions and intellectual scrutiny in reviewing the most relevant parts of this book. We greatly appreciate his friendship and the manner in which he shared his in-depth knowledge of orang-utan ecology and other relevant information involving conservation of the ape, in order to complement the information in this report. In the same vein, the support and contributions of Ahmad Yanuar, who has been involved in the searches for the legendary *orang pendek* of central Sumatra, is also kindly acknowledged. We thank Anne Russon as well, who shared with us some new insights into the psychology of orang-utans.

We gratefully acknowledge Ibrahim and Idrusman, both of Leumbang (Aceh Selatan), who efficiently conducted the surveys in Aceh under the guidance of Michael Griffiths, co-director of the Leuser Development Programme. To Mike we owe very special thanks for his friendship, generous hospitality and creative as well as logistic support, and we may perhaps also convey to him and to his colleagues at the Leuser Management Unit and the Leuser International Foundation the gratitude of many orang-utans in Sumatra for their efforts to conserve the ape's habitat in the largest rainforest reserve in the Old World. We gratefully acknowledge the efforts of the team of Universitas Indonesia, Yossa Isiadi, Suroso Mukti Leksono and Guritno Djanobudiman, who, under the guidance of Noviar Andayani, conducted surveys south of Lake Toba, which yielded important additional information for our understanding of the situation. With regard to the surveys and studies whose data have contributed to this report, we also particularly appreciate the supervision and support of the Indonesian Institute of Sciences (*Lembaga Ilmu Pengetahuan Indonesia* – LIPI).

We are indebted to our sponsors, above all the Institute for Forestry and Nature Research (IBN-DLO), the Tropenbos Foundation, the Netherlands Foundation for International Nature Protection (also known as *Van Tienhoven Stichting*), the World Wide Fund for Nature (*Wereld Natuur Fonds*) Netherlands, the Lucie Burgers Foundation, the municipality of Zoetermeer and the Golden Ark Foundation in the Netherlands. We particularly thank Niels F. Halbertsma, former director of the WNF, for his consistent support and friendship, and we gratefully acknowledge Siegfried Woldhek, present director, and Herman Eijsackers, of the board of WNF, who had the vision to prompt the organisation to provide essential support for this survey at a time when donor support turned away from species conservation. We would also like to acknowledge the fact that the Red Alert Program of the Flora and Fauna Preservation Society sponsored a preliminary 1991 survey of orang-utans in Kalimantan by Ahmad Yanuar and his colleagues, and note that the L.S.B. Leakey Foundation sponsored one of the preliminary 1992 surveys in Aceh (northern Sumatra) conducted by Carel van Schaik.

We are deeply indebted to the International MoF-Tropenbos Programme at the Wanariset Samboja Forestry Research Station in East Kalimantan, for providing logistic support and advice in organising the Kalimantan surveys. Very special thanks are due to the Programme's team leader, Willie T.M. Smits, who was receptive to revolutionary ideas and had the creativity and skills to establish, on a shoe-string budget, and in collaboration with the headmaster of the Balikpapan Expatriate School, Jonathan Cuthbertson, a new Bornean quarantine and reintroduction facility for the Orang-utan Survival Programme at the Wanariset Samboja location. The facility and organisation could well become the model for ape rehabilitation in Asia.

We are thankful as well for the support of the European Commission: Through its Leuser Development Programme, which was designed by the first author of this book, in collaboration with M. Griffiths, S. Poniran and several Indonesian colleagues, the European Union is at present the largest contributor to the implementation of the Orang-utan Survival Programme.

On behalf of the orang-utan we also gratefully acknowledge the precocious use of much of our information by the Environmental Investigation Agency for their powerful pamphlet entitled *The Politics of Extinction; the Orangutan crisis, the destruction of Indonesia's forests*, by the World Conservation Monitoring Centre (WCMC) for revising their publicly available thematic maps with reference to rainforest and ape conservation, and by the World Wide Fund for Nature Indonesia Programme for boosting a unique cyberspace workshop to once more discuss the plight of the ape.

The Netherlands Foundation for the Advancement of Tropical Research (WOTRO) also deserves to be mentioned with special gratitude because it has, since 1971, continuously sponsored research on Sumatran orang-utans at Ketambe,

> thereby safeguarding a significant local population of apes and other wildlife. It is also gratefully acknowledged that the Wildlife Conservation Society supports ongoing Sumatran orang-utan research at Suaq Balimbing; the World Society for the Protection of Animals has initiated a campaign to sponsor the Balikpapan Orang-utan Society which has so far provided the bulk of the support for the maintenance of the Wanariset Samboja orang-utan reintroduction facilities; the township of Zoetermeer in the Netherlands sponsors the establishment of a new orang-utan reserve in Sumatra; the International Federation for Animal Welfare has recently initiated sponsorship for orang-utan reintroduction in Sumatra, and The Netherlands' WNF is developing effective ways to sponsor orang-utan conservation in the near future.

We greatly appreciate the voluntary contributions of Rona Dennis and R. Tarigan of the ODA / PHPA GIS Unit, and we are indebted to Aart van de Berg and Arjan Griffioen of IBN-DLO, who added some final touches to an impressive GIS basis for Borneo and Sumatra and helped prepare the GIS basis of many of the maps. Special thanks are also due to Peter J.M. van Bree, curator of the Institute for Taxonomic Zoology of the University of Amsterdan, former chairman of the Golden Ark Foundation and former secretary of the *Van Tienhoven Stichting*, and to Herbert H.T. Prins, professor of international nature conservation at Wageningen University, for their constructive suggestions concerning improvements of earlier drafts of the manuscript.

The orang-utan in its natural surroundings, the Asian tropical evergreen rainforest.

We thank Jito Sugardjito, Junaedi Payne, Jane M. Bennett, David J. Chivers, Nico J. van Strien, Juwantoko, Jack Rieley, Susan Page, Suci Sri Utami, Sarah Cunliffe, Ian Singleton, Christine Luckett, Frank P.G. Princée, Eric Wakker, Hasjrul Junaid and Hemmo Muntingh for their constructive support and information on specialist issues. We also acknowledge Kathy Oates-MacKinnon, and Birute Galdikas-Bohap who, over the years, and beyond the scope of the present project, have provided the necessary intellectual challenges for a balanced concept of ape conservation and an optimal design of the Orang-utan Survival programme.

We would also like to express our gratitude to the Tropenbos Foundation for inviting us to publish this report as a Tropenbos publication. And we greatly appreciated the skills of Donna DeVine in transcribing our continental Anglo-Saxon in English.

In conclusion, we are especially indebted to the anonymous orang-utans in Sumatra and Borneo, who, while rapidly dwindling in numbers, have inspired us to act as advocates on their behalf in the world of humans. We fervently hope that we will succeed in this undertaking, and will be able to help transform the current persecution and displacement of the ape into a realistic chance of survival.

Apeology

Almost a decade has elapsed since the first signs of a disastrous course of events for the orang-utan and its habitat became evident. A conservative estimate of the decline in the ape population during this ten-year period runs into the thousands, of which so far fewer than 600 individuals have been able to be retrieved (for rehabilitation and reintroduction into the wild), and even then with dubious success. As close relatives of the ape, and in light of the escalating emergency situation in which it now finds itself, we humans have call to feel ashamed that so little – and often such misguided – action has been taken. One may wonder as well why the information presented in this report has taken so long to be made public, and why it has been compiled mainly by expatriates.

It is important to realise that the imminent annihilation of a spectacular mammal due to a surge of development is by no means an event unique to Southeast Asia. Virtually all of the world's larger wild animals that have survived up to the present are under tremendous pressures threatening either their wildness or their survival, irrespective of whether they occur in industrialized or developing countries. Everywhere, the growing numbers of people, with their interlinked market systems and insatiable demand for land and resources, are restricting unique animals in their ranging, are usurping and destroying their habitat or are subjecting them to outright persecution, simply because the creatures have a certain (market) value or are considered a nuisance.

Conversely, there should be no doubt that a large number of people both within Southeast Asia and abroad are seriously concerned about the ape's chances of survival. However, many people are unaware that their decisions concerning wildlands often result in the death of formally protected orang-utans, both directly and indirectly, through starvation. Very few people take nature conservation seriously, let alone modify their decisions in accordance with a knowledge of ecology and concern for the survival chances of even those familiar organisms towards which they might feel sympathetic. In any case, it is evident that no human organisation has yet been able to prevent the catastrophic course of events now dragging the ape towards extinction. After all, the irony of conservation is that if something has been conserved by a thousand wise persons for over a thousand years, any fool is able to destroy it in a few hours.

The problems causing the imminent extinction of the orang-utan are complex and can be solved only in an integrated way, involving whole networks of well-informed people taking well-informed decisions. It is our profound hope that this report will help create such a network.

Because this study concerns the ongoing decline of a formally protected species, it should come as no surprise that some embarrassing facts and critical implications may emerge. Consequently, the authors wish to emphasise that it is not their intention to accuse, or to blame persons connected in any way with such facts, but by means of our analysis to foster improvement of the situation. If mistaken decisions have been made and if grievous events have taken place, it has often been by default. All such factors are undoubtedly the consequence of a complex network of causes in which ignorance and misinformation reign supreme. In reality the situation may well resemble a *Wayang* performance involving some metaphysical puppeteer who appears to be labouring under serious misconceptions, greed, and destructive illusions of having power over Nature. Thus, it is hoped that some insight into the problems, constraints and pitfalls related to the steady decline of orang-utan populations may lead to the concerted actions which will give the ape a chance of survival into the twenty-first century.

For the record, it must be emphasised that the views and conclusions expressed in this report reflect a rough consensus of opinion on the part of the authors, which is not necessarily shared or underwritten by the sponsors.

Finally, on behalf of our supposedly *sapient* species, we should apologise to our wild relative, the orang-utan, not only for the ultimately lethal conditions our fellow humans are imposing upon it, but also because it has taken so long to initiate any action and obtain any realistic overview concerning its ongoing, human-induced suffering: *Kami minta maaf sebesar-besarnya.*

Summary

The orang-utan is a superb representative of a major sector of the structure of biological diversity in Sumatra and Borneo. Conservation of the living conditions of the orang-utan implies maintaining the integrity of the entire natural ecosystem of indigenous plant and animal species known as the West Malesian rainforest, i.e. the natural tropical evergreen forests of distinctive floral composition which stretch from the isthmus of Kra, in Thailand, across peninsular West Malaysia, south and eastwards, including Sumatra, Borneo and Java.

The main question behind this study is: What is the current status of the orang-utan? Or in other words:
- what is the current geographical distribution range?
- what have been the trends in the size of its range and numbers?
- to what extent is this range covered by (a) conservation areas, (b) timber concessions (i.e. modified habitat) and (c) plans for conversion (i.e. obliteration of the habitat)?
- what is the current quality of habitat in this range and what is the prospect for conservation or restoration of such habitat?
- what is a plausible average density of the ape in such habitat?
- what are the prospects for protection of the ape?
- what should and can be done to give the ape a chance of survival?

Several actions were undertaken to find answers to these questions, and the major results are:
- In Sumatra the orang-utan has a much more extensive range than was hitherto believed; small relict populations have been found in several locations in the Barisan mountain range and along the west coast of Sumatra; reports of the legendary *orang pendek* are currently considered to refer to orang-utans.
- The total accumulated expanse of land area where orang-utans are still found in Borneo covers approximately 150,000 km^2 of what is designated 'forest' on State Forest land, and in Sumatra it is considered to cover approximately 26,000 km^2. In Borneo this distribution range is fragmented into more than 61 major isolated sections; in Sumatra into at least 23 major sections. Of this area anything between 1-35% is estimated to be habitat for orang-utans.
- Virtually all the prime habitat, namely the alluvial and flood-plain forests, in which orang-utans can congregate in high densities, has been degraded by timber exploitation, while large sectors have been converted and settled by humans. The average carrying capacity of habitat in the said range of fragments is therefore reduced by anything ranging from 50% to total obliteration. This has serious negative consequences for the social organisation, reproduction and survival of the ape.
- Of the Bornean orang-utan population at the beginning of the twentieth century, no more than some 7% survives today; of the Sumatran population no more than 14% survives.

- No more than 16% of the distribution range of the orang-utan in Borneo has a protected status; more than 60% of the relict range in Sumatra is protected.
- The interlinked reasons for the deplorable current state of affairs are:
  (1) exponential human population increase; (2) people are focussed on domestication and exploitation while not taking nature conservation seriously; (3) massive destruction and degradation of the habitat of the ape, due to timber exploitation, forest conversion, transmigration, encroachment and arson; (4) poaching due to deficient law enforcement and a dramatic increase in potential ape-human conflict situations; (5) deficient protection of conservation areas; (6) national and international neglect involving the extermination of wild organisms.
- Because virtually all orang-utan habitat has been subjected to timber exploitation or encroachment, the issue for the ape's survival is now whether the wildland forest estate will be allowed to regenerate, and whether sufficient protected habitat area can be added to the current conservation area network. The possible belief that better regulation of the timber trade through certification is a solution may soon prove to be false, as wildland forests in Indonesia and Malaysia will soon be formally replaced by timber plantations, thus demolishing wildlife habitat in the process.
- Considering the current density and exponential increase in the human population in Sumatra and Borneo, in addition to the fact that most people are busy during the daylight hours scrounging about in their immediate surroundings for anything edible or tradeable, and given the government's poor commitment to conservation, it is highly likely that the orang-utan is doomed unless immediate large-scale integrated action is taken. Despite its formal protection status, the ape may be expected to be eradicated by the second decade of the twenty-first century, as a result of displacement by increasingly impoverished human masses occupying an increasingly devastated, desiccated desert-like environment, once characterised as being one of the richest tropical rainforest areas on earth.
- Nevertheless, the conclusion is that the orang-utan can be saved if:
  - authorities acknowledge that decisions concerning State forest land can result in protected orang-utans being sacrificed, and that decision makers can be held accountable for the demise of local orang-utan populations;
  - persecution and killing (poaching) is stopped entirely, through law enforcement and education, and if rehabilitation projects are run professionally and are forbidden to the public;
  - several large new conservation areas are established, and existing reserves are expanded and redesigned to cover the current ranging areas of significant relict populations, and if unauthorized access to protected areas is forbidden, and protection enforced;
  - timber concessions are given the responsibility and mandate for protection of their concession area against encroachment and against trespassers scrounging the forest for 'non-timber forest produce';

- timber logging concessions plan their operations according to reduced impact and selective felling protocols, acknowledge respect for the habitat needs of orang-utans and submit their operations to government control;
- logged-over forest can regenerate, and habitat can be restored in order to allow subpopulations to build up to 'normal' population densities again for optimal breeding success;
- conversion of the wildland forest estate is banned, and new plantations and transmigration projects are established on the millions of hectares of derelict grass steppes in Borneo and Sumatra;
- the organisation of nature conservation is made more effective and efficient;
- significant national and international support is raised for the Orang-utan Survival Programme, and
- the human subsistence pressure on forested areas is reduced, while socio-economic development is shared equally.

Section I: the orang-utan

Section I
The orang-utan

Illustration of the first orang-utan described for western science; caught in the county of Angkola in northern Sumatra and described by the Dutch physician Nicolaas Tulp in 1641.

Section I: the orang-utan

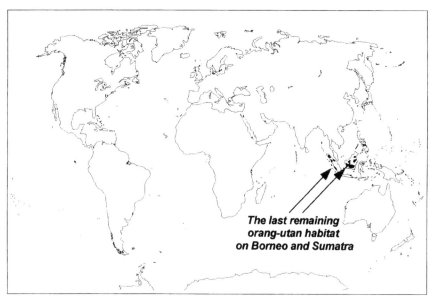

The presence of wild orang-utans in the world at the close of the twentieth century;

INTRODUCTION

Identity of an ape in danger of extinction

This report describes the situation of one of the closest living relatives of our own species, whose Malay name *orang-utan* appropriately means 'forest people'. Despite the good intentions of a few authorities and a formally protected status, the orang-utan has become a critically endangered species, the extinction of which within two or three decades is to be feared if no effective counter-action is taken.

The scientific name of the orang-utan has been contended (Röhrer-Ertl, 1983) (see Appendix 2). The ape was first described for western science in the early seventeenth century by two Dutch physicians, Jacob de Bondt, and Nicolaas Tulp, and subsequently assigned the taxonomic name *Simia satyrus* by Carl von Linné (Linnaeus, 1758). However, its formal name was changed for dubious reasons in 1927 by the International Commission on Zoological Nomenclature in *Pongo pygmaeus*[1]

The orang-utan is nowadays found exclusively on the islands of Sumatra and Borneo, and over 90% of its habitat occurs within the territory of the Republic of Indonesia. The main reason for the survey project that gave rise to this report was a deeply felt concern about the plight of the orang-utan, and the perceived neglect of this species by national authorities as well as the international conservation community. The survey was a first logical step towards the aim of developing an effective Orang-utan Survival Programme, commissioned by the Minister of Forestry of Indonesia.

Since the end of the nineteenth century, Indonesia's human population has increased from little more than ten million to a current population of over two hundred million people, while the orang-utan population has dwindled to a fraction of its nineteenth-century numbers. The reason for the continuous decline in orang-utan numbers is that humans and ape favour the same habitat, namely, the bottomlands (i.e. alluvial flood plains, peat-swamp forests and valleys), and the latter cannot defend its existential rights in any way. The outcome of the inevitable conflict is invariably fatal for the ape. Once land is cultivated and settled by people, it is usually unequivocally lost for the aboriginal ape inhabitant. Formal land allocation is predominantly concerned with the human condition, and almost completely ignores both the ecological functions of wildlands which are the basis for sustainable development, and for the survival needs of our closest relative in Asia. For a growing number of people the issue may be that of welfare temporarily increased, but for the orang-utan the issue is sheer survival.

---

[1] The taxonomist Chasen (1940) continued to use the original name ('Contrary to my usual practice' because he also disagreed with the Commission's Opinion (No 90).

The orang-utan has survived on the islands of Sumatra and Borneo, in the states of Indonesia and Malaysia.

The orang-utan has fascinated people throughout the ages in both the East and the West, but that fascination has rarely, if ever, had a positive effect on the ape's survival chances. This report presents the history of human – orang-utan relations, as well as a summary of knowledge on the ape's natural history. Against this background it presents the latest data on the distribution of the orang-utan in Borneo and Sumatra, based on special surveys conducted between 1990 and 1997. The calamitous drainage of peat-swamp forests and massive arson, which destroyed large tracts of hitherto prime ape habitat in Kalimantan during the prolonged drought of 1997-98, made a significant last-minute revision necessary. The report also provides spatial information on the complex of major threats to the survival of the ape and discusses the conservation measures taken so far. Finally, it presents the details of the Indonesian Orang-utan Survival Programme.

Although estimates of the total numbers of surviving orang-utans have been inaccurate, and any attempt to estimate the numbers of such an elusive creature may in fact be misleading, it is felt that authorities who make decisions about land use and the possible conservation of biological diversity need figures. Hence this report will also present figures for the estimated number of orang-utans surviving at present. However, it should be stressed that such figures are entirely meaningless when presented out of their context. The context is an inexorable trend: Continuous despoliation of the orang-utan's forest habitat, increasing fragmentation and reduction of its range, and deficient protection. Even if this report contained an error of 100% for the estimated numbers of apes surviving today, it would make no difference whatsoever for their survival chances. For it is the direction of development that will cause the numbers of apes to rise or fall within a decade. By modern and powerful technical means, an area of one-hundred thousand hectares of what has been called 'impenetrable virgin jungle' can be cleared and converted into an oil-palm plantation within about half a year. The fires of 1997-98 demonstrated that the destruction of both habitat and large numbers of apes can proceed much faster than had been hitherto believed. In any case, for the orang-utan, effective protection could cause a rise of up to 20% in numbers in ten years, while continuation of the present trend of forest conversion and over-harvesting will result in extinction in the wild state, no matter how high the numbers of wild apes still surviving today.

The protected status of the orang-utan

Most of the world population of wild orang-utans are to be found in the Republic of Indonesia; the remainder occur in the Eastern (Bornean) states of the Malaysian federation, Sabah and Sarawak. As a consequence the emphasis in this report will be on the Indonesian situation.

Protection of wild animals in Indonesia goes back to the colonial era, when, in 1909, private initiative induced the government to issue a Wildlife Protection Ordinance. It covered all wild mammals and birds, except those which were generally deemed harmful. Remarkably, the orang-utan and gibbons were not specifically mentioned and, among the 'monkeys' in general, were considered noxious and hence exempted from protection. This capital error was rectified by the Netherlands Indies Society for Nature Protection, which was established in 1912. In 1924, after a decade of lobbying, the Society eventually persuaded the colonial government to revise the first Wildlife Protection Ordinance so that the orang-utan came to feature prominently on the list of protected species.

After 1924, the ape became, theoretically, the best and most strictly protected species in Indonesia. Its protected status was reacknowledged and expanded in the *Peraturan Perlindungan Binatang-binatang Liar* of 1931 (*Staatsblad* 1931, No 266 jis. 1932 No 28 and 1935 No 513), and was as such recognised in Article 21(2) of the Act of the Republic of Indonesia No. 5 Concerning the Conservation of Natural Resources and their Ecosystems (1990). Yet, in December 1996, this comparatively safe administrative position was undermined by a new Decree of the Minister of Forestry (No 771/Kpts-11/1996), stating that all protected species could be sustainably harvested and used. Fortunately this decree lasted only a few months, until August 1997, when it was revoked and superceded by a new Decree of the Minister of Forestry (No 522/Kpts-II/1997), which restored the fully protected status of the ape, as well as other wildlife species on the original list.

The article with relevance to protected species, including the orang-utan, states that 'Any and all persons are prohibited from:
a. Catching, injuring, killing, keeping, possessing, caring for, transporting and trading in a protected animal in a living condition.
b. Keeping, possessing, caring for, transporting and trading in a protected animal in a deceased condition.
c. Transferring a protected animal from one place to another, within or outside Indonesia.
d. Trading, keeping or possessing the skin, body or parts thereof, of a protected animal, or goods made of parts of the animal, or transferring such parts and/or goods from one place to another, within or outside Indonesia.'

The Provision of Criminal Punishment (Art. 40, 2) states that 'Whosoever intentionally violates the provisions pertaining to Par. 1 and 2 of Article 21 (...) shall be liable to punishment by imprisonment up to a maximum of five years and a fine up to a maximum of one-hundred million Rupiahs (i.e. approximately US$ 5,000 in 1997; and US$ 1,250 in 1998). Further, Par. 4 states that 'Whosoever, through negligence, violates the provisions pertaining to Art. 21 (...) shall be liable to punishment by imprisonment up to a maximum of one year and a fine up to a maximum of fifty million Rupiahs.

Further, Art. 24 (1) states that 'Protected animals or parts thereof, which are found in illegal custody of a person shall be confiscated', and the live animal shall be 'returned to its habitat or (if release into its habitat is not possible, because the animal is no longer adapted to its habitat condition (...)) be handed over to institutions dealing with wildlife conservation, except in the situation that its condition is such that it is not likely to become fit, then it should be destroyed' (Art. 24,[2]; and Clarification of the Act).

The Act also notes that exceptions to Article 21 are possible. Thus orang-utans may be caught and kept in custody for research, and for such management as may increase the species' survival ('captive breeding'). However, a special permit was required for such exceptions, which, until 1996, could be issued only by the President of the Republic.

In addition, the Act's Article 22 (3) stipulates that in the event that the member of a protected species 'threatens or endangers human life', Art. 21 may be overruled, supposedly implying that even members of protected species may be destroyed whenever somebody feels 'threatened' by it.

Finally, the official Clarification of the Act concerning Art. 24 (2) states, 'protected (...) animals must be protected in their habitats'. However, the legal framework nowhere stipulates that the protection of a species should extend to conservation of the ecological integrity of its living conditions, i.e. its habitat.

In 1991, the Minister of Forestry issued a decree (No 301) which stipulates that people shall register endangered species which they 'have domesticated' (which supposedly meant that they are kept in illegal captivity). In a press article (*Jakarta Post*,

August 29th., 1991), the Directorate-General PHPA interpreted this decree so 'that animal lovers would be able to apply for permits to keep the endangered species' (...) supposedly until such time as the government would confiscate and rehabilitate them. Both the Indonesian Environmentalist Forum (WALHI) and the Indonesian Legal Aid Foundation (YLBHI) publicly protested the decree because it not only violated Art. 21 (and Art. 40), but even 'would encourage people to keep and domesticate protected species.' However, no further formal action was undertaken, except that on a later occasion the Directorate-General publicly stated that the orang-utan was, and remained, a strictly protected species for which no (temporary) permits for keeping by private persons would be issued.

> The peculiar new Decree (771/Kpts-11/1996) of December 1996 relegated the ape back among the common pool of 'natural resources' in general and all other protected species in particular. The decree stated that 'protected plant species and wild animals can be utilised sustainably by means of hunting and collecting from the natural state as well as by means of breeding.' The resourcist essence of Act No 5, of 1990, finally seemed to have overruled any further need for protection with the magic word 'sustainable'. Fortunately the Decree was revoked again in August 1997, restoring the fully protected status of the ape.

The orang-utan has been protected in Sarawak since 1957 (Wildlife Protection Ordinance 1957, 1990, 1995) and in Sabah since 1965 (Fauna Conservation Ordinance). In 1988 the Federal State of Malaysia ratified the protected status. It indicates a penalty of two years in prison for violation and an additional fine of up to approximately 15,000 US$. Both Indonesia and Malaysia are signatories to the Convention on International Trade in Endangered Species of Wild Fauna and Flora.

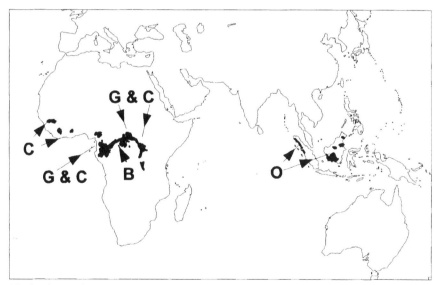

The distribution of the great apes in the world is confined to the tropics of Africa and Asia; C=chimpanzee, G=gorilla, B=bonobo and O=orang-utan. Note that the human distribution covers all of the indicated landmass, and currently overlaps the range of the great apes.

The orang-utan's place in nature

The Pongids

Our species, *Homo sapiens*, belongs to a taxonomic group which currently still includes four other great apes; together with the African bonobo (*Pan paniscus*), the chimpanzee (*Pan troglodytes*), and the gorilla (*Pan gorilla*), the Asian orang-utan (*Simia satyrus* Linnaeus 1758) is one of our closest extant relatives. According to genetic and biochemical similarities, these Pongids may have evolved from a common ancestor over a time period of less than ten million years (Sarich and Wilson, 1967). There is no scientific justification for designating the human species as a separate family (Margulis and Sagan, 1986).

The orang-utan, gorilla, chimpanzee and bonobo each have 48 chromosomes, while our own species has 46. However, the difference is readily explained as being the result of a mutant fusion occurring in the human lineage; evidently two pairs of subteleocentric chromosomes have fused to form the large second pair of metacentric chromosomes so typical of the human karyotype (De Boer and Seuanez, 1982). Schwartz (1987) found that of the four great apes the orang-utan shared the greatest number of identical DNA banding patterns with human chromosomes. However, in terms of temperament, social skills and behavioural expression, the central African bonobo is obviously closest to the modern human (Savage-Rumbaugh and Lewin, 1994). Be that as it may, the majority of the varied members of the Pongid radiation became extinct during the last 100,000 years, after the modern human form (*Homo sapiens*) emerged. Only the four above-mentioned typical forest relatives are currently known to have survived in the more remote tropical areas. At present all are facing imminent extinction at the hands of humans.

The orang-utan's anatomy, physiology and behaviour are so similar to our own that when the existence of the ape became widely known it caused considerable confusion among people convinced of the divinity of human uniqueness. Some early scholars grouped the ape among what were then regarded as 'exotic races' and mythological beings. Indeed, it is still a source of embarrassment to many people[2] who fail to appreciate our common evolutionary origin. Whenever the talents and characteristics of apes are seen to be comparable to similar, yet allegedly unique, human traits, e.g. language ability, reflection, consciousness, creativity, politics, and so forth, quasi-scientific opposition is extremely fierce, especially in the USA.

The indigenous peoples of Borneo were said to believe that transformations from human to ape, and *vice versa*, were possible (Brooke, 1866). It was recorded that with respect to their women the native men feared the sexual assertiveness of the male ape, and accorded it a status on a par with – or even higher than – other tribal people in

---

[2] A detailed description of the biology and behaviour of the orang-utan is beyond the scope of this report; for more information the reader is referred to Brandes (1939), Jantschke (1972); MacKinnon (1974), Rijksen (1978), Maple (1980), De Boer (1982) and Schwartz (1986; 1988).

their religious (head-hunting) rituals. To the first Europeans they allegedly described the species as an inland race of *people* (i.e. *orang*) who, it was said, refused to speak for fear of being enslaved and forced to labour (Bontius, 1658; Beeckman, 1718).

For some people the orang-utan's anatomy and behaviour are embarrassingly similar to that of our own.

> Psychological research has demonstrated that the orang-utan, like the bonobo, the chimpanzee and the gorilla, has a distinct consciousness regarding the 'self' (Lethmate and Ducker, 1973) and the value of kindred (Lippert, 1974), and it possesses an extraordinary ability for learning, deduction and invention (Rumbaugh and Gill, 1970; Lethmate, 1977), which allows it to make and to deploy tools (Lethmate, 1977; Rijksen, 1978; Bard, 1993; Russon and Galdikas, 1993; van Schaik et al., 1996). It is also gifted with a powerful long-term memory as well as an amazing talent for reading and comprehending environmental signs (Rijksen, 1978), including the understanding of spoken sentences in a language to which it has become familiarised[3] since its early infancy. Moreover, the orang-utan demonstrates a clear appreciation of aesthetic values, and an accompanying predilection for playful adornment.
> These mental qualities allow the Hominids, including the orang-utan, to develop a complex social organisation based on long-lasting relationships, of which the mother-child bond is the pivot. Hence, there is no scientific ground to deny the apes any of the moral rights and ethical values which civilised humankind has reserved for itself. In spite of a paucity of complex vocal communication, there is no doubt that orang-utans have the ability to think and to reflect upon their own position in relation to their social and physical environment, i.e. what in Islam is understood as aql and in Christianity as the 'soul', and consequently no civilised culture can deny an ape, like the orang-utan, 'personhood' (see also Cavalieri and Singer, 1993; Vines, 1993).

Arriving from the Asian continent, the orang-utan moved into the Sunda landmass expanse at least 1.5 million years ago, well before the first members of the now extinct *Homo erectus* entered the area during the middle Pleistocene epoch, estimated as having been some 400,000 years ago. To deprive the orang-utan of its birthright to the areas in which it has managed to survive for many millennia under persecution from tribal people is an ethical offence against the human concept of civilisation. Even if, due to more recent religious beliefs, modern Southeast Asian people nowadays find it difficult to recognise and acknowledge the orang-utan as a relative, it is nevertheless a true sign of civilisation if a nation is committed to protect it. Indeed, both in Indonesia and in Malaysia the ape is protected by law. Hence, because citizens are expected to obey the law, the ape can be seen to have 'the right' to live.

Two subspecies?

Taxonomy currently distinguishes two extant orang-utan subspecies, namely, one from Sumatra and the other from Borneo (van Bemmel, 1968; Jones, 1969). The subspecies have been geographically isolated since at least 10,000 years ago – and possibly much longer[4]– when the rising sea level divided Sumatra and Borneo into

---

[3] Language ability in orang-utans has not yet been studied to the extent it has been done with chimpanzees, gorillas and bonobos, yet there are indications that in terms of understanding language the orang-utan is on a par with the bonobo. Orang-utans seem capable of linguistic and cognitive abilities (Maple, 1980; Miles, 1993; Miles and Harper, 1994).

[4] The mitochondrial DNA analyses by Lhu Zi et al. (1996) on phylogenetic divergence between Sumatran and Bornean orang-utans suggest that the two (sub) species have been separated since some 1.5 million years ago, i.e. long before the islands separated. Note that this implies that, despite the possibility of a long-lasting genetic exchange across the Bangka-Belitung-Karimata range of southern Sundaland between the south Sumatran and southwest Bornean (and west Javan) populations before the Holocene, such an exchange may in fact not have taken place.

islands as it inundated the former Sunda lowlands. This isolation has resulted in some genetic and morphological differentiation, but when specimens of the two subspecies are brought together, e.g. in captive conditions, they readily breed and produce fertile offspring. The behavioural repertoire of both subspecies is virtually identical, although a difference in sociability can be distinguished (Markham, 1980).

However, phenetic analyses indicate that, on the biochemical level, the two subspecies are as divergent as are the chimpanzee and bonobo; differences in mitochondrial DNA between the subspecies are much greater than within any other great ape species (Ferris et al. 1981; Wilson et al., 1985; Janczewski et al.,1990; Karesh et al., 1997). In terms of karyotype, the difference is most markedly expressed in the Bornean subspecies having evolved a consistent pericentric inversion of the second chromosome pair (Seuanez et al., 1979). These characteristic differences would justify differentiation on a species-, rather than subspecies, level (Karesh et al., 1997)

Morphological differentiation is recognisable in the habitus (van Bemmel, 1968; MacKinnon, 1975; Mallinson, 1978; Markham, 1980) and, according to MacKinnon (1973), in the structure of the hair: 'Viewed under a microscope, the Bornean hair is flattish, with a thick column of black pigment down the centre; Sumatran hair is thinner, rounder, has a fine and often broken column of dark pigment down the centre, is usually clear near the tip and sometimes tipped externally with black' (MacKinnon, 1973; p. 239). Bornean orang-utans are more robust and have a darker skin and hair colour than the common Sumatran type.

However, it must be realised that neither the Sumatran nor the Bornean orang-utan can be easily characterised in a general, uniform type (Courtenay, et al., 1988). Indeed, Groves (1992) has noted that 'an arrangement into two subspecies (...) underestimates the complexity of geographic variation in the species.'

It has been known since the end of the nineteenth century that the habitus and physiognomy of orang-utans are almost as variable as those of humans. This is true even if it is taken into account that the ape passes through four major life-stages (infancy, adolescence, sub-adulthood and adulthood) which are reflected in a remarkably distinct appearance (Brandes, 1936; Rijksen, 1978). The early specimen collectors perhaps did not fully realise that these life-stage changes took place. For instance, in 1898 the taxonomist E. Selenka described as many as eight 'different races' from the water catchment of the upper Kapuas river alone. Hornaday (1885) believed that the Bornean orang-utan could be differentiated into two separate, but apparently sympatric, species.

More sophisticated methods of differentiation have revealed, however, that the orang-utan population in Borneo can indeed be realistically differentiated into at least three quite distinct types corresponding to their geographical distribution, namely a western (north of the river Kapuas), a southwestern (south of the river Kapuas, and in Central Kalimantan) and an eastern 'race' (in East Kalimantan and Sabah). Although they all appear to have the characteristic Bornean pericentric inversion of the second chromosome, it is interesting to note that in physiognomy the southwestern type is more reminiscent of the ('typical') Sumatran orang-utan than of the other Bornean 'races' (Courtenay et al., 1988). In terms of genetic diversity, however, the types or races of Borneo are not different enough to warrant subspecies classification (Karesh et al., 1997)

> Considering the most likely pattern of expansion, distribution and possible genetic exchange in relation to the geography of the Pleistocene Sundaland mass, it is presumed that until some 10,000 years ago the Bornean population of the southwestern type had possibilities for a continuous exchange with the Sumatran population along the approximately 400 km expanse of what may be called the Karimata – Belitung – Bangka watershed of Sundaland. Assuming that the Sundaland extension of the (Thai) river Chao-Phraya was so wide that orang-utans could not possibly cross except at its source, then the Bornean orang-utan must have entered the island exclusively along this watershed. Further expansion into Borneo must also have been constrained by the course of the large rivers, in particular the Kapuas and the Mahakam. One can imagine that for expansion to reach the northern side of the Kapuas River, the apes which currently constitute the northwestern type could make use of a relatively narrow habitat corridor in the upper course of the river along and across the watershed (see also Banks, 1931). The same applies to the eastern type apes branching off from the northwestern type to cross the watershed into the northern Mahakam catchment. The entrance to southeastern Kalimantan was possibly unsuitable for crossing (see p. 36). Both these currently surviving types thus ended up at the apex of their expansion, so that genetic exchange with the population of the remote Sundaland 'centre' has become impossible over the last several millenia.

The Sumatran orang-utan population north of Lake Toba does not seem to have such a distinct geographical differentiation. Nevertheless, two orang-utan morphs may be distinguished (even) within any one local community, i.e. deme, namely, a robust and a delicate form (Rijksen, 1978). The former is characterised by its cinnamon or orange-red hair colour, its comparatively light skin pigment and coarse, stubby hands and feet which often feature a degenerate or rudimentary thumb (and 'big-toe') lacking a nail. It may be called the light, stubby-fingered morph. The latter is characterised by its dark chocolate or maroon hair colour, a very dark, blackish skin pigment and delicately built, long, well-developed hands and feet, on which the nails of the thumb and 'big toe' are invariably present. Hence, it may be characterised as the long-fingered morph. The two morphs seem to interbreed freely in the wild state, and the relatively rare occurrence of the dark type suggests that it represents a recessive genetic factor in the population north of Lake Toba (Rijksen, 1978). If it was formerly assumed that the 'typical' light-haired Sumatran orang-utan was invariably of the 'robust' stubby-fingered morph (Rijksen, 1978; Courtenay, et al. 1988), the present survey found a light-haired 'delicate' (long-fingered) type of orang-utan in the coastal region of northern Sumatra, i.e. in the Kluet area (also van Schaik, pers. comm.) and on the northern side of the river Batangtoru near Sibolga Bay.

Research into the genetic diversity of Sumatran orang-utans (Karesh et al. 1997) has revealed that two distinct matriarchal lines of orang-utans exist sympatrically in Sumatra, their origins dating back some 600,000 years. It is therefore probable that the two morphs have in the distant past been representatives of two geographically distinct 'races', originating from different locations on the Asian mainland. It is tempting to speculate that one migrated west of the Malaysian watershed, entering Sumatra somewhere at the latitude of the present province of North Sumatra, while the other migrated east of the Malaysian watershed, entering Sumatra along the watershed of what is now the Riau archipelago. If so, at least some demes of both populations eventually got mixed up in recent centuries due to migration or expansion across former boundaries. It may then be surmised that the light-haired, robust morph is descended from a northwestern ('Indian') stock, and the dark-skinned, slender morph descended from an eastern Sundaland ('Indo-chinese') stock, which ostensibly also gave rise to the Bornean subspecies. Scanty reports on observations of individuals of the Angkola orang-utan population suggest that they may be predominantly a delicate type, having a remarkably dark physiognomy (Tulp, 1641, Hagen, 1890, van Balen, 1898, Borner, 1976), while demonstrating a much higher incidence of ground dwelling.

> It is perhaps of interest to note that the legendary yeti, mi-chom-po, mechume, mirgula, migöi, hlo-mung, nyalmu, ban-manche or gredpo of the evergreen forests of the Himalayan Range is consistently described by local inhabitants as a very strong, large, robust, reddish-brown ape, with long arms and human-like hands and feet, whose footprints rarely yield an imprint of the big toe (Choden, 1997).

Section I: the orang-utan

A HISTORY OF ORANG-UTAN DISTRIBUTION

Prehistoric expansion

From the fossil and semi-fossilised finds of orang-utan-like remains (Hooyer, 1948; von Koeningswald, 1982), it appears that the ape originated on the Asian mainland, somewhere along the developing Himalayan Range. Whatever the ancestor of the orang-utan looked like (see e.g. Koeningswald, 1982) when it emerged as a distinct species in the (sub) tropical southern forest fringe of the Asian mainland approximately two or three million years ago, some ape probably with a physiognomy similar to the current form dispersed southwards into Southeast Asia. Whether its social organisation then was the same as it is today is questionable (MacKinnon, 1972), but it must have facilitated the ape's expansion over more than 3,000 km southwards into the vast expanse of the Sundaland region during the

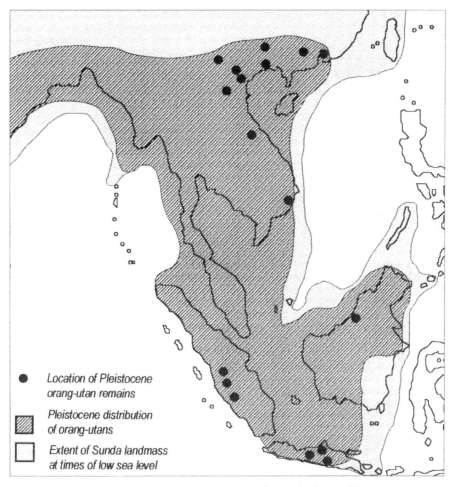

Records of Pleistocene orang-utan remains and the presumed prehistoric distribution of the ape.

Pleistocene. Considering the present distribution range and the ape's preference for alluvial forests and (coastal) lowland swamps, as well as for foothills of mountain ranges, it is probable that the expansion pattern was concentrated along the major rivers and the foothills of the ranges.

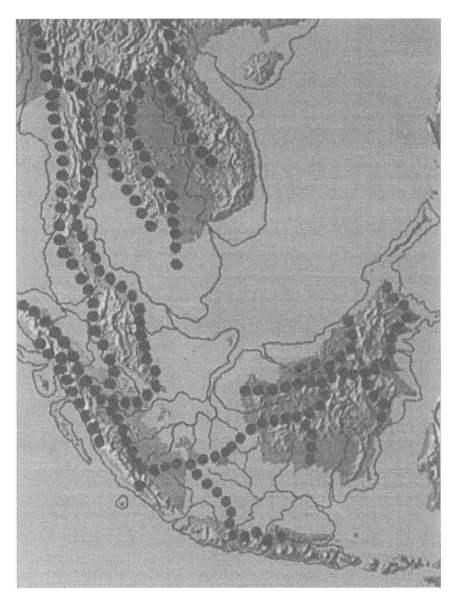

The apparent expansion of the orang-utan into the Malay peninsula, Sumatra, Java and Borneo, showing the Belitung – Bangka – Karimata link; the eastern (Indo-Chinese) branch was isolated at an early date and was probably driven to extinction.

Since the ape cannot swim, and can cross rivers only where the banks are somehow connected by tree canopies, a fallen trunk, or boulders large enough to use as stepping stones, the dispersion must have led wandering apes all along tributaries upstream into the foothills where the headwaters were narrow enough for them to reach the opposite side.

Until the end of the Pleistocene, the orang-utan occurred in most of the lowland forest areas of the vast Sundaland plateau which straddled the Asian continent south of the Himalayan Range. In terms of latitude, its distribution range stretched from the Siwalik plateau in northern India and the foothills of the Wuliang Shan mountain range in Yunnan, southern China, all the way south to what is now the island of Java, covering an expanse of at least some 1,5 million km^2. It is believed that three factors have determined the dynamics of the distribution pattern of the orang-utan, namely, (1) the land form and river courses, (2) the distribution of forest in general and habitat in particular, and (3) human invasions into the ape's domain.

The dispersal resulted in the orang-utan eventually reaching the eastern end of what is currently Java, as well as the northeastern end of the landmass currently known as northeastern Borneo (Sabah). Considering the reconstructed river courses along the bottom of what are currently the shallow South China- and the Java seas (e.g. Haile, 1973; Verstappen, 1975), the migration must have led the ancestors of much of the present orang-utan populations along the western side of a mighty river which extended from the current Chao Phraya River in Thailand southwards along the massif of the current Malay Peninsula and curved in a northeastern direction, where it then met its major tributary from the Sumatran massif, the confluence of the current rivers Tembesi and Musi. From that confluence the river followed a course northeast to discharge into the China sea somewhere between the present island archipelagos of Anambas and Natuna.

It is probable that Pleistocene orang-utans from the rainforests of the current southern Chinese provinces Kwangsi, Kwantung, Szechuan, Kiangsu and Yunnan, as well as from Indo-China, also migrated southwards along the eastern bank of this great river, and along what is now known as the Mekong River, into the eastern section of Sundaland. Such a population then became extinct, because it could never have reached the west bank and Borneo across the immense river(s).

During the Pleistocene the Sunda landmass was inundated several times, but major land bridges between the continental mainland and the islands of Sumatra and Borneo often remained. In such periods peninsular Malaysia was frequently still connected to Sumatra along the Batam-Riau-Lingga chain, and Sumatra was connected to Borneo along the Bangka-Bilitung-Karimata-Schwaner divide.

Thus, it can be surmised that the ape must have dispersed in Borneo along the mountain chains, starting in the foothills of what are now the Schwaner mountains, and fanning out into the lowlands along the rivers. In the lowlands the rivers are too wide for wandering apes to cross, but in the water catchments of the headwaters of the major rivers the ape may colonise from one water catchment to the next.

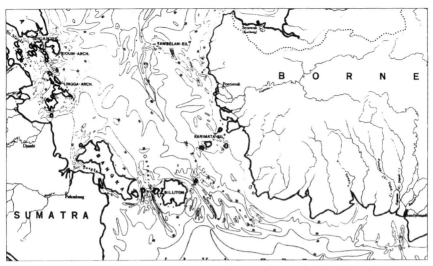

Geomorphology of the shallow sea between Sumatra and Borneo, showing the island chain which was once a land bridge when the sea level was lower.

In the following it will be demonstrated that the geomorphological and ecological conditions of some headwaters may well be unsuitable for ape dispersal. There the forest type is too uniform and too dry and unproductive, and the climate is inclement due to the high altitude (i.e. over 1,000 metres), so that no ape in its right mind would be tempted to venture forth. In Borneo one can clearly distinguish several such major 'bottlenecks' for migration even under the current conditions. One can imagine that these areas were even more effective barriers for dispersal when the sea level was lower and the climatic conditions in the mountains above 500 m altitude more severe.

> Smit-Sibinga (1953), in his study of the geomorphology and origin of the drainage system of Borneo, proposed that at some time during the Pliocene (i.e. 10-25 million years ago) the island was formed by the emergence of a long mountainous range which stretched from Sumatra, along the Bangka-Bilitung-Karimata chain northeastwards into the Schwaner mountains, the Müller mountains, and the Iran mountains, and on to the Apo-Duat and Crocker mountain chain to reach the northern tip of Borneo, extending into Palawan. Later, at the beginning of the Pleistocene (some 10 million years ago), the transverse range of the Semitau-Kuching-Tinda Hantung (Sambaliung) mountains was pushed up to form a cross-like frame along which accretion and erosion created flood plains and added what are currently Borneo's lowlands.

The Pleistocene era came to an end some 12,500 years ago (Hapgood, 1970) with the sudden geophysical upheavals that abruptly changed polar glaciation patterns. The post-glacial sea-level rise inundated all of the lowland, breaking up the landmass into the islands as we know them today. The once continuous ape population also became fragmented. It must have suffered the effects of forced migration to higher land (i.e. the current Malaysian mainland and Sunda islands), subsequent isolation and exponentially increasing hunting pressure by ever more humans trespassing into

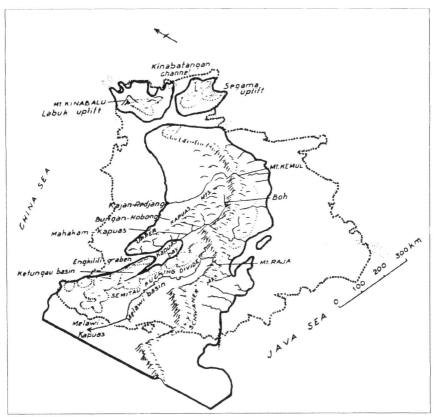

Development of Borneo as postulated by Smit-Sibenga in 1953.

its realm. The mainland population was possibly extinct before the first western explorers charted the coasts; the Javan population may have vanished as late as the 17th century (Dobson, 1953-54). All that was believed to have remained at the beginning of the twentieth century were two significant subpopulations: one – the largest – on the island of Borneo, the other mainly in the northern and western part of the island of Sumatra.

Fossil finds in Java (Dubois, 1922) indicate that at least two distinct waves of archaic people entered the Sundaland region during the middle Pleistocene. First, people of a remarkably early *Homo erectus* type who may have arrived as early as 1,600,000 years ago, and second, people of the Australoid *Homo sapiens* type who may have arrived close to some 100,000 years ago; the skeletal remains of this type in Java seem to be about 80,000 years old (Storm, 1996). Orang-utan remains found in caves indicate that the apes were hunted in central Sumatra (Hooyer, 1948) and the central region of what is now Sarawak, perhaps as early as some 35,000 years ago (Hooyer, 1961)[5].

---

[5] In Sumatra, E. Dubois found 3,170 orang-utan teeth in the Lida Air, Sibrambang and Jambu caves on the Padang highlands in 1889 (Von Koeningswald, 1979); in Borneo, T. Harrisson discovered 111 orang-utan teeth in the Niah cave complex in eastern central Sarawak (Hooyer, 1960); in Java, both Dubois and Von Koeningswald found orang-utan teeth, the former at Trinil, the latter at Gunung Kidul (W. Java) and Sangiran (C. Java), all in lower and middle Pleistocene deposits.

> Since the fossil finds from the middle Pleistocene *Tham Khuyen* Breccia cave deposits in Vietnam are coincidental with those of *Homo erectus* (Schwartz et al. ibid), it is possible that both belonged originally to the same prey-set of early modern humans.
>
> Most of the current human populations in Sumatra and Borneo are descendants of recent migrants from the Asian mainland who did not arrive before 3,000 years ago (Bellwood, 1985), mixed with remnants of the older (aboriginal) human types who were supplanted. The archeological record indicates that waves of tribal people from the region extending from along the foothills of the Himalayas to southern China moved southwards and settled in Sumatra and Borneo at such a relatively recent date, but a major cultural influence from Indo-China spread over the islands between 600-1200 (Cheng, 1969).

TABLE I

Overview of land area, forest cover and potential orang-utan habitat of Sumatra and Borneo after the Pleistocene Sunda landmass had broken up into islands and a peninsula: (*) forest cover minus mangrove and mountain forest over 1000 m alt.; (**) excluding South Kalimantan province; for the whole of Sumatra (including the extensive lowlands) orang-utan habitat is estimated to be 30% of the forest cover, for Borneo this is said to be 25%; geographic data from Collins et al. (1991).

	Total land area	Forest cover*	Potential habitat
Sumatra	473,000 km^2	430,000 km^2	129,000 km^2
Borneo**	740,000 km^2	640,000 km^2	170,000 km^2
Malay peninsula	340,000 km^2	330,000 km^2	115,000 km^2
Java and Bali	138,000 km^2	130,000 km^2	40,000 km^2

On the basis of the estimated forest cover of both islands during the early Holocene, it may be surmised that Sumatra has been the home of a meta-population of at least 380,000 orang-utans, while Borneo may have harboured a meta-population of at least 420,000. The Malay peninsula was probably home to some 340,000, and Java may have had a population of at least 100,000 orang-utans in early pre-historic times.

In early Pleistocene times this population, covering the Southeast Asian landmass, may have amounted to at least 2 million apes, discounting the population which may have occurred on the eastern side of the river that divided the Sunda landmass (i.e. the Indo-Chinese side).

Geomorphological barriers and bottlenecks for expansion in Borneo

Orang-utans can and do wade in gently flowing chest-deep water (e.g. in swamps), but they do not enter fast-flowing rivers to attempt fording, except where they are able to make their way across by securing a foothold on exposed boulders. Their method of traversing rivers is commonly along the crowns of trees which are growing on opposite banks and whose branches almost touch above the water, or along large trunks that have fallen across the river. This implies that rivers wider than some 10m are usually effective barriers for orang-utan dispersal.

However, rivers often become crossable in the headwaters where the stream narrows, and the crowns of trees and boulders in the stream bed provide ample opportunity for two-way traffic. Commonly, however, this narrowing occurs at higher altitudes, that is, in the more gently undulating country of Borneo, at an altitude of well over 300 m. Orang-utans do not favour conditions at altitudes over 1,000 m, and consequently the combination of conditions provided by rivers and high altitudes may also be a serious barrier for the expansion of an orang-utan population.

Five major areas in Borneo can be distinguished in which orang-utans are absent nowadays. In four of these areas, one may at least presume that human interference has been the major factor in the disappearance of the ape, namely, in the Kapuas basin, the centre of Sarawak, the water catchment of the river Kayan, and the coastal zones of Brunei and Sabah. Once the ape was eradicated locally, some major geomorphological bottlenecks and persistent persecution seem to have prevented repopulation of the areas, although the forest habitat remained.[6]

If rivers as well as the climatic and ecological conditions at higher altitudes can restrict orang-utan population expansion, then two important more or less contiguous areas can be distinguished which have played an important role in the geographical distribution of orang-utans in Borneo, namely, the cross-shaped central mountain ranges of the island (in particular the Semitau-Kuching range or Bentuang-Karimun area and Müller mountains) dividing West and Central Kalimantan from Sarawak, and the Schwaner and Müller range extending into the Pegunungan Iran (or Kayan-Mentarang region) and Apo Duat mountains, also called the Embaluh geological zone (Van Bemmelen, 1949).

The geomorphological maps of the island reveal that the soils in these two areas are a mosaic of Eutropepts, Kandiudults and Dystropepts on Plutonic basalt, and are relatively poor in nutrients. These mountainous areas support either somewhat uniform Dipterocarp types of forest which are relatively poor in tree species, or even more impoverished 'heath' forest (*kerangas*). The relative uniformity of these two predominant forest types may not allow for the necessary variation in temporal and spatial food supply to provide year-round subsistence for the apes. It is therefore not surprising that the historical record of orang-utan distribution seems to be extremely patchy along this mountain chain, even in areas of relatively low altitude, indicating that the ape may be concentrated in the main valleys.

Where much of the cross-shaped central mountain ranges of Borneo may well support habitat of poor quality in general, in particular their junction seems to be forbidding country for orang-utans, because it constitutes the highest mountain

---

[6] Payne (1988) noted that 'several aspects of the distribution of orang-utans point to the likelihood of a relationship with the distribution of minerals' his empirical findings, however, do not support this hypothesis. The present survey also yielded no indications that mineral deficiency was important enough to be taken into account.

complex of Borneo. It is significant that at this complex of the Gunung Liangpran (2240 m), Gunung Lapa (2246 m) and Gunung Batutiban (1880 m) all the headwaters of Borneo's major rivers originate, namely, Rejang, Kapuas, Barito, Mahakam and Kayan. These rivers, and the mountain ridges, divide Borneo – like a wedding-cake – into at least four separate quarters.

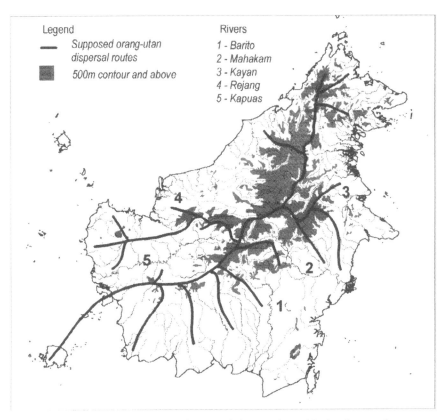

Central Borneo geology showing the estimated west-east habitat corridors and bottlenecks across the mountain ranges.

When the orang-utan dispersed northwards from the entrance along the Bangka-Bilitung-Karimata divide, it must have traversed at least one complex of altitudinal bottlenecks, at the centre of the Müller mountain range. More precisely, the ecological bottlenecks are formed by watersheds comprising several peaks of some 1,700 – 2,100 m in altitude (e.g. Gunung Pancungapung, Gunung Kerihun, Gunung Cemaru) that divide the rivers Ulu Busang, Ulu Kapuas, Ulu Mahakam and Ulu Baleh. Actually, on a more recent 1:1,500,000 map (1996; Nelles Verlag München, Germany), the pivot of Borneo is an unnamed mountain of 2,073 m in altitude (possibly Gunung Batutiban, or Tiban) from the flanks of which spring the two main headwaters (Oga, Mahakam) of the river Mahakam, the headwaters (Baleh and Ulu Balui) of the river Rejang, and the headwater of the river Ulu Kayan. After scaling this mountain complex from the south, the apes must have descended northwards into the headwater valleys of these major rivers. To the west (Kapuas, Rejang) and the northeast (Kayan) the valleys and ridges invite expansion, but to the east, along the southern banks of the river Mahakam, a wandering ape would be confronted by a maze of river obstacles and ridges forming the dissected water catchment of the upper reaches of the river Barito.

In many of the headwaters of this pivotal mountain complex in the Müller range the habitat is confined to the river banks and becomes poor and unattractive forest on the steep slopes. However, just north of the complex, the ancient mountain range is interrupted by a lower area of young volcanic origin, cutting through in an eastern direction. The area is known as the (fertile) Apo Kayan region.

This medium altitude divide (300 – 1000 m) must have enabled the orang-utan to disperse from the Ulu Busang valley into the rest of Borneo, that is the area between the river Kapuas and the river Rejang, to the west, into the central Sarawak plains to the northwest, and into the eastern and northeastern half of the island. The Apo Kayan corridor must have boasted excellent orang-utan habitat before humans occupied and converted it. It was and still is certainly the major corridor for people travelling between Sarawak and East Kalimantan.

From the Apo Kayan, the alluvial inroads into the zone, like the river Kayaniut (i.e. the headwaters of the Kayan), and possibly the rivers Bahau and Mentarang to the north, may have provided the corridors towards the gradually widening valleys, and flood plains of the east and northeast. The early *Mededeelingen* of the Netherlands Commission for International Nature Protection (1930) report on the occurrence of the ape in the upper reaches of the river Kayaniut (*Kajan Ioet*). That there is no historical record of the ape for the upper Kayan valley suggests its local extinction, which may well be due to the early inhabitation of this fertile stretch of river by peoples who were originally typical hunter-gatherers and who may have given rise to the Punan tribe.

> People in the southern sections of the Kayan Mentarang conservation area referred to the Apo-Kayan as a historic hunting ground of mainly Kenyah and Punun tribal peoples, where their grandfathers used to hunt orang-utans. The practice was allegedly still common until the Pacific War (1942-1945).

In the same series, A. Heynsius Viruli and F.C. van Heurn (1935) reported that an apparently isolated population of orang-utans occurs east of the river Barito, in the swamp forests west of the township of Tanjung, between the river Barito and its tributary, the river Negara. Being the only population of apes in South Kalimantan province, an explanation of its occurrence may be that the mighty river Barito and its tributary must have comprised a dynamic, bridled complex of meandering watercourses cutting through the very extensive lowland flood plains and swamps. Some of the former oxbows of the main course can still be discerned on satellite imagery, and it is probable that some part of the lower stretches of what is currently the river Negara was a main streambed of the river Barito in the distant past. Land accretion on the western side of what is now the river Negara, and subsequent cutting through of a new main channel, may, in the flood plains and swamps, have separated a small eastern population of apes from its main body. It is probable that a comparative analysis of DNA of the populations east and west of the current course of the river Barito could shed some light on this hypothesis.

Why the ape has been absent, or has disappeared, from the rest of the originally forested province of South Kalimantan has long been an enigma. For other areas the absence of the ape may be the result of persistent persecution by aboriginal hunter-gatherers and more recent forest scroungers, when considering the pattern of human invasions in the island (Bellwood, 1985; King, 1978); (see chapter Hunting). After all, the finds of prehistoric ape remains in the Niah caves indicate that the orang-utan was once, as early as at least 12,000 years ago, already a major prey for early nearby invaders (Medway, 1979). However, to explain the absence of orang-utans in the Meratus mountain range of South Kalimantan as being a result of traditional human persecution is perhaps not so easy. Whether traditional aboriginal gatherers ever occurred in the region is unknown. Yet, the area certainly has a very long history of human disturbance, especially by gold-prospectors and collectors of precious gemstones, and it is also possible that in South Kalimantan, subsistence hunting played a major role in the disappearance of the ape from that region.

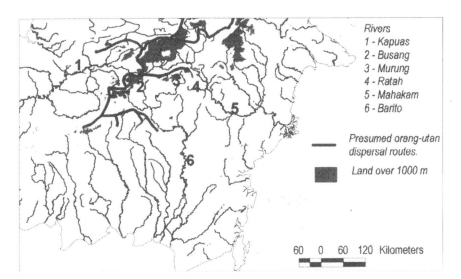

Geography of the 'entrance' to South Kalimantan province.

Can the absence of orang-utans from the largest sector of South Kalimantan east of the river Negara perhaps be explained by reference to other factors? If one assumes that human persecution can be ruled out, then there must be an ecological reason;

> Historically, the area of South Kalimantan was renowned as the hinterland of the major medieval trade centre of Banjar, from where gold, precious stones and forest produce were shipped to China, South Asia and the Middle East. This implies that from an early age, perhaps well over a millenium ago, all the eastern tributaries of the river Barito into the mountainous range must have been scoured by generations of gold-prospectors managing to live from the forest. And although the region came under the influence of Islam at some point during the 14th century (which might, as in Aceh, have suppressed local hunting of the apes for food), it is a fact that the region has suffered long and intensive exploration, mainly by non-Muslim Chinese and Dayaks, and encroachment, and is currently traversed by the densest network of roads and tracks in Borneo.

either the orang-utan never reached the forests of the Meratus mountain range, or it died out. In the latter case one can imagine two major options, namely, that the habitat could not support migrants, or the (relatively small?) population was wiped out by an epidemic.

The forest composition and ecology of the foothills of the Meratus mountain range does not differ significantly from the regions where the ape is found. It is therefore unlikely that an orang-utan population could not be sustained. Whether a possible former population died out due to an epidemic can unfortunately not be verified, but the ape's widely scattered social structure seems to render this unlikely.

Be that as it may, it is possible that the ape never reached the area. The entrance to the Meratus mountain range, from the northwest, along the headwaters of the river Barito and the river Negara, may well have represented a barrier for the dispersal of the ape. In particular the highlands between the upper reaches of these rivers and the river Mahakam can be considered an ecological barrier which, if not being virtually impassable for orang-utans, will certainly have precluded any regular expansion, or recolonisation, from northwest to southeast. During the supposedly drier periods of the Pleistocene these adverse conditions may have been even more pronounced.

Interestingly, the Sungei Busang is the dividing river between the two different species of gibbon in Borneo, i.e. the originally Sumatran *Hylobates agilis* in the southern part of the island between the rivers Kapuas and Barito, and the truly Bornean *Hylobates mulleri* in the rest of the island (Mather, 1992). Apparently the agile gibbon never crossed the entrance into southeastern Kalimantan in significant numbers. Is it possible that the orang-utan has been confronted with similar constraints there?

In any case, it is evident that in order to expand its numbers all over Borneo, the orang-utan must have dispersed along the major river systems and up into the headwaters to cross major geomorphological barriers, i.e. both rivers and a few crucially important watersheds at altitudes between some 1,300 and 1,700 m. And although orang-utans are known to favour lowland alluvial habitats up to some 500 m, it is certainly not uncommon for an orang-utan to venture into higher altitudes if and when the conditions are favourable. After all, even in Borneo, Whitehead (1893) found an orang utan alive and well at an altitude of over 2,000 m on Mount Kinabalu.

> First, the climatic pattern in southeast Borneo is considerably more bi-seasonal and generally drier than in the other sectors of the island. Second, the narrow valleys and high ridges separating the headwaters form a zig-zag, maze-like pattern of narrow, high-altitude ridge corridors covered with heath forest in their upper reaches, which may well preclude the crossing of all but the occasional ape.
>
> The entire Barito Ulu region comprises four major headwaters, the Joloi, Busang, Murung and Maruwai, and is virtually met by the river Mahakam and several of its tributaries, across a narrow, rugged and serrated ridge pattern which rises from about 500 m in its bottomlands up to some 1700 m at the ridge tops. The western water catchment of the Busang Ulu is contiguous with the Batikap Protection Forest complex of the northern sector of the Schwaner mountain range. Only in the western valley of the S. Busang, which is largely below 500 m, have people reported that the ape is occasionally seen. Elsewhere in the vast mountain complex hardly any reliable orang-utan sightings have been reported, and the ape is believed to be rare in most of the headwaters of the river Barito (Bodmer et al, 1992). The present survey also revealed that local people reported the ape to be absent or extremely rare. In the 7 years of field-research at the Barito Ulu Project on the S. Busang, orang-utans were encountered only twice. Heath forest (kerangas) is the predominant type growing in the nutrient-poor soils covering the plutonic basalt mother-rock in this region. Because the area is nowhere below approximately 500 m and largely covered with poor Dipterocarp and heath forest (Mather, 1992) it is probably too unproductive for Bornean orang-utans in any case.

Many of the original major corridors of expansion through Borneo are nowadays unlikely to have any remaining populations of orang-utans. Either the habitat has been converted for slash-and-burn agriculture and, more recently, plantations, or the ape's place in the habitat has been usurped by the archaic gatherers, the Penan, where the forest still remained. The rapidly expanding populations of shifting cultivators, notably Iban, Kayan, Kenyah and Kelabit, have all exerted a considerable impact on these inroads during the ongoing mass movements of people. One now finds cultivated gardens, grass steppes or, at best, degraded secondary forest there. Thus, it may be expected that orang-utans have been displaced and eradicated from the valleys of the Busang Ulu and the Kayaniut, from much of the flood-plain area of the rivers Kapuas and Rejang and from the foothills all along the border between Sarawak and Kalimantan, except perhaps in the inhospitable country where the Bentuang-Karimun and Lanjak Entimau areas meet. Due to human expansion and influence, the ape is also likely to have disappeared from the fertile valleys of the entire Kayan-Mentarang water catchment, and from central Sarawak. And although it is unknown whether archaic gatherers ever scoured the forests of the Meratus mountain range in southeastern Kalimantan, it is not impossible that more recent migrants (e.g. gold-prospectors) eradicated the ape, if it ever reached this region.

Considering the geomorphological barriers and bottlenecks of expansion, it becomes clear why the Bornean orang-utan may have begun to evolve into at least three, and possibly as many as five fairly distinct races or types. The populations of northern Borneo, eastern Kalimantan, western Borneo between the rivers Rejang and Kapuas, Central Kalimantan and southwestern Kalimantan between the river Kapuas and the Schwaner mountain range must have been more or less isolated from each other from the time the population expansion reached the opposite shores of the island. It is unlikely that any northern population individual had the opportunity to convey its genetic essence back, along more than a thousand kilometres of maze-like terrain, to the apes that remained close to the original entry gate in Sukadana county. The same must have been true for the other populations entering what were essentially ecological *cul de sacs*.

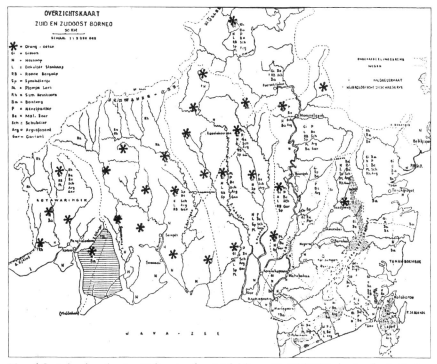

Map of Central and southeastern Kalimantan from Mededelingen No 10 (1935) showing the distribution data of the large mammals; the orang-utan is indicated by asterisks.

Historic distribution

Kalimantan

The first accurate field record describing the orang-utan, was written by Salomon Müller during his travels in 1836-37; it states:

> (...) the ape 'appears to be distributed over the whole of this vast island, with the exception of the high mountainous regions and more populous lowlands. For that reason it will be sought in vain in the environs of Banjarmasing, and northwards from there along the river Duson [i.e. Barito] it makes only rare appearances at certain times of the year in secluded places. It is less elusive a few days' journey further westwards, in particular in the surroundings of the Sungei Kahajan, and along the river Sampiet, near Kotaringin, and some other remote areas of the south and west sides of this island. The broad stretches of low and level alluvial terrain along these two coastal areas, broken occasionally (...) by minor mountain ranges or outcrops of high ground with difficult access (...) provide him with a spacious and secure abode.' (Schlegel and Muller, 1839-44).

A more comprehensive record of the distribution of orang-utans was published during the early twentieth century in the renowned magazine *De Tropische Natuur*. It concerned an article covering the distribution pattern of several large mammals in southern and eastern Borneo (Zondag, 1931), based on a widespread inquiry among colonial authorities (civil and military) in the region. In the 1930s, the ape reportedly occurred in the subdistricts *Kotta Waringin*, *Sampit*, *Boven Dajak*, *West Koetai* and along the river Sangkulirang, as well as along the river Kajan or *Boelongan*.

A comprehensive addition to this record was subsequently published as a Supplement to *Mededeelingen* No 10 of the Netherlands Commission for International Nature Protection in 1935. There A. Heynsius-Viruli and F.C. van Heurn noted that the orang-utan was most common in the very extensive, inaccessible swamp and flood-plain forests between the rivers Barito and Kahajan (*de drassige landstreek Loeau*).

The orang-utan was also reportedly common in the uninhabited area of *Kajan Ioet* in the district of Apo Kayan, and in the peat forests along the road between Mempawah *and Pemangkat (Si Boekit)*[7], in what was called the *Westerafdeling*, as well as in the region due north of Samarinda as far as the Bay of Sangkulirang, and on the border between the subdistricts of *Kloeat* and N. Tabalong (notably in the area Hapau Ondan). There are very few data on the distribution of orang-utans in the remote, and in those early days almost inaccessible, mountains and hills.

Finally, in 1938, the earlier reports were compiled in the article *Voorkomen en Verspreiding van eenige belangrijke dier- en plantensoorten* in the eleventh report of the Netherlands Indies Society for Nature Conservation, entitled *Drie Jaren Indisch Natuur Leven* (1936-38). In this article it is stated that the ape 'occurs everywhere in the swamp forests of what is now Central Kalimantan province and the adjacent drier

---

[7] During the present surveys these two populations were found to be extinct.

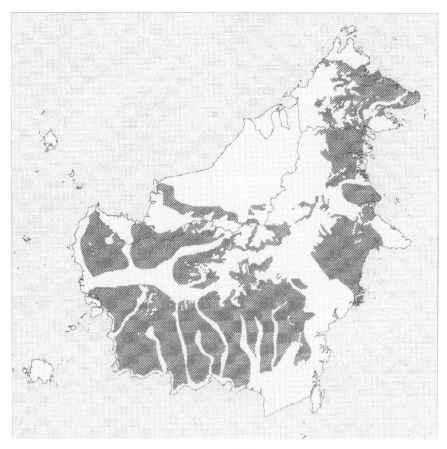

Recorded distribution of the Bornean orang-utan in the 1930s.

foothill forest areas but it is rare in the northern mountain region. In eastern Kalimantan it is to be found mainly in the area which is bordered by the lower Mahakam, the Sungei Telen, the Sungei Sangkulirang and the coast.' Yet, it 'seems to be absent entirely in the *Tidoengse Landen, Boelongan, Apo Kajan* (excepting *Kayan Ioet*), and *Beraoe*, as well as in Central and South *Koetei*.' The ape also used to be found in the county due south of Samarinda, in the water catchment of the upper Mahakam (upstream of Long Iram), and in the area of the lower Kayaniut – albeit in 'small numbers'[8].

The area south of the river Mahakam and east of the river Barito has usually been excluded from the historical range of the ape (Zondag, 1931; Rijksen, 1978; MacKinnon and Ramono, 1993). However, both Witkamp (1932) and Westermann (1937) have reported occasional sightings (and in particular the killing) of orang-

---

[8] It is not surprising that these early data correspond with the type localities of museum specimens collected between 1780 and 1941 throughout the lowlands of Borneo; data were kindly provided by M. Jose Braga of Bordeaux University (France) in 1995.

utans in the swamps surrounding the township of Tanjung, along the river Negara. They noted that the ape is absent in South Kalimantan province east of the river Negara, in the Meratus mountain range and the coastal lowlands to the east.

Sarawak, Sabah and Brunei

In 1931, E. Banks published *A popular account of the mammals of Borneo*. It is virtually the only comprehensive early twentieth century overview of the orang-utan's distribution range in the colonial states of Borneo under British rule. It reveals that the range was already much reduced during historical times. The nineteenth century explorers, like Wallace and Hornaday, were after great numbers of orang-utan specimens, and the measure of abundance was in how many one could shoot in a day. The favourite hunting area was in the peat swamps between the rivers Batang, Sadong and Batang Lupar, just west of the town of Kuching.

It is interesting to note that the historic record contains few sightings of orang-utans north of the river Rejang. Yet, the ape must have been common in this area until quite recently (Medway, 1977), considering the prehistoric remains discovered in the Niah caves, between Bintulu and Miri. The remains date back to at least some 12,000 years ago (Hooyer, 1961; Bennett, 1993).

In his 'Mammals of Borneo', Medway (1977) mentions that H. Low noted, in 1848, an irregularity in the distribution of orang-utans. The apes were not found 'in the

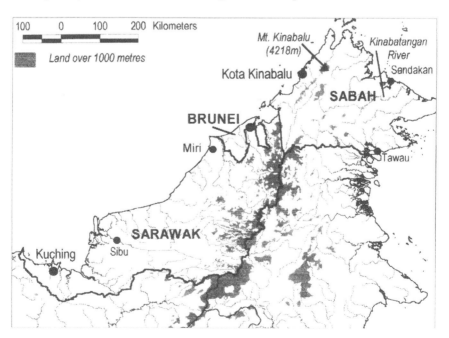

Major geographical features of Sarawak and North Borneo (Sabah).

Sarawak territory, nor in that of Sambas.' However, 'the tract of country between the Sadong and Batang Lupar rivers extending across the island in the direction of Pontianak, are the favourite haunts of this strange animal.'

Medway continues to note that 'collections made in the latter half of the 19th century at Simunjan, the river Sadong, (Sarawak; e.g. Wallace, 1865) and in the upper river Kapuas and its tributaries (West Kalimantan; e.g. Selenka, 1896) confirm this distribution, although Beccari (1904, relating events of 1865 – 1867) did report on sporadic occurrences in the upper reaches of the river Sarawak.' Brooke (1841) comments on the distribution pattern of the ape:

> (...) they are found both in Pontianak and Sambas in considerable numbers, and at the Sadong on the northwest coast, but are unknown in the intermediate country which includes the rivers of Sarawak and Samarahan. I confess myself at a loss to account for their absence on the Sarawak and Samarahan Rivers, which abound in fruit, and have forests similar and contiguous to the Sadung Linga and other rivers. The distance from Samarahan to Sadung does not exceed twenty-five miles, and though [orang-utans are said to be] pretty abundant on the latter, they are unknown on the former river. From Sadung proceeding northward and eastward, they are found for about 100 miles, but beyond that distance do not inhabit the wood.

Nevertheless, some recent records are available of specimens from north of the river Rejang and from the Moh tributary of the river Baram, close to the border with East Kalimantan (Harrisson, 1960). As the types of habitat north of this river, notably the

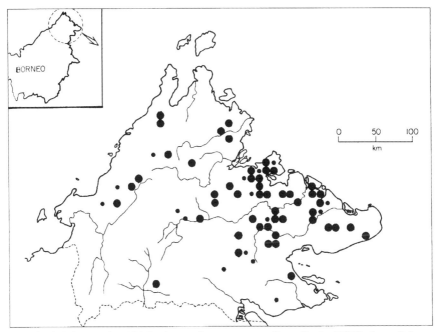

Map of Sabah showing the 1980 distribution of orang-utans according to Davies (1986); the small dots represent records of sightings prior to 1978.

extensive peat forests, seem to be suitable for orang-utans, it is hard to explain their absence, other than being the result of human persecution.

A look at the map of the distribution of the aboriginal hunter-gatherer tribes (known as Penan) reveals that precisely this area where the orang-utan is reportedly absent is the heartland of Punan distribution (Bugo, 1995).

Since the early 1960s, logging, forest management (i.e. poison girdling of unwanted species and liana destruction) and associated poaching has apparently caused a reduction in orang-utan numbers to a fraction, even in areas where orang-utans were once found in abundance, e.g. in the peat swamps. Reynolds (1967) provides an overview of all records in Sarawak, showing that in the 1960s the ape still occurred all along the boundary with Indonesia up to the headwaters of the river Baram. Sadly, however, in 1993, E.L. Bennett estimated that of the 350 orang-utans which Schaller believed to have occurred in the region during the early 1960s, no more than 30 had survived.

According to Medway, 'old records of Sabah cover the state from (...) the southwest (St. John, 1862) to the east coast (Allen and Coolidge, 1940; Davis, 1962), including the highest recorded altitude of 8,000 ft (2,400 m) on G. Kinabalu (Whitehead, 1893)'[9].

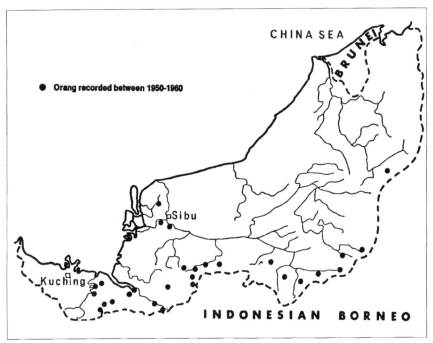

Map of the orang-utan distribution in Sarawak in the 1960s, after Reynolds (1968).

---

[9] According to Payne (1988) this record is supposedly an error; one should, however, not discredit too readily the observation of a respected naturalist, because it is certainly not improbable that orang-utans are able to reach such altitudes during day trips, in particular when much sought-after strangling-figs (e.g. *Ficus deltoidea* var. *kinabaluensis*) and a wealth of oaks (*Lithocarpus, Castanopsis, Trigonobalanus spp.*) and berries (*Rubus spp.*) occur in the 1,000-2,500 m zone (Corner, 1978; Cockburn, 1978) on mount Kinabalu.

Many of these data are corroborated by Reynolds (1967), which suggest that orang-utans lived all over the forested expanse of north Borneo, except perhaps in the southwestern corner near Brunei.

Sumatra

In 1826, the conservator C. Abel published an account of a large, preserved male orang-utan specimen which had been sent to the British Museum by Commander Cornfoot. The ape had been shot in 1825 in the environs of Trumon in Aceh, when it was encountered while walking along the ground in the swamp. Even up to the present, the English-speaking world wrongly believes this to be the very first record of the Sumatran subspecies, because Tulp's record of the first specimen from 'Angola' has consistently been misrepresented. The county of 'Angola', which in modern Indonesian is spelled Angkola, lies in the southern Tapanuli (North Sumatra) regency, south of the town of Padangsidempuan.

In 1878, the Italian explorer Count Odoardo Beccari spent several months in 'the province of Padang' from where he reported that orang-utans occurred in Tapanuli province at Rambung, and in the hinterland of Sibolga (Beccari, 1904: p. 205). He also noted that the skeleton (which was kept at the Zoological Museum in Florence) of an orang-utan, with remarkably long fingers and toes, originated from the area around Palembang.

A note on the distribution of the orang-utan in Sumatra was first published by the German explorer B. Hagen in 1890, although much of northern Sumatra at the time was still dangerous *terra incognita* for Europeans. In 1905, another German explorer, Gustav Schneider, again reported sightings of the ape in the hinterland of Sibolga (i.e. Anggolia) and at the mouth of the river '*Badiri*' (i.e. probably the river Batang Toru). Interestingly, these early reports suggest that these southern orang-utans are very dark ('black') in appearance. In *Mededeeling* No 5 of the Netherlands Commission for International Nature Protection (1928), W.A.J.M. Waterschoot van der Gracht reported on unsubstantiated information concerning the occurrence of orang-utans due west of the Indrapura volcano (Mount Kerinci), i.e. the Way Hitam water catchment.

In 1934, *Mededeelingen* No 10 of the Netherlands Commission for International Nature Protection contained a review published as a result of an inquiry in 1932 among a number of colonial contacts, which tried to establish the distribution of certain unique species and to explore the need for their protection in 'reserves'. In a special supplement to these *Mededeelingen*, published in 1935, Heynsius-Viruly and van Heurn presented an update of these results, including a map of the historical and current distribution of the red ape.

According to van Heurn, the orang-utan survived in two separate areas during the 1930s: (1) the large contiguous mountain forest areas of Langkat, Aceh's east, north and west coast, and including the central Gayo- and Alas *landen*, and (2) a smaller stretch of hill forest in the county Habinsaran, parallel to the east coast of Sumatra southeast of Lake Toba, comprising the upland area of the river Asahan, and the water catchments of the rivers Bilah, Panah and Rokan, at the current provincial boundary between North Sumatra and Riau. Van Heurn mentions that orang-utans used to be common in the forested region between Deli and the *Battaklanden*, but had probably become extinct during the first decades of the 20th century. The inquiry also revealed that orang-utans still occurred in the Rokan district[10] as well as in the Silindoeng district, i.e. the mountainous area due east of the coastal township of Sibolga.

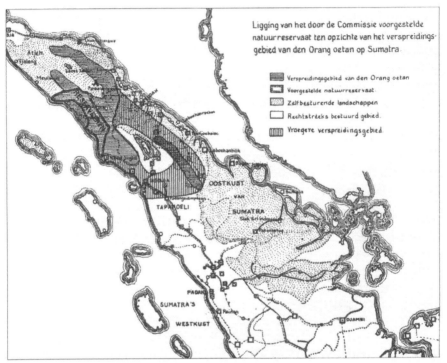

The distribution of Sumatran orang-utans as published by Heynsius Viruli and van Heurn (1935); the horizontal hatching indicates the (confirmed) 1930s' range, the vertical hatching indicates the then presumed historic range.

In the 11th report of the *Nederlandsch-Indische Vereeniging tot Natuurbescherming (Drie Jaren Indisch Natuurleven; 1936-38)*, de Voogd and Rengers Hora Siccama reported that orang-utans were commonly observed in the Protection Forests of Dolok Sembelin (south of the river Renun, in Dairi), near Sipirok (35 km north of Padangsidempuan) and along the river Batangtoru (Central Tapanuli). It is evident that in the late 1930s the Habinsaran relict of the distribution range was still

---

[10] The headwaters of the large east coast river Rokan in Riau province leave the Bukit Barisan mountain chain in the district Pasaman as the *Batang Sumpur*. The Rimbo Panti wildlife reserve is traversed by this river, and a major township at the headwaters of this river Sumpur is Lubuksikaping.

A history of orang-utan distribution

connected along the southern and western sides of Lake Toba with a heartland of orang-utan distribution around the Trumon-Singkil swamp and the adjacent Bengkung-Kaparsesak lowland plateau between the west coast and the river Alas.

In 1937, the American Committee for International Wildlife Protection commissioned Clarence R. Carpenter to conduct part of a US wildlife survey of Sumatra. He was to revise in particular the reports on the distribution range of the orang-utan. Carpenter spent almost one month of field work, during which he claimed to have travelled 5,500 km by car (i.e. an incredible average of 183 km/day), and some 50 km by foot, and to have conducted 38 interviews. In addition he was able to draw on the results of a new inquiry made among all the military commanders in northern Sumatra, organised for him by the Netherlands Indies Society for Nature Protection.

It is unlikely that Carpenter could accurately read the handwritten Dutch reports, but one wonders whether he had even informed himself sufficiently by reading the available English language literature. After all, he believed that orang-utans were restricted to Aceh and Langkat north and west of the river Wampu, whereas the literature of the time reports a much wider distribution. Carpenter emphasises a 'centre of distribution' in the watersheds of the rivers Simpang Kanan and Peureulak, i.e. the environs of the village of Lokop in the direction of Blangkejeren (west) and Takengon (northwest). His reports of orang-utan sightings on the western side of the Barisan range indicated another 'centre of distribution' in the extensive coastal swamp forests between Lami and Blangpidie, up along the water catchment of the river Teripa, and extending southwards to Tapaktuan and Bakongan and the swamp forest around Trumon.

According to Carpenter, the main regions which were excluded in Aceh are 'the high central districts above 1500 m, the cultivated and densely populated districts on the east coast, the grasslands (*blangs*) of the north in the region between Sigli and Koeta Radja (Banda Aceh), the rough mountains north of Lamno and cultivated sections of the West Coast, especially around Meulaboh' (Carpenter, 1938).

Carpenter accepted the report of Captain B. Veth, which stated that the southeastern boundary of distribution could be imagined as 'a line drawn from Singkil on the west coast to Pangkalan Brandan on the east coast.' He considered the northern boundary to be an imaginary line between Lamno on the west coast and Sigli on the east. It is curious that Carpenter decided on this line, which had been suggested to him by a Lieutenant Bloem, because, according to the collection of original reports which he consulted, Captains L.H.J. Koprogge and A.G.W. Navis had reported orang-utan sightings in the upper valley of the river Atjeh and in the northern environs of Sigli[11].

Carpenter was apparently unaware of the publication of Heynsius-Viruly and van Heurn, and overlooked the large southwestern extension of the range on the eastern bank of the river Alas (Simpang Kiri) south to the coastal township of Barus, as well

---
[11] The original reports are preserved in the library of the Van Tienhoven Foundation.

as the isolated relict range in the southeastern section of the Tapanuli district (i.e. Angkola). In hindsight, it seems that Carpenter's limited view of the Sumatran orang-utan distribution range was to set the scope of international and national nature conservation with reference to orang-utans for many decades to come.

In the early 1970s, the Indonesian forester K.S. Depari reported that orang-utans still occurred in the forests along the river Batang Toru, near Hutabaru and in the Rimbo Panti wildlife reserve (Rijksen, 1978). Borner (1976) noted that a sub-adult male had been shot in the reserve (in 1973), and that people reported having seen at least two other such apes in the region[12]. Interestingly, the newspaper reports at that time insisted on 'black' orang-utans which had been seen 'running along the ground', like the one that was shot.

The primate survey conducted by C. Wilson and W. Wilson in 1972 not only confirmed the presence of orang-utans in southern Tapanuli, but also added two remarkable records to the possible distribution range of the ape in Sumatra (Wilson and Wilson, 1973). Local informants reported that orang-utans occurred in the swamp forests near Kota Badak (north of Pakanbaru) and near Sungei Kampar (south of Pakanbaru, in Riau province).

In 1986, the French forest ecologist Y. Laumonier sighted an orang-utan and numerous nests in the forests on the slopes of Gunung Talamau (Gn. Ophir) (pers. comm.). The forest complex on this, and the surrounding mountains, was, until recently, contiguous with the Rimbo Panti wildlife reserve and adjacent Protection Forests in the regency Pasaman (i.e. the northwestern section of West Sumatra province on the equator), as well as with the Baruman reserve along the border with Riau province. It is the habitat of the southern extension of the Angkola population, and was formerly famous for its *orang pendek* sightings.

In conclusion, until Indonesian Independence (1945) the Sumatran orang-utan was believed to have survived in four isolated populations:
(1) the largest contiguous population was to be found in Aceh province, from Singkil on the west Coast, along the river Alas (Simpang Kiri) north as far as the upper Alas valley (near Aguson) and contiguous with the eastern Barisan mountain chain in Langkat north of the river Wampu;
(2) a much smaller population survived on the eastern side of the lower river Alas, along the headwaters of the river Simpang Kanan, southwards along the western slopes of the Toba plateau from the coastal town of Barus on the west coast, and northwards to the highlands of Sidikalang and along the western side of the river Renun (including the Dolok Sembabala I and II – the Dolok Sembelin – and the Batu Ardan or Dolok Sopolaklak Protection Forests);

---

[12] In 1996, this event was still fresh in the mind of the people in the village of Panti, which indicates the tremendous impression this rare occurrence must have made. Although a relatively high incidence of nests indicates that orang-utans still occur in the area, they are apparently never encountered by people entering the forests.

A history of orang-utan distribution

(3) a relatively small population in the mountainous area of Habinsaran county, between the river Asahan[13] and the current Padangsidempuan – Gunungtua – Langgapayung-Rantauprapat road or the upper reaches of the river Barumun;

(4) a small population (?) at the headwaters of the river Rokan, i.e. the area currently known as the Baruman wildlife sanctuary and Protection forest, although this record was not indicated on the accompanying map. This area also featured prominently in stories of sightings of the legendary *orang-pendek*.

Since the 1970s a surviving Angkola population has been added to this list, while several new records indicated that the ape still occurred much farther south (east) than had been previously assumed (Wilson and Wilson, 1973; Rijksen, 1978, 1982).

The most devastating impact on the historical range of the orang-utan in Sumatra probably took place during the first two decades of the twentieth century, when the vast lowland forests in northern Sumatra east of Lake Toba and the Serbolangit range,

A dark, slender-type orang-utan adolescent walking along a flat surface; note the open position of the left hand and the plantigrade feet. The youngster was confiscated near Tarutung, south of Lake Toba, in 1973.

---

[13] A male orang-utan was seen in the forest near the Asahan waterfall as late as 1971.

in Deli and Langkat, were converted, first into tobacco-, then into rubber- and, more recently, into oil-palm plantations. Well into the 1930s the majority of orang-utans captured for international trade originated from Langkat.

Other major impacts have been large human influxes into the Alas valley (in the 19th century, and again in the 1960-70s), and the massive forest conversion, reclamation and transmigration projects in the South Aceh regency east of the river Alas, and the Bahbahrot – Meulaboh swamp forests on the west coast, dating from the late 1980s. In this period much of the forests in the eastern Tapanuli (Habinsaran) region and the extensive lowlands of Riau province also became subjected to conversion. The full extent of the massive destruction of wildlife in these areas remains unknown.

The orang-pendek enigma

During the first decades of the twentieth century, when interest in wildlife gained momentum in colonial circles in Sumatra, reports of sightings of a mysterious forest-dwelling creature, named *orang pendek* (i.e. short person), began to emerge in local newspapers, as well as in the Natural History magazine *De tropische Natuur*.

Also known in other regions as *letjo* (Rokan district), *umang* (Karo and Pakpak district), *manteu* (Singkil, westcoast Aceh), *si-bagau* (Painan, Barisan), *sedapa*, *sebaba* (Banyuasin, north Palembang), *se-goegoe* (south Bengkulu, Ulu Ogan), *slimor* (Barisan selatan), *atoe rimboe*, *atoe pandak* (Rawas district) or *li koyaoe* (upper Batang hari area), the *orang-pendek* was commonly described as being of short stature (between 80 cm and 1.5 m tall), with brown, dark brown or black shortish hair on a tan or dark brown body, and the hair 'of the head or the shoulder blades hung down long along the sides of the back.'

It was said to have no tail, long arms which, when the creature was in a standing position, hung down to the level of the knees, and short legs. Its face was 'naked', of a brown colour, with a small nose and a high forehead; its ears were human-like, and its wide mouth had remarkably large canines and large teeth of a yellowish or whitish colour. Its hands were human-like, long, and covered with short hair on top. It was usually encountered on the ground, near shallow streams, searching for grubs in dead logs or for 'shellfish' (probably snails) in the shallow water. Its diet supposedly also consisted of 'vegetables, fruit, snakes and worms.' It was said to be most commonly encountered wherever durian trees (*Durio spp.*) had ripe fruits, deep in the forest along water courses, although the creature was also reported to steal sugar cane and bananas from gardens near the forest edge. One reporter insisted that the orang-pendek could best be met near a durian grove during a moonlit night (Dammerman, 1932); another described how one had stolen a bowl of rice from a shed, ate the contents and then returned the bowl.

When encountered, it appeared to be extremely shy; it was seen standing upright, with 'extraordinarily long arms, holding on to a small tree' (Westenenk, 1962). One observer subsequently saw it climbing up, others usually reported that it fled along the ground, to disappear in the dense undergrowth; during its retreat it made a hissing sound or something which sounded like 'hoe, hoe' (Westenenk, 1919; Jacobson, 1917; Maier, 1923; van Herwaarden, 1924; Dammerman, 1924).

All reports stress the very remarkable feet, although it is unclear whether anyone actually described his or her observation of the actual feet, rather than an interpretation of the alleged 'footprints.' The feet, then, were said to be turned backwards with the heel to the front, leaving a print which is human-like but *has the big toe on the outside*, rather than in the normal position. Some reports of plaster casts also mention *both* a 'remarkably curved sole' *and* 'flat-footed', and the different footprints of one set were once even supposed to be mixed from two individuals, one small, the other larger.

> In cases where only plaster casts of alleged *orang-pendek* footprints were presented for scrutiny, they had sometimes actually been taken from tracks of a Malay sun bear (*Helarctos malayanus*) or a human. As head of the Zoological Museum in Buitenzorg (Bogor), Dammerman (1932) even once received a series of photographs and the skeleton of an alleged *orang-pendek* which had been shot and collected in the Rokan district (east coast of Sumatra). Analysis of the skeleton and pictures, however, revealed that the specimen was a hoax, prepared from a leaf monkey.

Although many authors explicitly denied the possibility that the creature they described could be an orang-utan, it is difficult to find better accounts of early twentieth-century meetings with the red ape, when one considers the descriptions of the physiognomy and behaviour of the alleged *orang-pendek*. Even the descriptions of the sounds closely match; the common 'squeak' vocalisation (i.e. 'hoe, hoe') and the 'kiss sound' (i.e. the 'hissing') are both typical 'contact vocalisations' of orang-utans (Rijksen, 1978). Furthermore, some of the locations where the creature reportedly occurred concur with what is now known to be within the historic or present distribution range of the orang-utan, e.g. Rokan, Karo and Pakpak districts. The only facts which may seem to be at odds are the names, the ground-dwelling habit, the peculiar footprints and the locations far beyond the conventionally known distribution range. After close scrutiny of plaster casts of such footprints, some zoologists actually concluded that if the *orang-pendek* were perhaps a mythical memory of the orang-utan (Dammerman, 1938), its prints were most probably made by the Malay sun bear (Dammerman, 1932).

Before the 1930s, when the German zoo director G. Brandes provided an accurate description of the normal physiognomy and behaviour of the wild-caught adult orang-utans that he kept in special facilities at Dresden, very few people, both colonial and local, had ever seen a full-grown orang-utan or knew anything of its normal physiognomy and behaviour. If western people ever sighted an orang-utan, even in Indonesia, it was usually a juvenile which was either kept as a pet or caged in a zoo, and all such apes died before they reached sub-adulthood. Such young apes usually had short hair due to poor conditions in captivity, and were rarely in a position to demonstrate normal behaviour.

It is probable that those colonial people who described an encounter with *orang-*

pendek had never seen a wild orang-utan, but did have a reference image of a caged ape. Hence, they had conventionally learned that it was an obligatorily arboreal creature of such fierce temper that it was likely to attack upon encounter. Such a false conventional image hardly meshed what they came upon in the wild: a short creature scurrying away along the ground. Thus they denied the possibility that what they in fact could have seen was a ground-dwelling orang-utan.

It is perhaps even more important to realise that, until recently, indigenous tribes in Borneo and Sumatra had never used the name '*orang utan*' for the red ape (Dammerman, 1938; Rijksen, 1978). Their name for the ape was quite different from that of western convention: Indeed, the name by which the red ape is known in the western world was a concoction dreamed up in the seventeenth century to explain something of the identity of the Bornean *mias, keo, kuyang, kaheyu*, or whatever, to the Dutch physician Jacob de Bondt. Since *orang-utan* in Malay parlance is usually applied derogatorily to indicate a person's rude, if not bizarre, attitude, it is likely that *orang pendek* is a less offensive transcription of a local name for the ape. Now that many of the more educated Indonesians employ the modern, conventional name, it is striking that some reports from timber concessions in the remote regions where the *orang pendek* allegedly occurred record in a straightforward manner nothing but the presence of orang-utans, e.g. from southern Bengkulu province.

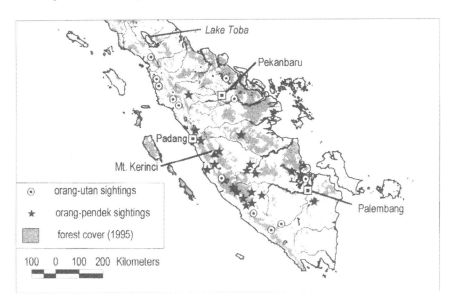

Reported sightings of orang-utans and orang-pendek in Sumatra south of Lake Toba; data based on the present surveys, and information from the literature since 1917.

Nevertheless, the enigma of the lopsided, backwards footprints remains. There is no mammal known which has a foot structure that deviates in the manner described for the *orang-pendek*. As a matter of fact it is highly unlikely, from an evolutionary perspective, that such a limb structure would have evolved. However, the prints are clear. The question, therefore, is: How can a normal limb structure produce such a print?

The answer is readily given by some orang-utans, in particular those of the delicate, dark, long-fingered type. What is considered to be the footprint, is in fact probably the imprint of the hand, which is kept flat with folded fingers, but turned backwards with the wrist pointing in the walking direction, the bent fingers pointing obliquely backwards, and the thumb pointing outwards. Delicate-type orang-utans often walk that way, especially on soft soil and mud. The robust-type Sumatran orang-utan (as well as its Bornean kin) more commonly walks with a clenched fist on the ground, which allows the wrist to remain in a normal position, with the thumb inwards[14]. The feet of all orang-utans produce a hooked or curved image (see MacKinnon, 1974), and since much of the ape's weight falls upon the arms, the feet may in fact leave a much weaker imprint in soft soil, so that such prints may be readily overlooked. Perhaps most important, however, is that the recent searches for orang-pendek have also recorded footprints which are well within the usual range of differences in orang-utan prints.

Footprint of an alleged orang-pendek from Central Sumatra as collected by the team of Debby Martyr and Achmad Yanuar, and which was published in the Sunday Times of Oct. 12th, 1997.

Thus, considering that the presence of Sumatran orang-utans in areas hitherto unknown, i.e. at least as far south as the equator, was confirmed during the present surveys, and that these apes reportedly have a higher incidence of ground dwelling,

---

[14] Perhaps the difference in the hand position, either open with the palm touching the soil, or closed with the backs of the bent fingers touching the ground, is primarily an adjustment to increase support; an open hand does prevent the hand from sinking into the (swampy or muddy) soil; note that the Himalayan 'Yeti' or *mirgula* also produces back-to-front footprints.

it is presumed that most of the more accurate reports of *orang-pendek* in fact refer to orang-utans. It is intriguing that such orang-utans may well have a higher incidence of ground dwelling, may be predominantly of the delicate, long-fingered variety and have a more crepuscular existence than their brethren in the north.

ECOLOGY AND NATURAL HISTORY

Feeding ecology

Since the earliest reports, it has been known that orang-utans are fruit eaters. Frugivory has important implications for many aspects of an animal's biology and in particular its way of life. Thus, the quantitative and qualitative distribution of the ape's fruit diet in time and space will be a major determinant of its ranging behaviour, population density and, consequently, social organisation. Field researchers during the 1970s habituated for the first time small populations of resident apes in particular, well-studied areas, and since then biologists have acquired some insight into the complex lifestyle of the orang-utan.

TABLE II

A comparison of field data on feeding behaviour in feeding-time percentages. B = Borneo; S = Sumatra.

Field researcher	Isl.	Fruits	Flower	Leaf	Bark	Insect	other
Rodman 1973/'77	B	54	2	29	14	1	0
Galdikas '78	B	61	2	21	11	5	0
Wheatley '82	B	59	1	19	18	3	0
Suzuki '89	B	61	0	25	13	1	0
MacKinnon '74	B	62	1	24	11	1	1
MacKinnon '74	S	85	-	10	4	1	-
Rijksen '78	S	58	1	25	3	14	1
Sugardjito '86	S	65	1	16	11	7	0
van Schaik (unpubl.)	S	68	0	13	1	17	0

Thus, the main staple of the orang-utan diet is fruit. In good quality habitat, between 57% (males) and 80% (females) of all recorded feeding time is spent feeding on fruits (Sugardjito, 1986). The lowest recorded foraging time for fruit, during periods of extremely low fruit availability, is still 16% of the total (Galdikas, 1978). Although a wide variety of species (e.g. about 200 at Ketambe: Djojosudharmo, pers. comm.) of fruits is eaten, some types of fruit are over-represented in the ape's diet: Orang-utans have a clear preference for fruits with soft pulp, arils or seed-walls around the seed, including drupes and berries (Dammerman, 1937; Djojosudharmo and van Schaik, 1992). Orang-utans also prefer fruit trees with large crops (Leighton, 1993). Both preferences mean that figs, especially those from large-crowned stranglers, comprise the staple food wherever and whenever they are available.

However, the fruit diet is supplemented by considerable quantities of leaves, both young shoots and mature leaf material. Taking together all of the data, it can be seen that feeding on leaves constitutes 25% of the feeding time on average. When fruits become scarcer, the incidence of leaf eating increases. Still, even in good quality habitat, during the fruiting season between 11-20% of feeding time (Sugardjito, 1986) is spent every day in the consumption of leaf material.

In some places leaf (and stem) material appears to be the survival food for some orang-utans during periods of low fruit availability. For instance, they often eat the growth shoots of rattan species and the stems of creeping Araceae, which are chewed into a wadge. In Borneo, the young leaves, top shoot (or "heart") and inner meristematic tissue of the mature leaf stem of the locally abundant palm *Borassodendron borneensis* are an extremely important survival food source during periods of fruit scarcity (Dransfield, 1972; Rodman, 1977; 1988; Leighton and Leighton, 1983; Frederiksson, pers. comm.). Local people in the northern section of Central Kalimantan insist that orang-utans can only survive in areas where the *Bendang* palm (i.e. *B. borneensis*) occurs in considerable numbers (Dransfield, *ibid.*). After a lean fruit season the stock of *Bendang* palms is seriously damaged by orang-utans (Rodman, 1977).

In addition to the consumption of fruits and leaves, an average of some 6% of the feeding time may be spent on collecting insects (ants, termites, leafhoppers, crickets, bugs, etc.), every day; adolescent and sub-adult individuals may even spend as much as 24% of their daily activity in insect hunting (Rijksen, 1978). During certain periods, foraging for insects seems to gain in importance and may constitute much of an ape's daily activity pattern. This may well be related to a relatively greater availability of insects, rather than a shortage of fruit.

In times of fruit scarcity, and as a minor part of the regular diet as well, the ape may spend up to 18% of its feeding time (average 12%) eating the growth layer under the bark of some particular trees, notably *Ficus*, but also other Moraceae and (e.g.) *Payena* spp.

Certain flowers may also feature as an irregular item of diet, notably *Bombax valetonii* (in Sumatra) and *Madhuca sp.* (fam. Sapotaceae, in Borneo).

An occasional bird's nest may be looted for its eggs, and a small vertebrate (e.g. geckos, squirrels, and the slow loris[15]) may be added to the diet whenever it happens to be encountered in such a manner that it can be easily grabbed.

Orang-utans are fond of honey and, despite the inevitable attack by angry bees, will try to rob any bees' nest they can reach, often shielding their face with a leafy twig picked for that purpose. In conclusion, the red ape is a typical gatherer, or opportunistic forager (MacKinnon, 1974; Galdikas, 1978), readily using tools

---

[15] In 1981 Sugarjito and Nurhada observed an orang-utan devouring what they thought was a dead infant gibbon, but which was probably a slow loris. In the 1990s the capture and eating of slow loris individuals was observed several times by Ms. Sri Suci Utami at Ketambe (Azwar: *Orangutan kanibal di Gunung Leuser*, Mutiara 03.12.1993) and by Mr Bahlias P.G., at Suaq Balimbing, in Sumatra in 1994 (van Schaik, pers. comm.).

whenever required[16] (Rijksen, 1978; Van Schaik *et al.*, 1996). Yet, unlike the chimpanzee, there are so far no indications that the orang-utan is an active hunter.

The orang-utan diet may vary markedly from month to month (MacKinnon, 1974; Galdikas 1988; Mitani, 1989), but in some places there is an almost continuous availability of some high-quality fruit and the variation is remarkably small (van Schaik, pers. comm.). Indeed, the diet as well as the variation is significantly different in different research sites (Rodman, 1988; see also Table II). Whereas figs of at least eight strangling-fig tree species were the main staple food and available during at least eight months of the year (Rijksen, 1978; Sugardjito, 1986) in one area, in other areas figs are an insignificant food source (e.g. Tanjung Puting, Galdikas, 1978; and Suaq Balimbing, van Schaik, pers. comm.), and strangling-fig trees are actually rare.

In some regions orang-utans also occasionally ingest quantities of soil, sometimes eating sections of the tubes of soil produced by termites along tree trunks, at other times descending to the ground in order to pick up and eat clumps of uprooted earth. The apes may also frequently visit the typical 'mineral licks' or *uning* where many other mammals convene to eat of the clay or sandstone-like walls (MacKinnon, 1972) of a cliff or earth depression. Apparently these soils contain either some important minerals (Payne, 1988), or a high concentration of kaolin, which is, at times, an essential requirement to neutralise the high quantities of toxic tannins and phenolic acids in the vegetarian diet. In other regions, however, such as swamps, orang-utans have not been observed to ingest soil (van Schaik, pers. comm.).

Table II suggests that Sumatran orang-utans tend to eat more fruit and insects, and less 'bark' than those in Borneo. Other differences between the two subspecies will be discussed shortly.

---

[16] Tool use by captive orang-utans is well known, and in experimental situations the red ape deploys tools as intelligently as, if not more so, than the chimpanzee (Lethmate, 1977); tool use by wild orang-utans has been observed during locomotion ('fishing' for a branch or liana which was out of reach), during insect foraging (Rijksen, 1978), honey collection (Van Schaik *et al.*, 1996), protection against bees, and for flushing out of tree-holes young squirrels and geckos (Rijksen, unpubl.) which can be caught and eaten. Insect foraging and the opening and processing of ripe *Neesia* fruits by orang-utans in the swamp forest of the southwestern Leuser Ecosystem (Van Schaik *et al.* 1996) has also been recorded as a cultural routine.

Patchy orang-utan distribution

Altitude

The orang-utan is a lowland species, and the local communities or demes of apes can be found in their highest densities between sea level and at altitudes of some 200-400 m (Payne, 1988; Van Schaik and Azwar, 1991). In Sumatra, however, orang-utans, and in particular adult males, may occasionally be encountered on mountain slopes as high as over 1,500 m but, again, the occurrence of demes at altitudes over 500 m is increasingly rare. In Borneo, the altitudinal limit of orang-utan demes seems to be at about the 500 m contour (Groves, 1971).

Nevertheless, it is noteworthy that the limit is not simply altitude. Some highland plateaus of around 1,000 m in Sumatra (e.g. the Kappi area in the Leuser Ecosystem) seem to support a population of apes, including reproductive females. And in Sabah, Payne (1988) reported 'two mountain populations concentrated between 700-1,300 metres above sea level.'

It is possible that this altitudinal limit is a reflection of the presence of preferred food types rather than of climatic factors. The abundance of fleshy-pulp fruits declines strongly with altitude (Djojosudharmo and van Schaik, 1992). Yet, the incidence of nutritious acorns and chestnuts (*Lithocarpus, Castanopsis, Trigonobalanus spp.*) increases sharply up to an altitude of 2,000 m, and even some important strangling-fig trees (e.g. *Ficus drupacea*) still grow in this zone (Corner, 1978). It is important, however, to note that the higher altitudes on a single mountain slope may well have wetter and colder nights than those in larger mountain complexes, due to the *Massenerhebung* effect. Evidently, cold and frequent rain are environmental conditions disliked by the ape.

Even in the lowlands the orang-utan is not evenly distributed. A review of the available literature on orang-utans, supplemented by survey data from many different sites in Sumatra and Borneo (Rijksen, unpubl.), indicates that the ape is more commonly found close to streams, rivers and in swamps; it is found in greatest densities in (alluvial) forest patches in river valleys, and in the (flood-plain) peatforests near swamps, or between rivers. One is less likely to come across the ape in any numbers further away than 10-15 km from a watercourse or a swamp with open water. No doubt the main reason for such preference is the higher incidence of preferred fruit trees close to rivers, but an additional reason may be that rivers and streams are the best geographical landmarks for spatial orientation. It seems that the ape is as rare or absent in extensive, relatively uniform lowland forests on well-drained 'dry' flat ground, as it is in some of the mountain complexes above a certain altitude.

Habitat distribution

Contrary to a first general impression, the tropical rainforest is not of uniform composition. It is perhaps best described as an immense patchwork of plant communities of different composition, comprising many thousands of species. Some patches of a particular composition are large, others small, the patches being imaginary entities in the eye of the beholder. Within this patchwork, an orang-utan is estimated to be primarily interested in plants which provide food, especially trees and lianas with the 'right' kinds of fruit. Such food sources are obviously not randomly distributed over an area of forest.

Orang-utans appear to be unevenly distributed in an expanse of forest, both in time and location. Yet they seem to be most often present in alluvial forest and other ecotones[17]. Why they display temporal preferences for certain areas is probably explained by the quantitative and qualitative distribution of their main food sources in time and space. In particular the time factor is of crucial significance here: For survival an orang-utan population is dependent on a composition of trees and lianas which provide food during a continuous sequence of productive seasons the year round, and within a reasonable travelling distance. For a frugivore like the orang-utan, a forest area can therefore be differentiated into a number of patches of different quality. The largest expanse of forest may in fact hardly be of interest in this respect, because it probably has a low density of food sources, which are widely spaced and commonly produce food during one season exclusively. The patches of greatest interest are undoubtedly those characterised by such a diversity of food sources that fruit productivity is sequential, covering most, if not all, of the year.

The composition of forest patches is related to ecological factors and history. Long-term ecological research in untouched primeval rainforest (Bruenig, 1996) revealed that plant-species richness and the composition of a particular plant community fluctuates. It was found that a current richness of species in such rainforest is the result of the specific history of locations and is closely related to contemporary edaphic, climatic and atmospheric conditions. Plant-species richness is related to, and limited by, soil conditions along a gradient of declining rooting depth[18], aeration, moisture, nutrient status and humus quality. The ecological gradients – i.e. 'ecotones' – between water, dry land and (the foothills of) mountains are sites where conditions for the diversity of fruit-bearing species in particular are highest.

The drier conditions away from the rivers seem to prevent the development of species richness. Here the forest is commonly dominated by extensive aggregations of long-lived, high-rising mature hardwood trees (Dipterocarpaceae). The regular deposition of floodwater-borne nutrients is virtually absent, and many of the trees are dependent for their survival on atmospheric deposits and on symbiosis with fungal mycorrhizae, thus efficiently recycling the nutrients lost by the periodic shedding of leaves. These trees typically bear wind-dispersed seeds, produced during

---

[17] An ecotone is an interface where two areas of different ecological (e.g. geomorphological, hydrological) conditions meet; a 'border zone between two habitats' (Fitter and Fitter, 1967); alluvial areas, such as flood plains or a river valley, are also ecotones.

[18] The 'best' soil types, supporting the highest species diversity, are deep ultisol/acrisols and humult ultisols; diversity declines when a humus podzol becomes shallower (Bruenig, 1996).

characteristically erratic ('mast') fruiting seasons which occur at very long intervals (between 4 and 15 years). In this way the trees apparently conserve reproductive energy over several years before producing massive quantities of fruits that are less attractive for animals like the orang-utan (Ashton, 1964; 1976). In a comparison of the West Malaysian and the Central Bornean (mountain) rainforest, Mather (1992) attributed the significantly lower primate biomass in the Bornean forest to the high incidence of Dipterocarps in relation to other tree families.

In a typical Malesian rainforest, which contains patches of orang-utan habitat, any one square kilometre may harbour up to a thousand different tree species[19], in addition to over a hundred species of woody lianas, and a large number of herbaceous plant species. MacKinnon (1974) noted that tree-species diversity appears to be greater in valleys than on hills.

> If certain minerals are in short supply, and there is no water-borne deposit every now and then, the competition will be so severe as to allow only a limited number of well-adapted species to establish themselves. The typical symbiotic communities of Dipterocarps and Mycorhizal fungi are fascinating examples of such adaptation (Smits, 1994). Moreover, some extensive dryland areas of Borneo are so poor in soil quality, e.g. almost pure sand (silica) outcrops, that the (primary) forest is typically stunted. Such forest has hardly any edible fruits and typically reproduces after bouts of several years (i.e. mast), when sufficient resources have been conserved. The Iban Dayaks call such forest *kerangas* (Whitmore, 1982), and never cultivate it because they know it to be unsuitable for agriculture and horticulture. In such deprived forests an orang-utan is not likely to survive. Still, it is perhaps not so readily understood that from the viewpoint of the orang-utan, the extensive stands of commercially valuable Dipterocarp trees on dry land can also be 'poor' forest habitat.

Of the tree species in the patches of interest to human forest-dwelling gatherers, approximately 10-19% bear fruits of nutritious (and commercial) value (van Valkenburg, 1996). Such patches are also of prime interest to orang-utans. In the alluvial swamp forest of Tanjung Puting (Central Kalimantan) between 54 and 60% of all trees (> 10cm DBH) reportedly are potential food sources for orang-utans (Galdikas, 1978), although no more than 8 – 17% of the identified trees are mature enough at any time to actually provide fruit in significant quantity. In general, good-quality orang-utan habitat comprises an association of trees and lianas of which 30-50% provide fruits of nutritious interest to orang-utans. However, where the variety of species is high, the number of individuals of any one tree species per unit of space may be limited; indeed, some tree species may be found in such low densities that the distance between individuals is quite commonly over five kilometres (e.g. *Heritiera elata, Bombax valetonii*) (Rijksen, unpubl.).

Of 28 major fruit tree species important for orang-utans in the Segama forest (Sabah), MacKinnon (1974) found four to occur 'frequently' (i.e. up to eight/ ha), while eighteen occurred in a density of less than two/ ha. In alluvial orang-utan habitat

---

[19] The 50 ha permanent plot at Pasoh in West Malaysia contains 660 species (224 genera in 67 families) of trees with a diameter at breast height (DBH) of 10 cm or larger (Soepadmo, 1995); the forests of Sumatra and Borneo are usually slightly richer in species; any sample plot has between 100 and 250 species of trees/ha of > 10cm DBH (M.M.J. van Balgooy pers. comm.; van Valkenburg, 1996) and between 250-350 species of smaller diameter (Whitmore, 1995; Bruenig, 1996).

at the Ketambe area (Sumatra), some 56% of the food tree species were 'dispersed', that is, occurring in a density of less than one tree/ha, while four (of the 52) species were 'common' (i.e. more than nine trees/ ha) (Rijksen, 1978). Interestingly, also in chimpanzee habitat in Africa, Ghiglieri (1985) has found in the Kibale forest (Uganda) that nine of the twelve most important fruit tree species ('bulk resources') were distributed in a non-random, irregular pattern, more or less grouped together.

Thus, in the amazingly diverse tropical forests of the Old World, some patches contain a number of tree species which occur in a non-random and irregular distribution pattern, and remarkably high densities. Hubbell (1979: 1304) noted that 'when total population size [of trees] is considered, the trend is towards increased clumping with decreasing abundance.' In Sarawak, Newberry *et al.* (1986) found that almost 50% of the most abundant (64) tree species in their sample of some 16,000 trees occurred in a 'clumped' pattern, the scale of which matched the common size of gaps (Bruenig, 1996). Virtually all these species bore fruits which were dispersed by animals. Indeed, in some areas of rainforest one can find groves of anything up to approximately one hectare in which fruit trees are so predominant that they appear to have been propagated by a gardener. No doubt the seed-dispersing effect of orang-utans (and other large animals) gives a selective or 'horticultural' vector to the natural succession – making them in effect 'garderners' or 'cultivators' of much of their own provisions.

> There is evidence that orang-utans, over several generations, maintain a high incidence of certain fruit trees (e.g. *Aglaia* spp., *Aporusa* spp., *Baccaurea* spp., *Dracontomelum* spp., *Elaterium* spp., *Ficus* spp., *Mallotus* spp., *Mezettia leptopoda*, *Tetramerista glabra*) in their habitat (MacKinnon, 1974; Rijksen, 1978; Galdikas, 1982; Adriawan and Senjaya, unpubl. report 1986). Galdikas lists twenty-three species which were effectively dispersed in faeces, and another twelve which were carried over some short distance and as intact seeds were discarded from the mouth (Galdikas, *ibid.*). It is not coincidental that all the recognised bulk resources are among these species; these fruit trees have adapted to produce seed of the 'right' size (1-2 cm diameter) with a nutritious fleshy arillus or fruit wall, to be ingested undamaged and consequently dispersed by orang-utans. Indeed, many of the seeds will not germinate unless their seed-wall is scarred or they have been subjected to the chemical impact of the alimentary tract. In more generalized evolutionary terms, such fruit trees seem to represent a typical adaptation to medium-quality environmental disturbance, where wind, water and large animals may cause calamities[20] on a localized scale.

In more general terms, one may expect up to no more than 15% of the most important food tree species to be 'common' in good-quality habitat. That is, such tree species occur in an incidence of nine or more trees/ha.

It is believed that when in permanently wet (swampy) conditions such habitat must have at least 40 species of food trees, and at least 60 when in dry alluvial conditions. However, a good habitat for orang-utans is not simply composed of trees: a considerable number of the preferred fruits (and leaves) in the orang-utan's diet are provided by lianas, a form of vegetation which can constitute up to 10% of allflowering plants in the Malesian region (Richards, 1952). In some areas the fruits and

---

[20] For a fruit tree, a harvesting visit of orang-utans may indeed be a calamity because the apes are usually very destructive gatherers. At Ketambe the highly preferred *Durio oxleyanus* and *Heritiera elata* trees required at least six years to recuperate and bear fruit again (see also Leighton, 1982. 145). Under domestic conditions a durian tree produces fruit annually or once in two years.

leaves of lianas comprise as much as some 17% of all food plants in the diet (Rijksen, 1978). Moreover, lianas are a major means of arboreal transport for the orang-utan.

Habitat patchiness; every hectare (i.e. 100x100 m) has a different composition of plants, due to geomorphological and ecological conditions.

The best known staple resources for orang-utans are *Tetramerista glabra* (Meliaceae), *Sandoricum beccarianum* (Meliaceae) and *Neesia cf. glabra* (Bombaccaceae) (Van Schaik, pers. comm.) in peat- and flood-plain forests in Sumatra, while *Tetramerista glabra*, *Gironniera nervosa* (the rough 'laurel', related to the Moraceae and Urticaceae) (Galdikas, 1988), *Lithocarpus* sp. (Fagaceae) and *Nephelium* sp. (Sapindaceae) (A. Sebastian, pers. comm.) are likewise favoured in similar wet environments in Borneo. It is interesting to note that many of the fruits in flood-plain forests appear to have the appropriate qualities favoured by orang-utans: large fruit size, large crop size, highly nutritious and with sweet or fatty pulp. In the drier alluvial valleys the strangling-figs (*Ficus* spp.) and several members of the Meliaceae (notably *Aglaia* spp.) as well as Euphorbiaceae (notably *Baccaurea* and *Mallotus* spp.) are the staple resources.

In conclusion, the optimum habitat of an orang-utan deme comprises at least two major geomorphological landscape types, namely, a freshwater fringe and an adjacent dry upland region. The water fringe may be a flood plain, a swamp or an alluvial valley, the upland area is usually a foothill region. Both types must of course be sufficiently large, and accessible within a reasonable distance, i.e. less than five kilometres. Good orang-utan habitat is usually a mosaic of patches of different degrees of woody plant diversity in which some have a remarkably high density of fruit tree species (>20% of all trees).

It is, however, interesting to realise that once a forest is established its plant community as a whole, but especially its canopy structure, has a considerable regulating influence on edaphic and climatic conditions. One can even say that it 'nurtures' its own biotic structure, maintaining a narrow range of climatic variation, conserving nutrients as well as water, and providing locations and conditions where food for (e.g.) a frugivore can be found almost the year round.

Temporal variation in food productivity

Plants commonly produce their seeds and fruits during intervals which are related to seasonal – and/or ecological – conditions. This may imply that all the fruit trees in an area have a synchronised fruit production pattern. Indeed, such mass production may be advantageous in an evolutionary sense. Conversely, a frugivorous animal could enjoy a blissful abundance during this productive season, but would have to face starvation until the next. Fortunately for the frugivores, such a simple seasonal pattern is not common in tropical rainforests.

Under a regime of seasonal availability of food, the frugivore is obliged to move. Indeed, several animal species in the Malesian rainforest have a nomadic existence, and make long-distance movements from one seasonal crop to another (e.g. the bearded pig *Sus barbatus*, the pig-tailed macaque *Macaca nemestrina*, flying foxes *Pteropus vampyrus*, and several hornbill species, but in particular the gregarious wreathed-hornbill *Rhyticeros undulatus*). A considerable proportion of an orang-utan population may thus also be forced to migrate, although in some areas at least several individuals appear to settle as residents. It is also remarkable, however, that most other frugivorous primates, such as gibbons and long-tailed macaques, can live year-round in home ranges of less than 1 km^2 (e.g. Chivers, 1980; Palombit, 1992).

The (maximum) number of individuals of a wild animal species that can live in a particular area is determined by the carrying capacity of the habitat, that is, for orang-utans, the timely productivity of the vegetation in terms of food, and the availability of safe resting sites. A relative shortage of food will incite competition, and subordinate members of a deme will need to search for resources in other areas or to accept alternative resources, otherwise they will perish. Thus, when other basic requirements (e.g. water, resting sites, etc.) are in ample supply, the carrying capacity is set by the lowest sustainable level of food availability for a particular number of individuals at any period of time.

Fruit productivity is related to the density of fruit trees, the fertility of the soil, and the duration and frequency of climatic seasons. For the frugivorous orang-utan, its habitat should preferably have a considerable number of fruit trees with regular, disjunct and frequent fruiting seasonality, which are spaced within a reasonable travelling distance.

In the eastern sector of Borneo two seasons clearly predominate: one wet, the other dry. In Sumatra, however, and in the western half of Borneo, two to four annual seasons may be obvious in the open cultivated areas and urban centres, but these are usually not very pronounced in the rainforest areas. There the differentiation of seasonality is usually so subtle as to be almost indiscernible for the human observer, except in years with exceptional climatic patterns (e.g. 'El-Nino years'). Richards (1952: 199) drew attention to the fact that 'In evergreen tropical forest flowering generally extends throughout the year and there is no season in which a proportion of the species are not in flower, some of them blossoming almost continuously; but even in the least seasonal climates there are maxima of flowering at certain times of the year (...)' (see also Fogden, 1972: 340). Indeed, in the Sumatran and west Bornean rainforest the climatic conditions are commonly fairly equitable the year round, especially where (ground) water is plentiful (e.g. near rivers and swamps), so that the tree community follows a similar 'phenological anarchy,' as has been described for the habitat of chimpanzees in the Kibale forest (Uganda) (Ghilieri, 1985).

Seen from another perspective, one might also say that, in Sumatra and western Borneo, several species of trees, and even individuals within species, have their own independent fruiting periods. In particular many strangling-figs seem to have their 'private' productivity rhythm and may produce fruit in two or more bouts per year, unrelated to the seasonality of other fruit trees and other strangling-fig individuals, even of the same species (Rijksen, 1978; Van Schaik, 1986). Perhaps the most important staple fruit-bearing tree for the orang-utan in swamp forest, *Tetramerista glabra* (Tetrameristaceae), is known to produce fruit the year round (Soerianegara and Lemmens, 1993).

> The Ketambe area in Aceh (Northern Sumatra), which can be typified as a Meliaceae forest (Abdulhadi and Kartawinata, 1982), contained at least 91 different species of fruit trees and lianas, of which 16 were strangling-figs (*Ficus* spp.).
> Nevertheless, during the leanest period at Ketambe, i.e. November-March[21], anything between 20 – 30 trees of at least four species were found to bear fruit; at no time were fewer than 17 fruit-bearing trees discovered by the orang-utans. The strangling-fig trees at Ketambe did not follow the seasonal regimes of the other fruit-bearing species. The population of strangling-figs in fact displayed a wave-like productivity pattern in which some 2-3 months of low productivity levels (5-15% of the trees in fruit) alternated with months of high productivity (20-28% of the trees in fruit) (Sugardjito, 1986; see also Palombit, 1992).

One can imagine that a large diversity of fruit-tree species may increase the likelihood of disjunct fruiting, while the frequency of fruit production is probably dependent on soil fertility and availability of water. The conditions which allow for frequent and disjunct fruit productivity are apparently found mainly in alluvial areas, including swamps (Leighton and Leighton, 1983, van Schaik *et al.*, 1995) of a particular scale, i.e. river valleys wider than approximately 500 m. It is likely that in such valleys the permanent availability of water and a regular fresh nutrient supply, owing to flooding and run-off from the surrounding slopes, are the determinant factors for a regular, multiseasonal productivity of large quantities of fruits in a number of tree species.

---

[21] This lean period is not fixed between the months of November and March. In recent years it appears to have shifted so as to begin in ebruary and end in April (Van Schaik, in Sugardjito, 1987).

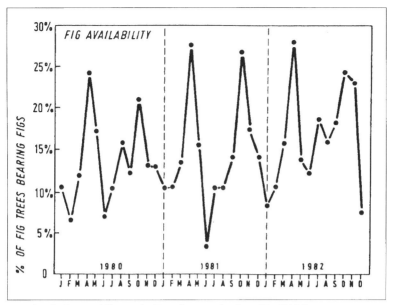

Strangling-fig fruit availablity in the Ketambe area; after Sugardjito (1986).

It is essential to realise that the 'ideal' habitat patches for orang-utans, where fruit is available in sufficient quantity throughout the year, are uncommon. If in some alluvial habitat patches fruit is available for up to some 5-6 individuals/km^2 the year round, the availability is usually more temporal still, as well as scattered in relatively small quantities. It must be borne in mind, however, that the productivity in an 'ideal' habitat patch may still occur in seasons when the availability of food is abundant, and exceeds the requirements of the few residents. Under such conditions of small 'good' patches of habitat in a wide forest expanse of generally low-quality productivity, most deme members are undoubtedly obliged to expend considerable energy in a nomadic existence, travelling from one source of temporary abundance to the next, as do the other nomadic frugivores such as flying fox and hornbills.

It may well be that in the large expanses of dryland forest of Borneo ideal habitat patches are rare and widely scattered. There the vast majority of fruit-tree species have a tendency towards interspecific and intraspecific fruiting synchrony. Leighton and Leighton (1983) found that the forest at Mentoko (East Kalimantan) provided clear annual peaks in food supply (February through May), followed by a long period of scarcity.

Not all the fruit trees in an area produce fruit regularly. For orang-utan habitat the percentage of species (trees, lianas and strangler-figs) to bear fruit annually should exceed some 35%, with at least 35% of the members of each species, on average, participating in the fruiting, while the diversity of species fruiting each month should exceed some 10% (data from the Ketambe area, northern Sumatra; see e.g. Palombit, 1992).

It is presumed that the shifts in phenological phases are determined by subtle differences in patch dynamics and altitude, that is, the interrelationship of a particular community of plant species, local climatic conditions, local hydrological conditions and soil fertility (Richards, 1952; Jacobs, 1988; Bruenig, 1996). In particular in undulating terrain, on a scale larger than one square kilometre, not only soil fertility but also climatic and hydrological conditions may vary considerably in space. For instance the conditions on the western slope of a mountain can induce a composition and seasonality in plants which are different from those on an adjacent northern or southern slope. The microclimate in alluvial forest (along a river) is markedly different from that in a large stretch of dry lowland forest.

It was noted in northern Sumatra that the intra-population fruiting synchrony (or seasonality) of a particular tree species could 'move like a wave' over tens of kilometres[22]. In the same region, S. Orbons (in Te Boekhorst *et al.* 1990) found that fruit production at intermediate and higher altitudes, respectively, peak up to one and two months later than in the bottomlands. Thus, fruits of one particular type which were available for one month in one particular area, or at a particular altitude, became available during the next month in an adjacent area, or at a higher altitude, and a month later in an area further away.

The long-term observations concerning habituated orang-utan demes indicate that while some individuals remain in an area during temporary low food availability, others readily move. One can imagine that an orang-utan would stay if it is confident that it has exclusive access to the remaining scarce resources, and the expectation that the scarcity will only be temporary. Such confidence can be drawn from its social status within the deme and may even be boosted by the expectation that, in another place, the availability of food, as well as socio-ecological constraints (e.g. competition, lower familiarity with the area and its possible residents), are doubtful. If the ape decides to stay and the primary resources dwindle below subsistence level, then its only option is to alter its diet. Indeed, resident orang-utans are able to remain in a small area by switching to 'fall-back' resources (also called 'keystone' resources, Terborgh, 1986) that allow them to weather the lean period. Such resources are leaves, growth shoots of Araceae, the cambium layer of some trees ('bark'), palm 'heart' and rattans. Since the ape develops a reserve of fat during periods of food abundance, this enables it to survive the leaner periods for some (considerable) time.

Yet for the majority of individuals in a deme, their social position apparently does not allow them to take the risk of losing the competition for scanty resources. They are obliged to spend energy expediently by moving, either following the 'waves' of fruiting seasons from one patch to another, migrating over larger areas, or by moving into the regions with a very dispersed and generally low supply of food items. Both

---

[22] The cultivation of rice and marketing of fruit (e.g. for durian, rambutan) also moves like a wave over large geographical regions in western Indonesia (Bottema, 1995).

> After a poor fruiting season, orang-utans at Kutai (East Kalimantan) were in a remarkably lean and hungry condition, their ischial bones protruding from their sides (Rijksen, unpubl.). The young orang-utans which were illegally caught and confiscated during a poor fruiting season are also in a markedly less healthy condition than those confiscated during and after a good fruiting season (Smits, pers. comm.).
>
> Most illustrative, however, of the remarkable resilience of orang-utans in the face of famine is the report by Suzuki (1992): Several of his research subjects survived the extensive drought and subsequent forest fires that razed the Kutai wildlife reserve in 1982-83. Despite the prolonged unavailability of fruit for half a decade, and marked changes in the regenerating forest, several adult apes survived on bark and vegetable matter.

these options bear a serious risk; the former concerning aggressive encounters with possibly unfamiliar residents, the latter involving chronic starvation, as more energy may be spent in searching and collecting than can be gained from low-quality food items.

One would thus expect that when availability and/or quality of food drops below a threshold in a patch of good habitat the individuals would prefer to stay put and face some competition if there is a reasonable expectation of sufficient food in the near future. Only those of lower social status in the deme must decide in time to move away in search of new provisions[23]. It appears that many apes efficiently utilise the spatial variation in the timing of fruit abundance. Their powerful memories undoubtedly play a role in their ranging decisions, and the observations suggest that the apes can read the signs (e.g. the mass movement of hornbills, and flying foxes) indicating an abundance of fruit elsewhere. That such decisions are taken at the proper moments is indicated by the fact that orang-utans rarely travel much further distances during periods of fruit scarcity than during periods of fruit availability.

The long-term observations in both Sumatra and Borneo indicate that most orang utans in an established deme appear to make use of at least two habitat segments of good quality. Such patches typically provide an excess of fruits during their fruiting season, and may temporarily support extremely large numbers of apes. In view of the topography and common patterns of patch distribution in Malesian forests, it is probable that in many regions such segments are separated by quite some distance of low- to extremely low-quality habitat. Consequently, individual orang-utans may be expected to be familiar with the topography and ecology of a large (i.e. > 30 km^2) range.

---

[23] As noted earlier, this is not what Suzuki (1989) found in the burned forests of the Kutai area in East Kalimantan. However, it is not unlikely that the resident orang-utans in the Sanggata study area realised that there was no chance of finding a better spot for survival under the circumstances; after all, the forest had been burned over enormous areas. Such a hypothesis is corroborated by the observations concerning orang-utans in the open-pit KPC concession mine area in East Kalimantan, who apparently saw no options for moving in the midst of the boundless devastation.

The within-patch distribution of food in quantity, quality, time and space must somehow 'fit' the energetic requirements and physical constraints (i.e. ranging distance) of the ape. Under usual conditions an orang-utan covers an average distance which does not commonly exceed some 1,300 metres each day[24]. This distance apparently reflects the average limit of daily energy expenditure in travel. Considering such a constraint, one can imagine that there is a limit of habitat patch distribution for orang-utans. If patch distribution exceeds that limit, as may well happen in the extensive dry lowlands of Borneo, then the area is obviously unsuitable for orang-utans, even though all the known fruit-tree species may occur there.

Thus, an orang-utan deme occupies a range that typically includes a variety of habitats, some comprising patches of high quality, where a base line of fruit production is available throughout the year, others where an abundance of fruit is available only temporarily, interspaced within large sectors of poor nutritional quality at any time. Some of the best habitat patches can support up to five apes the year round, while other sectors are too poor to sustain more than a single ape over a one-week period. This fact has important implications for the design of conservation areas.

Representing the structure of biodiversity

The orang-utan has an extraordinary range of food possibilities, and appears to make adaptive choices for its subsistence in different regions and types of natural forest. Indeed, when one takes into account the ape's former migration into the Sunda landmass region, it may be surmised that the ape can survive in all types of natural forests in the vicinity of open watercourses. The major limiting factors for the ape's subsistence are a particular diversity of tree and liana species, as well as a sequential seasonality of its diet's keystone species such that food is (almost) continuously available within its ranging capacity.

Considering the places where orang-utans naturally occur, the ape's existential requirements (i.e. the minimum size of its range) and the diversity of species of animals with which it is commonly associated, one may consider the orang-utan to be the best *representative* of the highest quality structure of the Malesian rainforest's biological diversity. Moreover, the fact that its demography is extraordinary sensitive to both hunting (Soemarna *et al.* 1995) and forest disturbance (see Chapter The impact of logging) make it an 'umbrella species' for conservation of the rainforest. An umbrella species is one whose home range is large enough, and its habitat requirements wide enough, that when it becomes the focus of protective management the entire structure of the original biological diversity of its range is automatically protected as

---

[24] Mothers with dependent offspring usually travel between 500-700 metres/day; adult males travel slightly further than adult females and often show crudely circular ranging patterns covering anything between 600-800 metres on average per day (except when being chased). Of all the age-groups, adolescent and young adult females without offspring range farthest, travelling in an erratic wide-ranging pattern while covering anything between 600-1000 metres/day, while sub-adult males commonly travel almost as far as young females, yet more directionally and often far beyond the range of their natal deme.

well (Stork, 1995). Its size, predominantly arboreal existence and generalised frugivority mean that no other species can match the orang-utan in functional representation. If an orang-utan deme is present in a regular density, i.e. between 1-10 apes/km^2, then the area is likely to harbour at least five other primate species as well, including at least one gibbon, at least five hornbill species, at least 50 different fruit-tree species, and 15 different lianas, the structural relationship of which reflect the delicate environmental conditions of the undisturbed Malesian rainforest.

Orang-utans are dependent on seasonally available fruits from a very wide variety of trees, requiring a large area of habitat, and therefore are among the best representatives of rainforest biodiversity.

Thus, the orang-utan's presence and population density may be deployed as the yardstick for conservation of a Malesian rainforest, without much further analysis of the structural diversity of plant and animal species in the particular area. It means that preserving (i.e. protecting) a viable wild orang-utan population is identical to conserving the most valuable areas of Malesian rainforest, with the most unique structures of biological diversity.

If one adds to this novel characteristic the fact that the orang-utan is a species which seems to have an extraordinary appeal for humans, it will be clear that this ape is one of the best 'mascots' or 'flagship species' for raising awareness and support for the protection of the endangered biological diversity of the Malesian rainforest.

Variable orang-utan ranging patterns

Empirical facts

Orang-utans display considerable variation in individual social behaviour and ecology, due to differences in sex, age, reproductive state and social status, as well as in talents. There is no 'standard orang-utan'.

Long-term observations of habituated demes of orang-utans reveal that some individuals, notably adult females (with infants), appear to live permanently in one particular area for several years. It is uncommon not to encounter these individuals over periods of more than a few weeks. Most deme members, however, seem to spend longer periods away from such a 'home base', while a few may be found there occasionally or only once.

The spatial and temporal patchiness of food and its variation in quality explains the finding that a considerable number of deme orang-utans display seasonal nomadism, as had already been noticed by Hornaday (1885) for the Bornean, and by Ruppert in 1936 for the Sumatran orang-utan (in Carpenter, 1938).

In general, long-term studies of habituated orang-utans suggest that one can distinguish three classes based on dispersal activity: (1) residents, who are found for many years to be present for most of each year in one particular area (Rijksen, 1978; te Boekhorst, et al. 1990), (2) commuters, who are seen regularly for several weeks or months each year for many years and appear to live a 'nomadic' existence, and (3) wanderers, who are seen very infrequently (or once) in a period of at least three years and may never return to the area. The residents are commonly a minority of a studied population. In a high-quality habitat patch of approximately two square kilometres, it is possible to find one or two resident males and up to three resident adult females (with offspring) belonging to a deme of at least 30 individuals. For the Ketambe deme, which enjoyed a good-quality habitat patch (of approximately 1 km^2) during the early 1970s, the relative differentiation between residents, commuters and wanderers was, respectively, approximately 30%, 60% and 10% (see also te Boekhorst et al. 1990); that is, the commuters comprised the bulk of the population. However, an overview of reports of other field studies indicate that some areas seem to have no resident orang-utans at all (see Table III).

A long-term study of interactions reveals that the residents are usually individuals of high social status. A resident occupies a home range of high-quality habitat which, in the case of adult female residents, may be as small as anything between 0.6 (Horr, 1975) and 1 km^2 (Rijksen, 1978; Suzuki, 1989), and is still shared with others. Resident adult males commonly range over larger areas, supposedly for socio-reproductive reasons, but their non-exclusive home range rarely exceeds 10 km^2 in high-quality habitat.

The commuters occupy much more extensive ranges, and utilise more than one major habitat site of high or reasonable quality. Such sites may be separated for quite some distance, i.e. > 5 km, by forest of lesser or poor quality, or by the home-ranges of conspecifics, in which they are barely tolerated.

It is considered that the development of many young orang-utans follows the sequence from dependent resident (during their infancy and juvenile stage) to (social) commuter (during their adolescence and sub-adulthood) to (quasi-solitary) resident (as adults), if and where the habitat and an individual's social relationships pattern permits. An individual may be obliged to remain a commuter during its adulthood – or to become a wanderer – if its relative social status remains low.

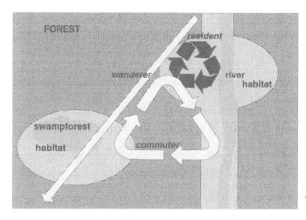

Schematic representation of the various ranging styles of orang-utans; good habitat is represented by the lightest shading.

TABLE III

Comparison of ranging data

Researcher	Isl.	Study area	Dur.	# Ind	Res.	Non res.	Range male	Range female
Horr 75	B⁰	24 km²	1 yr	27	?	-	5.2 km²	0.6 km²
Mitani 82	B⁰	3 km²	1 yr	>15	6	>9	-	1.5 km²
Rodman 73	B⁰	3 km²	1 yr	8	6	2	1 km²	0.5 km²
Suzuki 87	B⁰	50 km²	5 yr	30	12	18	4-8 km²	0.6-3km²
Galdikas 78	B	20 km²	3 yr	54	4?	50?	6 km²	2-6 km²
Rijksen 78	S*	1 km²	3 yr	28	5	23	6-10 km²	1-3 km²
Schurmann 82	S*	5 km²	4 yr	29	8	17	>8 km²	>3km²
van Schaik 94	S	3 km²	2 yr	45	8	37	>5 km²	> 3km²

The names with an asterisk refer to the Ketambe area in the Leuser Ecosystem, Aceh (Sumatra); the ⁰ refers to the Mentoko/Sangatta area in Kutai, East Kalimantan (Borneo); Galdikas' research was carried out at Tanjung Puting wildlife reserve (central Kalimantan); van Schaik's research was conducted at Suaq-Balimbing in west Aceh (Sumatra).

The wanderers seem to have no familiar range at all, but go *walkabout*, continuously trekking alone over large distances, in search of sustenance and perhaps sex. Wanderers may (try to) associate for some time with a commuter group. Since all wanderers seem to be adult or older sub-adult males, it may be inferred that they have failed to become accepted by, and felt anxious within, or were 'cast out' of, a group of commuters and residents. Considering the structure of observed orang-utan demes, it may well be that to become a wanderer is the ultimate fate of at least 20% of all males[25].

Socio-ecology

In view of these spatial dynamics it is not surprising that the orang-utan has remained an enigma in a sociological sense. Perhaps this is typical of the Pongid condition. For centuries the scientific world has had great difficulty accepting that even people within one ethnic group have a wide range of options for spatial movement and cultural diversity. Communities of people may adopt any social structure ranging from a solitary lifestyle to the commune, and from strict monogamy to extreme promiscuity, dependent on the ecological and cultural conditions. For the apes it was apparently hoped that they could be seen to conform to a standard model; thus *the* gorilla is often said to conform to a one-male group structure, *the* chimpanzee supposedly lives in a multi-male society and *the* orang-utan is known as the solitary ape.

Yet, there is no standard socio-ecological pattern for Pongids, if only because they live in – and maintain cultures adapted to – different ecological conditions. Some people spend most of their lives on open water, others inhabit territories on steep mountain ridges. Some chimpanzees live in woodland savanna, others in dense rainforest, while some orang-utans dwell in the dry foothills of mountain forest, others in swamps. If there is a common or basic pattern in the many forms of social organisation of *the* Pongid, it is that of 'an open and panmictic society across the contiguous distribution range of the species in which individuals may socialise because it is, under the given conditions, most convenient to do so' (Goodall, 1963; Sugiyama, 1968). An individual's decision to remain with or split from a party is probably determined by intuition reflecting (unconscious?) cost-benefit patterns of evaluation in which the expectation of benefits pertaining to socialising are weighed against the known and expected availability of food and the supposed risk of competition by companions (Ghiglieri, 1985).

As a consequence, it remains to be seen whether *the* orang-utan is a solitary ape, as it has so often been characterised. It is certainly strange that in particular the extraordinary intelligence of the orang-utan does not seem to fit a solitary image. Conversely, a supposed solitary nature seems to contradict the observations that the

---

[25] In view of this fate, it is perhaps understandable in ontogenetic terms why sexually mature (i.e. sub-adult) males actively enforce copulations (i.e. rape); it may be their only chance in life to reproduce, no matter how unlikely the copulation is to result in conception.

ape, especially under artificial group conditions, displays a clear talent for the establishment of a complex pattern of relationships, with dominant individuals assuming a controlling and hence a protective role (Rijksen, 1978). Recent observations suggest that this ability plays a major part in the social organisation of wild orang-utans, and that females in particular maintain a relatively high degree of sociality when conditions are favourable.

In fact, the raw description of the fusion-fission pattern of a chimpanzee community in a forest environment (Ghilieri, 1985) is strikingly similar to that of an orang-utan deme. Like orang-utans, chimpanzees combine temporarily into larger than average aggregations (and socialise more with one another) when feeding constraints are lifted (Wrangham, 1975), and demes of both species maintain a social network of relationships that persist over lean seasons at least. The differences between these apes are in degree rather than in quality.

To a casual observer it is far from obvious that orang-utans live in groups, in the sense that individuals are often within close range of and usually within view of one another. More careful field work, however, reveals that the same individuals are often seen in one particular area, while at other times most of them seem to be absent. Indeed, many field workers have struggled with their impression that a number of seemingly unassociated apes appear to have an uncanny synchronisation in their daily movements. After all, the members of an orang-utan deme often maintain large inter-individual distances, so that group formation can only be inferred when different members are followed simultaneously. The infrequent interactions then clearly reveal elements of familiarity or an established social status, if not a suggestion of bonding.

Group formation in orang-utans is most obvious in adolescents and sub-adults who may commonly move near to one another. The adults usually keep a considerable inter-individual distance, except when a male and proceptive female engage in a temporary sexual consort relation. Only occasionally do adults come within such close proximity to each other that clear interaction patterns can be discerned by the observer. Such instances usually, but not invariably, take place in large fruit trees such as strangling-figs, or in areas with high fruit-tree densities.

When following several individuals simultaneously, however, one can reconstruct a pattern indicating that they do travel synchronously, albeit dispersed over an area which may be as large as one or two square kilometres. The leading position is often taken by a dominant adult male, sometimes by an adult female. How an individual knows or senses the whereabouts of its group's members remains one of the great mysteries of orang-utan social behaviour. Indeed, for much of the day the members may seem to be well beyond one another's visual and auditory range. Perhaps intelligent anticipation, in combination with an amazing topographic memory, plays a role, but some students of orang-utan behaviour have come to believe that the apes possess a kind of 'sixth sense' (Rijksen, unpubl.).

The social networks of orang-utans consist of adult females and their offspring and may include a number of adult and sub-adult males. Superficially such open 'groups' are not unlike the social organisation of the sympatric pig-tailed macaque or the forest dwelling chimpanzee of Africa (Ghiglieri, 1985).

In any case, the nature of social interactions suggests that residents and commuters belong to a single social network, as they seem to know each other well, evidently having established particular social relationships, including what may be interpreted as 'friendly' bonds (Rijksen, 1978). The interaction pattern in encounters suggests that wanderers are usually strangers to the members of the social network.

The socio-ecological structure of the orang-utan is undoubtedly determined by female reproductive interest (Horr, 1977; Rijksen, 1978; Wrangham, 1979; van Hooff, 1995). Because the smaller, more agile orang-utan female could – in principle – easily avoid sexual abduction by adult males, the extreme sexual dimorphism of the male orang-utan is believed to be the evolutionary reflection of the female preference for mating with adult males who display exaggerated physical features and an assertive attitude (Rijksen, 1982; Utami and Mitrasetia, 1995), i.e., large size, long hair, wide face, loud and long calls.

It is assumed that orang-utan females strive to achieve the supposedly more comfortable strategy of being a resident during their reproductive years, and especially when their infant is between 0-4 years of age. It may depend on a female's social relationships[26], both with males and other resident females, whether she can rise to, and retain, such a status.

Despite frequent displacement activities in large fruit trees, it is virtually impossible to detect from the behaviour of wild orang-utans overt competition for food, even in periods of scarcity. The impression is that all such displacement as occurs within a group has little to do with food, but instead reflects the individual's perception of its relative position in the tree, plus simple assertiveness.

Yet, when the availability of fruit is low it is rare to find orang-utans other than residents in an area. The rare sightings of commuting or wandering adult females and sub-adults during such periods indicate that they are extremely cautious – if not reluctant – about approaching trees in which a resident female is feeding, and often wait in hiding. Nevertheless, an adult female wanderer may join the others without hesitation when an adult male is present in the tree. It is interesting that in Sumatra the assertive hierarchical, or territorial, attitude of resident females seems to be suppressed when an adult male is present. There is no sexuality apparent in such interactions.

> It may well be that competition for food is so difficult to detect because most of this ape's interactions either seem to be restrained due to its introverted nature, or take place at an extraordinary inter-individual distance. Thus the established dominant position of one individual may well have influenced the other before any potentially competitive situation could become evident to an observer.

---

[26] Social relationships and bonds with other females are developed mainly during the years of adolescence and pre-motherhood.

When considered on the level of (long-term) relations, however, it appears as though in the introverted orang-utan a continuous sexual (or reproductive) interest has a consistent (almost institutionalised, yet restrained) impact on all social interaction patterns. Among orang-utans it seems that sex, rather than exclusively serving reproductive purposes, is also deployed as an important social instrument in the establishment and maintenance of relationships.

Whereas the proceptive female vertebrate can generally remain passive, expecting the sexual services readily offered by males, a male must apparently actively pursue his reproductive interest in the face of overwhelming (sexual) competition. In orang-utans the sub-adult males travel widely and enforce copulations (*rapes*) at any given opportunity. Characteristically, the rapes occur when a sub-adult or (young) adult female is newly met, or met again after a period of separation. In particular, (hesitant) females of lower status are victimised. It is believed that these rapes function mainly to establish a relationship[27], as there is no evidence that such rapes have immediate reproductive significance (Rijksen, 1978).

The orang-utan male is unique in that the habitus of adulthood, and especially the development into large body size and the characteristic cheek pads, is not just related

Orang-utan male in the prime of his life (20 years of age), offering his sexual services ('presenting') to a female (outside the upper right-hand corner of the picture).

---

[27] High-ranking sub-adult females also can 'rape' lower ranking, newly encountered females, by manual intromission of the victim, usually while masturbating (see also Rijksen, 1978).

to age and sexual maturity; it can be suppressed or postponed for anything up to ten years (Maple, 1980), i.e. a phenomenon called bimaturism. There are clear indications that the ascent to a higher social status in the deme for (sub-) adult males is somehow associated with long-lasting sexual relationships with a number of females, and that the typical adult size and demeanour comes with his (perception of) socio-sexual status (Kingsley, 1982; te Boekhorst, et al. 1990).

When the sub-adult achieves the typical adulthood physiognomy, effective rapes become rare, if not almost impossible, due to the increasing discrepancy in agility between the adult male and the females. When a male has eventually attained a relatively high social status, it apparently makes rape unnecessary, even if his physique did not preclude the hazardous chases[28]. Having attained typical adulthood, the male orang-utan solicits sex with a proceptive female by means of posturing and a penile display[29] (Schürmann, 1982).

For males to achieve and advertise their resident status requires a great deal of assertiveness. Resident adult males usually appear to occupy a high status in terms of dominance with reference to other males, and there seems to be a continuous, albeit largely hidden, competition for the resident status by commuter (and/or wandering) adult and upcoming sub-adult males. Thus, an adult male's resident status is invariably bound up with a home range which covers a relatively high-quality habitat patch in comparison with other habitat patches in the surroundings, automatically including the home ranges of at least five resident females. It is of particular interest that females play an active role in the dynamics of resident social status among adult males in an area (see Utami and Mitra Setia, 1995).

Thus, though the status of a typical adult male has probably largely been established in long-term assertive competition, especially with other males, it must be secured by individual relationships with females. If the resident females within his home range have offspring, however, the resident male's sexual interest may remain unsatisfied for anything up to six years if he were to rely on these females alone. Indeed, a male of resident status will usually find that his sexual interest is unlikely to be satisfied[30] unless other, commuting or wandering, proceptive females are attracted to his home range. In the event that his home-range does not attract such females, he may in fact be forced by the lack of sexual opportunity to resume life as a commuter, or even to become a wanderer.

Hence, one will find the highest ranking adult males, at the peak of their assertive

---

[28] Under experimental conditions in captivity, however, fully adult orang-utan males, after separation from a female for some time, revert immediately to rape, enforcing copulations (Nadler, 1995).

[29] It is remarkable in this respect that in comparison to that of the bonobo and the chimpanzee, the erect penis of the orang-utan is small and inconspicuous.

[30] It is assumed that such a situation in e.g. langurs (Hrdy, 1977), humans (Devereux, 1976; Kempe, 1976) and gorillas (Fossey, 1983) can lead to infanticide; as yet there is no indication that infanticide occurs in orang-utan demes under natural conditions.

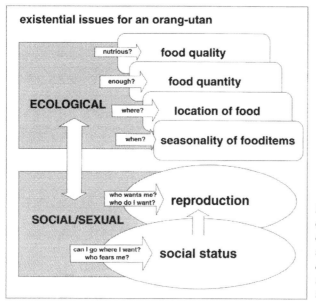

Ranging necessity is due to the outcome of a dynamic interplay of variables: social status, reproductive status, seasonal availability of quality food, topography of the area, immigration and death.

and reproductive vigour, as residents in the best patches of habitat having the highest incidence of different fruiting seasons for various favoured fruit trees. It is therefore hardly surprising that in the first place a resident adult male displays a protective attitude towards (any) females entering his home range, and regularly advertises his presence.

High-status orang-utan males react to the distress of a conspecific by threatening and/or assertively engaging the animal or person causing the distress (Rijksen, 1978). As a consequence, a higher ranking orang-utan male may demonstrate a control role in the presence of others. He does not tolerate distress in the interactions of two subordinates in his presence, and controls potential conflict situations by active interference. It is remarkable that this even applies when humans are victimised; a case is known in which a wild, habituated adult male resident reacted to the distress of a woman observer who was about to be abducted by a human male, by threatening and chasing the latter away.

Like other Pongids (including humans) adult orang-utans at times emit a loud, far-ranging call. The (long) call of orang-utan males undoubtedly has powerful assertive overtones, but it may initially be an indication of potential protection and reproductive interest for (commuter) females, rather than a demonstration of sexual interest for the resident females, or a threat to rival males. Indeed, the call may even have distinct qualities of announcing the availability of food – as with chimpanzees – although very little is understood of the message content of possibly variable loud calls in apes. In any case, protection against overt assertiveness by the resident females as well as sub-adult males undoubtedly gives a commuting proceptive female opportunities to both meet with a male of proven assertive quality, and to feed.

Indeed, under the male's protection she may even have a better chance to establish herself as a resident.

The arena: the reproductive market of the orang-utan

The orang-utan male shows the greatest degree of sexual dimorphism of all Pongids. In terms of mammalian (and avian) social structure, extreme sexual dimorphism in males is usually correlated with either a one-male group or a *lek*-type mating system. It is considered to be mainly induced by female choice for assertive quality in males, rather than for additional care of offspring.

> Two processes are believed to be responsible for the evolution of the adult male's extremely large body size: Primarily female choice and secondarily male-male competition. First, the female orang-utan seems to prefer to mate with males of the highest status accessible to her. Second, in orang-utans, adult males are highly intolerant of each other; usually confrontations quickly lead to one male chasing the other male, and sometimes very serious fighting ensues, causing injuries. To the extent that body size determines the outcome of these conflicts, larger size may be selected for. It is probable, however, that agility is another important factor, possibly also favouring younger over older individuals.
>
> However, the outcome of previous interactions may well play an overriding role in determining status among orang-utans. During the early 1970s at Ketambe, a visiting, delicately built adult male (Moses) appeared to consistently supplant the larger, heavily built (but younger) resident adult male (O.J.) during his brief visits. Yet the proximity and interaction patterns of potentially receptive females involving either male suggested a long relationship (i.e. bonding) rather than ad-hoc choice for size alone (Rijksen, 1982).

The observations of Utami and Mitrasetia (1995) on female orang-utans actively inducing competitive confrontations between resident males suggest that the spatial organisation of an orang-utan deme in some parts of its range conforms to a very complex *lek*-type mating strategy. It differs from the common *lek* systems of birds and ungulates, however, in that orang-utans seem to establish and entertain long-lasting relationships between adult males and females. We will henceforth denote the place where this organisation is evident as the 'social arena'. The males of highest status in a deme are residents in top-quality habitat patches. Such males occupy the central positions of the social arena, which attract many of the members of the deme when fruits are abundant there.

The large size of the composite range of a deme, however, makes it difficult for a human observer to form an accurate image of the range's social arena, especially if it also may shift over the years from one forest patch location to another. Complicating factors which have so far precluded its recognition are probably the extent of the overlapping resident ranges, the extreme length of the reproductive cycle and birth interval and the fact that individual relationships seem to play an important role in orang-utan mate choice (Rijksen, 1978). Thus a receptive female may in fact avoid a particular high-status male occupying a central position in her range's social arena.

> The lek or arena is a matrix of contiguous display sites of different males in the range of a deme where younger females in particular come to become impregnated by male(s) of the highest (territorial) status, usually male(s) occupying the central positions. Crook (1965) noted that the occurrence of a lek (mating) system in birds depends on a seasonally abundant food supply, and this also seems to apply to orang-utans – albeit in a slightly different sense.
>
> The social arena of the orang-utan can be defined primarily in terms of habitat characteristics; it is the area with the highest diversity and quantity of staple (and/or favoured) resources which provides the widest annual distribution of fruit availability, relative to other patches in the range of a deme. As a consequence, the social arena will be occupied by a relatively high density of residents. It is possible that the range of a deme has more than one social arena. Conversely, it is probable that if habitat conditions fail to allow for the establishment of a social arena, the viability of a deme is significantly reduced.

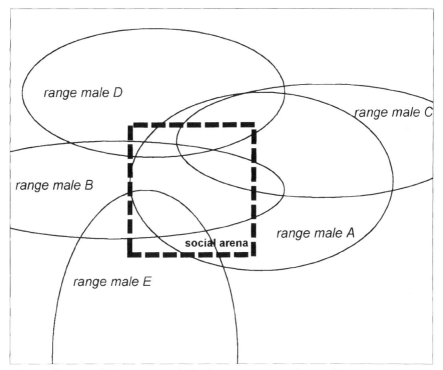

Schematic illustration of the orang-utan male distribution pattern, a social arena; the imaginary square represents the home range of at least three resident adult females (with dependent offspring).

Thus, a social arena is located in those areas of a deme's range where, relative to other patches in the range, quality fruit is available most often and in sufficient quantity. Hence it is the place where most individuals of the deme congregate most frequently, not unlike a village centre, a mall, a pub, a fair or the sports-stadium in human society.

If, for a female, the resident status is the most economical in terms of food availability in time, quantity and quality, being a resident in a social arena also implies that the

> In a social arena, the central positions that are occupied by the highest ranking males cannot be called territories in the ethological sense; a male of high status, by advertising his presence (e.g. through long calls), seems to be giving a warning signal, but observations indicate that any territorial space is not much larger than a sphere in which the high-ranking male can directly perceive the (assertive) presence of other (sub-)adult males. It must be realised, however, that a resident high-status male in the central position of one particular arena may not be the most dominant male in the deme: It has been observed that a dominant resident is temporarily supplanted if a commuting adult male, apparently from an adjacent arena, pays a short visit during a particular fruiting season (Rijksen, 1978). Since such events take place without overt interactions, it suggests that the males know each other well, and belong to one deme of mutually familiar individuals.
>
> Only other (sub-)adult males of lower status usually avoid entering the direct perceptive range of a high-ranking male (Rijksen, 1982). Nevertheless, some hitherto subordinate males may of course actively challenge a dominant male in the vicinity, seeking a confrontation, by uttering long calls. It is interesting that (sub-)adult females also have a very similar call (the lork call) reflecting their sense of status (Rijksen, unpubl.).
>
> Characteristically, some species subject to the lek mating system show bimaturism, with some males remaining less extravagant in size and demeanor while operating as 'satellites' next to the overtly competing top males. These smaller-sized males have adopted the strategy of 'stealing copulations'. Orang-utan sub-adult males may assume such a role for several years into maturity, retaining their sub-adult physiognomy. Indeed, many rapes in wild orang-utan demes appear to be perpetrated by males adopting just such a 'satellite' role.

resident male(s) may offer her some protection against harassment by sub-adult males. For a high-status male, being a resident within a social arena is the most economic position in terms of attracting (possibly receptive) females and demonstrating his power and qualities. However, competition for resident status will be high. For an ape to advertise his position within this arena by way of the typical long call demonstrates a daring prowess which conveys reliable information on the day-to-day state of competition among males, and at the same time advertises his availability for protective and sexual services among the females.

An orang-utan individual may change strategies expediently during its lifetime, ranging from the strenuous way of life of the commuter to that of the comparatively comfortable resident, if and when circumstances permit. Or, alternatively, it may be obliged to reverse the sequence and end up the dangerously unpredictable way of life of a wanderer. Virtually all orang-utans begin their adolescent careers as commuters. As juveniles they learn to appreciate the advantages of their mother's range while enjoying the benefits of their mother's status and experience. When they are being weaned, becoming adolescents and entering the social phase, they travel together with peers, and gradually extend both their ecological knowledge and their social network over the ranges of these peers.

The general picture emerging from field-research records is that the social arenas of an orang-utan deme are notably found in river-valley and alluvial flood-plain (peat-swamp) forest. These arenas can harbour up to thirty apes, including as many as seven adult males (van Schaik, pers.comm.) in times of fruit abundance, and may retain an average density of at least five – and sometimes well over fifteen – apes per square

> During sub-adulthood, both male and female orang-utans show a dramatic increase in assertiveness. Since orang-utans have excellent long-term memory and appear to be able to skilfully evaluate complex social situations, such assertiveness is not always expressed directly. Among females this is expressed especially in triadic relationships involving a male, and then seems to bear a strong resemblance to jealousy. It results in the break-up of previously strong social bonds, and in fact seems to force the individual into a state of greater solitariness. The ontogeny of this change is poorly understood and has never been studied under the experimental conditions of rehabilitation or zoo captivity. This ontogenetic change in the individual's nature, and the apparently increased need for female choice in the patchy habitat of the rainforest, dampens inter-individual tolerance, and seems to have facilitated male competition and female selectiveness to such an extent that it has resulted in the social arena type of organisation which we find today.

kilometre over a period of several months of the year. However, the most deme members are distributed over a much larger continuous area of forest in which the habitat patches can sustain a considerably lower year-round density of apes.

Density estimates

Assessments of orang-utan density are plagued by practical and conceptual problems. The practical problems concern the difficulty of estimating numbers per unit area, and of extrapolating this estimate to a larger region. The conceptual problem involves the difficulty of estimating a relevant minimum area of habitat required by a deme, and the spatial scale of habitat differentiation required by its individual members, some of whom adhere to the resident strategy, while others are commuters. The spatial patchiness of orang-utan distribution, in combination with the dynamics of temporal availability of food and the unique social organisation of the ape, presents formidable problems not only for an assessment of population size in a certain area but also for the conservation of the species. If estimated local population densities are extrapolated over the entire range, the neglect of spatial variation in habitat occupation may result in a tremendous error in an assessment of the population size (see e.g. MacKinnon, 1986).

> After a few months, research sites may become comparatively more attractive for habituated apes because of added security. The daily presence of researchers tends to deter predators, such as the elusive clouded leopard (*Neofelis nebulosa*) and (exclusively in Sumatra) the adjag (whistling dog, *Cuon alpinus*) as well as the tiger (*Panthera tigris*), as has been suggested by Griffiths and van Schaik (1993). Further, it is not unlikely that some habituated apes actually take social and territorial advantage of the presence of a human observer, by exploiting the awe-inspiring and intimidating effect the observer might have on less habituated commuters and unhabituated wanderers.

Until recently, most density assessments were based on long-term studies of known populations. However, it is impossible to extrapolate estimates derived in this way to larger areas, because long-term studies tend to be selected in habitat covering social arenas.

Alternative techniques are therefore needed to acquire a more accurate picture of densities in a range of habitats, including those forest patches not frequented by the ape. Although orang-utans are notoriously difficult to detect in the rainforest, one can readily verify their presence in an area by looking for the typical platforms or nests which are constructed daily for their repose at dusk, and sometimes for play or a midday rest. Such nests remain visible for a considerable period of time, and can therefore be the basis for an assessment of the number of apes which have made use of a particular area. The first attempts to assess densities by means of nest counts go back to the 1960s (Harrisson, 1961; Schaller, 1961; Milton, 1964). Van Schaik et al. (1995) have refined the nest-count line-transect method, which has been validated at two different sites of known densities and been found to give reasonably accurate results.

Table IV shows a compilation of all reported densities from studies on orang-utans over the last two decades. The findings clearly demonstrate that densities vary tremendously according to general type of forest. Habitats near rivers, notably freshwater and peat-swamp forests in flood plains and alluvial bottomlands, may, in an undisturbed state, yield mean densities of 2-7 individuals/km^2, and in exceptional cases reach well over 10 apes/km^2 (I. Singleton, pers. comm.). Dryland forests outside flood plains and valleys appear to be much less densely inhabited and seem to represent marginal or outright poor orang-utan habitat, with temporary densities up to a maximum of 2 individuals/km^2. Forests at higher elevations rarely attain densities over 1 individual/km^2, and usually far less. The differences between Sumatra and Borneo will be discussed below.

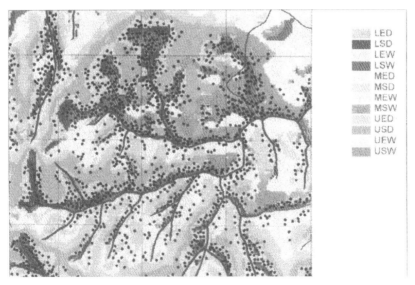

GIS model of the habitat distribution of an orang-utan deme; the spots are food trees. Note that the highest density of food trees is in the valley bottom and on the lower slopes; different shades in the legend indicate geomorphological conditions.

Table IV

An overview of the recent reports of orang-utan surveys applying the nest-count line-transect method gives the following range of results (dens = number of individuals per km2; ref = the reference as indicated below the table).

Habitat type	Borneo	Dens	Ref	Sumatra	Dens	Ref
Flood-plain and peat-swamp	Sebangau	2.2	1	Suaq-Balimbing	6.9	6
	Kulamba	3.0	2	Trumon	7.0	10
	Tanj. Puting	3.0	3	Bahbah Rot	4.5	10
	D. Sentarum	3.5	14			
Alluvial / bottomland forest	Gunung Palung	3.5	4	Ketambe	5.5	6
	Lokan	2.1	5	Kompas	3.0	7
	Ulu Segama	1.2	2	Low Mamas	3.2	7
	Mid Kinabat.	2.5	2			
	Low Kinabat.	2.3	2			
	Kutai/Sangatta	2.0	11			
Upland (hill and dipterocarp forest)	Ulu Segama	0.8	8	Renun	1.0	9
	Kawaq	0.3	2	Bohorok	1.0	9
	Tabin	1.1	2	Bohorok	2.2	6
	Danum Valley	0.3	12	Bengkung	2.0	6
	Crocker range	0.1	2	Manggala	1.2	6
	Meliau	0.8	???	Suaq B.-hills	1.0	10
Submontane and montane forest	–			Kapi	1.2	7
				Ket.-submont.	1.2	6
				Ket.-montane	0.7	6
				Mamas-subm.	0.7	6
				Dg. Megaro	0.4	6
Sel. logged / second. forests	Sebangau	1.0	1	Sikundur	1.1	6
	Katingan	0.5	13	P. Lembang	1.3	6

References:
(1) Page et al., 1995; (2) Payne (1988); (3) Galdikas (19778); (4) Leighton and Leighton (1993); (5) Horr (1975); (6) van Schaik et al. (1995); (7) Rijksen (1978); (8) MacKinnon (1971); (9) MacKinnon (1973); (10) van Schaik, Idrusman and Sitompul (unpubl.); (11) Suzuki (1992); (12) Johns (1992); (13) Rieley, et al. (pers. comm.); (14) Russon and Erman (1996)

Contrasts between the islands

As shown in Table V, point densities of orang-utans on Sumatra are roughly twice as high as in similar habitats on Borneo. Table V also shows that the altitudinal range appears to be much higher in Sumatra. It was noted earlier that Sumatran orang-utans tend to spend more of their feeding time eating fruits and insects and less on the growth layer (cambium) of some trees. Sumatran orang-utans are also believed to be more social (Markham, 1980; Courtenay et al. 1988) and gregarious (Rijksen 1982) than their Bornean conspecifics.

Table V

A comparison of average orang-utan densities in undisturbed sites of major habitat types in Borneo and Sumatra, as derived from nest-count surveys (Table IV).

Habitat	Borneo	Sumatra
Flood plains and peat swamps	2.9	6.1
Alluvial bottomlands	2.3	3.9
Uplands	0.6	1.4
(Sub) montane	-	0.8

All these differences suggest that from the orang-utan's perspective the average Sumatran forest is more productive than its Bornean counterpart. The ecological factors underlying this dramatic difference are varied, and may well be linked to the divergent geological history and the scale of land area units of these two large islands.

The geological history of Sumatra differs markedly from that of Borneo. Subduction of tectonic plates has produced the build-up of the Barisan mountain range and extensive volcanism in Sumatra, whereas the mountainous core of Borneo is of a more ancient origin. Hence, soil fertility is highest in the western half of Sumatra, where the best orang-utan habitat is currently found. In addition, the Barisan mountain range is subjected to a more equitable and generally higher level of rainfall, possibly resulting in less pronounced seasonality and shorter periods of drought.

During these last two decades of the twentieth century many of Borneo's and Sumatra's lowland forests have been systematically ravaged and destroyed in order to satisfy the demands of European, Japanese and American timber markets; a log-landing site in East Kalimantan (1993).

The greater altitudinal range in Sumatra may be due to the *Massenerhebung* effect caused by the Barisan mountain range. This effect results in an upward shift of altitudinal vegetation limits in Sumatra, relative to the conditions in Borneo. Thus, suitable habitat for the orang-utan has a greater altitudinal range in Sumatra.

The geomorphological scale which ultimately underlies the distribution of habitat patchiness also appears to be markedly different in the two islands. In Borneo's relatively flat topography, the scale of the patches is much more (3-10 times) extensive than in northern Sumatra's mountainous terrain. The condition in Sumatra is characterized by tightly folded terrain allowing for relatively small-scale habitat patches of high tree and liana species diversity. The trees and lianas in Sumatra are also influenced by a more patchy distribution in micro-climatic conditions, inducing a higher variation in seasonality on a relatively small scale. This could well allow for a higher carrying capacity and hence a higher density of apes. Finally, the spatial dynamics of the climate in Borneo are also different from those in Sumatra. Hence, the availability of fruits over time in the larger, comparatively uniform forest patches of Borneo is undoubtedly lower per square kilometre, so that seasonality is more spaced out over the year. One can therefore understand why it is generally believed that the Bornean orang-utan is less evidently gregarious and more nomadic than its Sumatran brethren.

The impact of logging

Introduction

In 1969, in a report entitled *On Game Conservation*, W. Stevens informed the Government of Malaysia that 48% of the mammal species in a forest seemed to have disappeared as a result of timber exploitation. Khan (1978) found that the six primate species in a logged forest in peninsular Malaysia suffered losses between 23% (long-tailed macaque) and 57% (siamang) of their original population as a direct consequence of logging operations.

One can imagine that the specific habitat demands of the orang-utan make it particularly vulnerable to the ongoing loss, degradation and fragmentation of its prime habitat. In particular the alluvial lowland and/or swamp forests, as well as the contiguous hill forest, all of which are crucial for orang-utan survival, are under increasing pressure of human exploitation and occupation. However, foresters commonly believe that as long as the apes are not directly targeted for illegal harassment, they will be able to evade the disturbance caused by tree felling. It is presumed that the apes find temporary refuge in adjacent areas, then eventually return to the logged-over forest. This of course presupposes that adjacent forest areas of suitable, subsistence-level habitat still exist for a compressed population of at least twice the number of its original inhabitants. There are no long-term studies to support this irrational supposition.

During surveying work in East Kalimantan, Wilson and Wilson (1975) concluded that orang-utans in fact did not seem to be affected by selective logging. It is questionable, however, whether their rapid survey technique was sufficient to uphold such a notion. In the early 1990s Payne and Andau reiterated this impression by concluding that 'while logging of natural forests undoubtedly has an impact on orang-utans, there is no evidence (...) that such logging has had significant effects on the survival of wild breeding populations' (Payne and Andau, 1992). Although data from the 1970s are scanty, the researchers apparently felt confident enough to claim that 'population densities in lowland Dipterocarp forests are similar before and up to 20 years after logging'; and all of this despite Payne's earlier findings (Davies and Payne, 1982; Payne, 1987) indicating the contrary.

In 1986, Davies noted that surveys in areas of logged forest had revealed that the ape occurred 'at much lower densities than in primary forests of the same region', while in some areas it had disappeared. The surveys by Rijksen in 1973 and 1978, and by Aveling and Aveling (1979) during and after logging operations in the Sikundur area of the Gunung Leuser wildlife reserve, had yielded results similar to those found by Davies. The survey data revealed a dramatic drop between 50-100% in orang-utan density in affected, supposedly preferred habitat patches in primary forest areas immediately after logging. It is certainly relevant to mention here that the orang-utans which were displaced by the logging operations at the time had ample opportunity to move into adjacent, unspoilt forest areas.

The occupancy rate of the logged-over forest did not seem to recover to more than some 40% of the original density, even after one decade. Van Schaik, Azwar and Priatna (1995) also reported that according to their comparative nest-count surveys in different types of (logged and pristine) forest, orang-utans had become rare in exploited forests. In all their rigidly standardized surveys in northern Sumatra they found that relative densities of orang-utans in selectively logged and secondary forest were 30-50% of the densities recorded in primary forest, even when the secondary regrowth was over 20 years old. This was found in all types of secondary forests alike. It is perhaps interesting to note that these findings are consistent with the recorded effect of logging on a chimpanzee population in Uganda (Skorupa, 1986).

The comparative study by Rao and van Schaik (in press) again corroborates these findings. Applying the same rigid methodology – including a control for habitat quality – they found that 'orangutan density in logged forest is (at most) 40% of that in equivalent untouched forest, five years after logging.' They concluded that 'selective logging leads to a reduced density of orangutans in logged forest' and recorded a marked difference in the behaviour of apes in logged forest: Resting was less frequent and of shorter duration than in untouched forest, while feeding bouts were markedly shorter. This corresponds to a reported increase in travelling; all orang-utan individuals were found to travel more frequently and over longer distances in areas disturbed by timber extraction. It was concluded that feeding efficiency is markedly reduced in logged over forest, also because a higher than normal proportion of leafy vegetation is taken.

Recent research within the framework of the long-term Kalimantan Peat-swamp Forest Research Project in the Sebangau catchment in Central Kalimantan (Page and Rieley, 1996) has once more reinforced these findings. Although it seems that, due to logistic constraints, logging in peat-swamp forest tends to have less, and quite localised, impact on the over-all forest structure, while regeneration may be more rapid and complete than in dryland forests, it must be realised that some of the prized timber trees in swamp forest represent the major food source for orang-utans, e.g. *Tetramerista glabra* (see Soerianegara and Lemmens, 1993).

> The Sebangau research area covers approximately 50 km^2 and comprises a range of swamp-forest types, including areas which have been logged over a period of some 20 years. The comparative density of orang-utans was estimated on the basis of the nest-count technique (van Schaik and Azwar, 1991). A clear correlation with degrees of disturbance and regeneration of the forest was found (Page *et al.*, 1995). While in undisturbed tall peat forest the nest counts indicated 2.2 ind./km^2, it dropped to 0.5 ind./km^2 in areas of high disturbance, and reverted to 1.2 ind./km^2 in (20 year old) regenerating forest, i.e. 50% of the original.

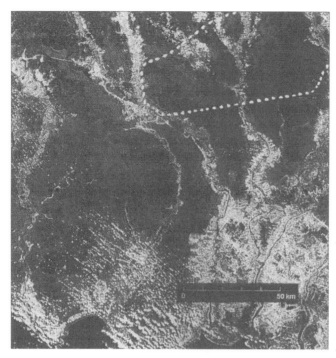

Composite satellite image of the Sebangau catchment in the left half of the illustration, and the KAKAB-PLG reclamation area in the right half – the township of Palangkaraya is seen in the upper middle section of the illustration (1997 image by Kalteng Consultants, Germany). Dark patches represent peat swamp forest.

The impact of logging on the ecosystem

A number of systems for harvesting timber from natural or wildland forests have been developed. These combine prescribed techniques for spatial information on harvestable trees (e.g. surveying or 'cruising') and rules for felling (species, size and age class of harvestable trees, technique, recovery of the logs, etc.) as well as for subsequent silvicultural practice, such as the eradication of climbers, hemi-epiphytes and non-commercial species, and the propagation of desired species.

What could cause the very considerable reduction in the numbers of orang-utans in logged-over forest habitat, even if a selective felling system is applied? The decline in orang-utan density observed by Rao and van Schaik (in press) was considered to be directly attributable to a recorded loss of fruit trees. This conclusion is supported by an evaluation of the most extensive study to date of the ecological effects of selective logging in a forest area in West Malaysia (Johns, 1983). Although orang-utans did not occur in Johns' study area, the ape once occurred there and the structure and composition of the forest is comparable, if not virtually identical, to that in Sumatra and (western) Borneo.

From Johns' list of 1138 trees (in a 2.08 ha plot) which were affected by the extraction of 38 of the 60 marketable tree trunks (i.e. 18.3 trees/ha or 3%), it is possible to deduce to some extent the effect of selective logging on food availability for the ape.

Two major groups of systems can be distinguished for Borneo and Sumatra:

(1) The monocyclic system, exemplified by the Malaysian Uniform System (MUS), where one logging operation harvests all marketable trees, and the cleared area is left to regenerate naturally until – in 60 – 80 years – a new cohort of mature trees is available. MUS caused extensive damage, which was aggravated because the surviving trees of no commercial value were usually eliminated by poisoning, and the cleared areas were often simply converted to other kinds of land use. Application of MUS was stopped in 1979.

(2) The polycyclic, or selective logging system, which is based on repeated logging operations on a rotation cycle of 25 – 45 years, relying on the growth of immature trees and the succession of saplings. In Indonesia, timber exploitation in the Permanent Forest estate is formally regulated according to selective felling guidelines (TPI – *Tebang Pilih Indonesia*) within the overall framework of the Basic Foresty Law and additional Acts. Major guidelines are that only trees over 50 cm diameter at breast height (DBH) can be harvested in Common Production Forest (HPH), while for Limited Production Forests (HPT) the limit is 60 cm DBH. For wildland forests of HPT status, at least 25 trees of 25-40 cm DBH must be left intact on each hectare. If fewer than 25 trees of the stipulated size-class remain, then enrichment planting must be carried out. The TPI system requires a full inventory (cruising) prior to logging, and is estimated to increase productivity.

Since the late 1980s, a lease contract for timber concessions must involve a total area in excess of 50,000 ha. The concession holder must present to the Regional Forestry Office for approval a five-year plan as well as annual harvesting plans based on accurate maps, indicating an assessment of the stock. Harvesting blocks of some 1000 ha are then formally allocated by the head of the Office. Felling near streams and rivers is not allowed and some 10% of the concession area must remain unaffected for 'biodiversity conservation'. Since the timber stock over 40 cm DBH is virtually depleted, and the timber industries in Sumatra and Kalimantan are collapsing, the regulations are under considerable economic and political pressures to be adapted. For instance, the range of species to be harvested may soon be expanded officially.

In Sabah three types of logging concession have been in force since the beginning of the timber industry, namely, full concession, special licence and form 1 licence. The full concession is valid for 21 years, allowing one selective harvest under a fixed annual quota according to an approved felling plan. The special licence is valid for ten years with similar constraints as the full concession, and the form 1 licence is valid for one year only, and has no restrictions. The Malaysian Uniform System has been applied since the early 1970s. There is no regulation of the extraction of timber and in Sabah the high-lead extraction method is commonly applied. This method, dragging logs along high-tension steel cables to a central spar tree (usually uphill) at a loading site, may be efficient, but it utterly destroys the forest so that regeneration is much delayed, if not impossible.

Of the 1138 trees tallied before logging started, some 657 were of 85 potentially edible species for an orang-utan. After the felling operations, 47% of these trees were left in a productive state, although the diversity was reduced to 23% (i.e. 20 species). The reduction appeared to be random; a similar percentage of all trees was unaffected (i.e. 554 or 48%), and even the target (Dipterocarp) trees suffered the same impact (42 of 87 i.e. 48%). Because the diversity of food types is as important for survival as quantity, these figures suggest that selective logging reduces the carrying capacity for orang-utans by more than 50%.

The harvest of some 3% of the trees in a plot falls well within the limits of the common interpretation of the selective felling concept. More common is the extraction of some 10% of the trees (Burgess, 1971); the lower economic limit of extraction is 8 trees/ha, a common range is 15 – 75 trees/ha (Burgess, *op. cit.*).

However, because this relatively low level of extraction had already resulted in the destruction of almost 51% of the trees, Johns concluded that the term *selective logging* is misleading in any case. So-called selective logging results in 'essentially random destruction' of the forest, especially when carried out in the absence of regulation enforcement. It causes not only more than a 50% loss of trees of all size-classes, but also some 68% loss of standing biomass. It is not uncommon that the extraction rate in the Dipterocarp-rich hill forests of Borneo is well over 25 trees/ha, with an additional 20-30% of the area being devastated and compacted for tractor (and skid) roads (Kartawinata, 1978; Brouwer, 1996). This implies that for orang-utans the habitat is virtually obliterated.

> It must be realised that despite removal of the standing tree crop, the Malaysian Uniform System, which was commonly deployed in Malaysia until 1980, was a typical sustainable forestry regime with a focus on timber production alone. It was a first step to transform, in a most economic way, a wildland forest into a timber plantation, deploying the natural recruitment of the largely removed forest. It had no concern whatsoever for any ecological function other than timber production, and considered all other species either as weeds or irrelevant. Thus, first of all climbers and lianas were deemed noxious and destroyed. Yet removing a major sector of the food species favoured by (e.g.) orang-utans, and then, after logging, killing the defective and non-commercial trees through 'poison girdling' (Whitmore, 1975) – also called refinement management. The application of MUS and its obligatory refinement management is believed to account for the fewer numbers of primates in even very old logged-over forest in West Malaysia (see e.g. Marsh and Wilson, 1981), and may well be a major cause of the demise of the orang-utan in most of the last forest refuges in Sarawak outside the Lanjak Entimau area.

It has been suggested that an initial loss of food-source trees due to extraction may be more or less compensated for by a greater incidence of fruiting in remaining trees, which would supposedly be stimulated by the opening up of the canopy (Chivers, 1973). Such a buffering effect has not been recorded for the common fruit trees favoured by orang-utans (Johns, 1992; Rao and Van Schaik, 1995; Rijksen, unpubl.). On the contrary, the record shows that because the diversity is affected through the disappearance of more than half the soft-pulp fruit-bearing species, there must occur a shortage of food, both temporally and qualitatively.

Johns (1983) recorded that the frugivorous white-handed gibbons (*Hylobates lar*) at Sungei Tekam in West Malaysia were obliged to shift to a more folivorous diet after selective logging. He also noted that the findings of the study by Davies and Payne (1982) in the same area demonstrate how the trophic structure of the rainforest avifauna changed after logging, with a reduction of more than 40% in frugivorous and insectivorous/frugivorous bird species. Both observations were clearly 'correlated with a lesser abundance of fruit and flowers following logging' (Johns, 1985: 361).

The dramatic decrease in food productivity relating to frugivores probably extends over long periods of time, since it has been demonstrated that 'selective logging disrupts the nutrient cycle (Herrera *et al.* 1981) to such an extent that in larger gaps

> A comparative study of the distribution of wild fruit trees of interest to indigenous people in untouched and logged-over forest in East Kalimantan (van Valkenburg, 1996) strongly supports the image of impoverishment due to commercial timber extraction. Several important soft-pulp fruit trees, such as (e.g.) *Mangifera spp.*, *Garcinia spp.*, *Lansium spp.*, *Durio dulcis*, *Durio kutejensis* and some *Artocarpus spp.* were found to be absent in logged-over forest plots and apparently had disappeared from the regenerating associations of forest trees. The results of the study, covering 11 different forest areas, suggest that 'the commercial fruit species are confined to the least affected logged-over plots and the primary forest plots' (van Valkenburg, *ibid.*).

(i.e. > 1000 m^2) some 30-50 years of atmospheric input of Ca, K, Mg and P, and about 260 years of atmospheric input of N is being lost' (Brouwer, 1996). Such losses will undoubtedly have a negative impact not only on fruit productivity but also on the establishment and development of new fruit trees, and hence on the long-term availability of food for frugivores.

Be that as it may, this complex of impoverishment with reference to habitat quality readily seems to explain why (infant) mortality among primates increased markedly after logging (Johns, 1983; 1985; see also Dittus, 1980), and makes it understandable why more than half the original orang-utan population disappears from a logged-over forest area: Sustenance is reduced dramatically for many years.

Thus, the exploitation of a natural or wildland forest has an impact which goes beyond the removal of a few individual timber trees; the major food sources of the orang-utan, i.e. the fruit-bearing trees and lianas, appear to be affected by collateral damage, ecological after-shock (e.g. exposure), or silvicultural measures ('liberation thinning') to stimulate the growth of commercial timber species. The degree of impact of such a primary logging effect is related to the number of trees harvested, the technique of felling and removal of the logs, the accessibility of the felling sites, the control of collateral activities (e.g. hunting and poaching for subsistence by the logging crews) and subsequent measures to prevent encroachment of the forest plot for slash-and-burn cultivation and arson.

> Bruenig (1996) emphasised that the essential steady-state or homeostatic conditions of a rainforest derive from the trees in what have been termed the A and B layers of the canopy; indeed the trees which are the target of logging and are most severely affected by timber extraction. The state of the canopy determines the quality and diversity of the environmental conditions within the forest structure and hence the regenerative capacity of all organisms, as well as the conservation of soil quality. Serious and extensive fragmentation – or removal – of the canopy therefore has a tremendous negative impact on both the living conditions of many organisms – other than typical pioneers and opportunistic herbivores – and the regenerative capacity of a diverse plant community.

In other words, the most crucial long-term primary effects of logging on a forest plant community due to canopy fragmentation are (1) the disruption of the hydrology of the area, (2) the exposure of the soil system to the vicissitudes of the climate (notably radiation/heat), reducing the number and function of soil organisms and allowing splash and sheet erosion of the humus layer and top soil, and (3) increased vulnerability to fire. This complex of impacts results in a serious delay in regeneration and succession, so that in terms of habitat quality the effects of prevailing logging practices are extraordinarily extended in time.

> The quantity of nutrients in a rainforest is the product of a complex living system encompassing the growth and decomposition of plant and animal biomass, especially the soil macro- and micro flora, and the hydrological conditions which may cause losses through leaching and denitrification, and gains through weathering and deposition. The topsoil contains virtually all dead organic matter and free nutrients (Lal, 1986). Hence, when the topsoil is exposed to erosion and mechanical displacement by bulldozers, nutrients and organic matter are readily lost (Gillman et al. 1985), while up to some 20% of the living biomass is removed in the form of timber in any case. Soil which suffered the impact of mechanical logging practice (especially on skid-roads and log landings) is compacted, has reduced infiltration and low water-retention capacity, is nutrient poor, and is prone to erosion; the soil organisms die off and few plants remain so that regeneration is barely possible. Uhl et al. (1982) estimated that it will take thousands of years to recover the original biomass at such sites.

The effect of poorly controlled selective logging immediately after log extraction; most of the remaining trees will die of exposure and desiccation.

The possible secondary effects of logging

If the primary effects of timber exploitation are serious in terms of forest and habitat degradation, the possible secondary effects are usually much more devastating. Cruising (i.e. inventory surveys) and logging operations open up hitherto uncharted and almost inaccessible forest areas. Then, in the absence of enforcement of the concession regulations, a whole sequence of destructive activities is usually initiated.

This chain of events often ends in the total obliteration of the forest ecosystem for the purposes of what is called 'conversion' or 'cultivation', and to the benefit of either an emerging industrial landlord or (a community of) small-holders. Arson – or the deliberate application of fire – plays a major role in this process.

The sequence of destruction, collateral to timber harvesting, is as follows: (1) trees are felled, their crowns removed and their trunks extracted from the forest along newly opened-up ski-roads, (2) the remaining crowns, as well as surrounding trees smashed in the process, and other vegetation, dry out due to exposure, (3) timber poachers use the skid-roads to extract more trees, (4) some gaps are occupied by squatters who clear a larger area around the gap, producing more drying debris, and, finally (5) the debris on the fringes of a cleared field are set ablaze. In times of a prolonged dry season, such arson creates fires that extend all along the skid-roads to adjacent gaps and cleared fields.

Owing to its ever-moist conditions, even at the end of a normal dry season a pristine rainforest cannot burn. There is not enough combustible litter concentrated in one place to generate sufficient heat to ignite surrounding vegetation. When such a forest has been affected as described, however, and a concentration of debris catches fire, for whatever reason, the flames are soon hot enough to affect the living (wet) forest in its surroundings, killing most of the remaining trees. Woods (1992) found that after the fires of 1982-83, the mortality of trees due to the combined effects of drought and fire in a selectively logged forest ranged from 38% for large trees to 72% for small ones (10-20 cm DBH). The regeneration capacity of burned logged-over forest is also affected dramatically; on average less than one Dipterocarp seedling per 25 m^2 survives, and instead of the normal colonization by pioneer trees such as *Mallotus spp.* and *Macaranga spp.* the regeneration is dominated by grasses (*Paspalum conjugatum* and, after regular fires, *Imperata cylindrica*) (e.g. Beaman *et al.*, 1985).

The same applies even more strongly to swamp forests where logging in combination with drainage can create enormous quantities of combustible material. Drained, dried-out peat will continue to burn and smoulder under the surface even in heavy rains, and the destruction of the vegetation cover is usually so complete as to entirely preclude regeneration of a forest cover. Burning rapidly removes tropical peat layers to a depth of several metres, exuding poisonous smoke and leaving a substrate which has markedly altered chemical and physical properties. A drained and burned area of tropical peat permanently loses more than 90% of its water retention capacity. After many years such areas may be colonized by only a dense mat of ferns (*Stenochlaena palustris*) and monotypic stands of *gelam (Melaleuca spp.)*. The ferns tend to die off during the dry season, creating a large mass of highly combustible material, which invites repeated arson.

All the fires which annually raze thousands of square kilometres of State forest land and farms in Borneo and Sumatra after a period of drought are deliberately set in logged-over and encroached forests. Such arson occurs with increasing frequency, and covers larger areas of State forest land every year. The fact that burning under such circumstances is usually uncontrollable, and causes poisonous smog, enormous economic losses and a considerable health hazard all over Southeast Asia, seems to be of little concern to those who set the dried-out vegetation ablaze. Estimates of the immediate economic losses range from US$ 26 million (official Ministry of Forestry assessment, *Jakarta Post* December 16th, 1997) to US$ 6 billion (*Int. Herald Tribune*, April 20th, 1998).

As long as law enforcement regarding forest exploitation and encroachment is absent, and scouring for non-timber forest produce as well as temporary settlement in forest areas is condoned, there is no hope that this disastrous malpractice can be contained. It is evident, however that the usual attempts to treat symptoms, such as training and equipping fire brigades, have no effect other than to add to the waste of money and resources. It is sobering to realise that in the period between 1982 and 1993 more than 22 international workshops and aid projects were devoted to forest-fire prevention in Indonesia, while – judging by the gargantuan extent of the arson in the prolonged dry season of 1997-98 – up to one hundred million US$ was wasted in assessments, talks, missions, training and the purchase of state-of-the-art fire-fighting equipment (see also Hartono and Sato, 1993).

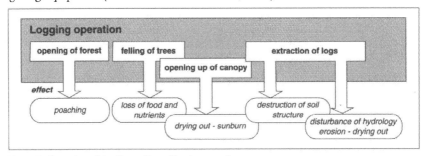

Schematic illustration of the direct effects of logging on a forest ecosystem.

After the 18-month drought in 1982-83 an estimated total area of some 33,000 km^2 of (regenerating and secondary) forest, *ladang* (i.e. a former forest area cultivated by means of slash-and-burn practice) and savanna was destroyed by fires in East Kalimantan, and some 10,000 km^2 in Sabah (Woods, 1992; Collins *et al.* 1992). After the drought of 1997-98, the most recent fires in Sumatra and Borneo affected a much larger area – i.e. at least 50,000 km^2 in Borneo alone. In Borneo all the affected forest areas were within the distribution range of the orang-utan.

Immediately after the fires, it became clear that all the trees in adjacent areas had died. It is considered that this was due to (heat) radiation and accelerated desiccation, as such areas had been unaffected by the fire itself.

If such extensive forest areas are destroyed, it is probable that the regional and continental climate of islands the size of Sumatra and Borneo is being permanently affected, leading to more pronounced seasonality, greater extremes in the periods of drought and more protracted torrential rains during shorter wet seasons. Such a climatic change will strongly facilitate further forest fires, and may eventually lead to a dry savanna-type habitat in Borneo and the eastern lowlands of Sumatra.

Ecology and natural history

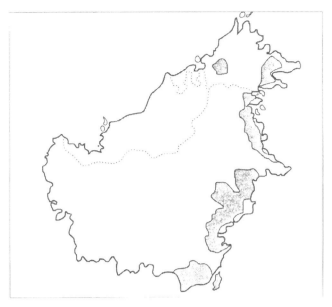

Map of Borneo showing the approximate extent of the fires in 1982-83.

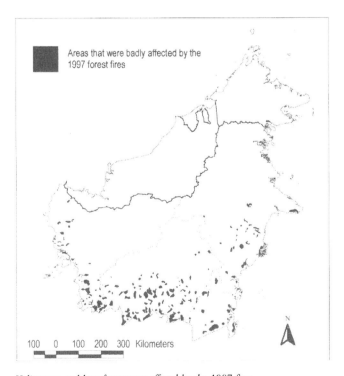

Kalimantan and how forests were affected by the 1997 fires.

The impact of logging on the socio-ecology of the orang-utan

Fluctuations in the trophic structure due to disruption of the forest structure are greatest among the mammals. The data show that especially the primate density decreases abruptly (Khan, 1978; Johns and Skorupa, 1987; also Dittus, 1980). Johns (1983) concluded that in particular the animals with a specialised diet are the most seriously affected by logging, and although for some opportunistic feeders (e.g. pigs, deer, rodents) the disruption appears to be a tremendous advantage, allowing them to increase in numbers, for many others, and in particular the orang-utan, it appears to be an 'ecological crunch' (Wiens, 1977). Six years after logging, the inconsistency of species composition is still apparent (Johns, *ibid*).

Thus, most fieldstudies both directly and indirectly support the common belief among naturalists that orang-utans are seriously disturbed by logging operations. The apes usually flee in response to the noisy operations, in particular the crashing down of trees. However, of the three social classes – residents, commuters, and wanderers – it seems that residents may often try to stay put. Indeed, their resistance to abandoning their range may be so strong that such apes are sometimes killed because they remained in the trees during the actual felling operations. Such cases may be at the basis of the rare reports that orang-utans do not flee logging operations (Davies, 1986).

It should hardly come as a surprise that logging crews and foresters consider that the density of apes has increased. The markedly higher incidence of 'long calls', given in response to the noise of crashing trees, often gives foresters and inexperienced naturalists the false impression that the numbers (of adult males) of a hitherto cryptic deme have increased. And in adjacent areas within hearing distance of timber operations, the density may in fact increase when residents flee and enter the ranges of their neighbours (Davies, *ibid*.; Rijksen, unpubl.). In this way the numbers of orang-utans in some relict patches of undisturbed forest may suddenly rise to several times the usual density of apes[31].

Evidence collected during the present study indicates that the initial disturbance is often exacerbated by illicit hunting, or by the deliberate harassment of any ape detected by the logging crew. The creature is then chased, cornered and isolated in a tree which is subsequently felled. If the ape survives the crash to the ground it is usually clubbed or slashed to death, to be butchered and eaten if it is an adult. If it is an adolescent or a juvenile, it has the dubious fortune of being captured alive, for 'amusement' or for sale as a pet. It should be comparatively easy to prevent this kind of serious mismanagement by having Timber Associations urge their members to abide by the regulations, punish any harassment of wild apes and supply timber crews with sufficient protein from domestic stock.

---

[31] Such an instance of temporary 'refugee crowding' due to logging was reported from conversion forest near Moara Wahau (East Kalimantan).

If dispersal away from the logging operations into adjacent ranges is possible at all, the majority of apes move (Davies, 1986; Rijksen, unpubl.). On such occasions the commuters may first try to retreat to one of their several major habitat patches or arenas, which, in the absence of seasonal abundance, soon causes what may well be called 'refugee crowding', i.e. an overshoot of the carrying capacity and considerable social unrest. Such disquiet is unlikely to be readily recognised by an observer, however; not only because the introverted and cautious nature of the ape would render any overt actions unlikely, but also because any increase in ape activity and social interaction would be so welcomed in the usually uneventful routine of orang-utan observation that an observer might easily accept it as a new norm.

In any case, the generalised famine in what used to be excellent habitat for anything up to a dozen residents, combined with the now permanent addition of several commuters, quickly reduces the carrying capacity of an area. This implies that eventually all members of the deme ultimately face starvation and are obliged to become wanderers, moving along into increasingly less familiar territory. Such a refuge is either occupied by unfamiliar orang-utans (up to carrying capacity level) or is covered with forest of such poor quality that no ape could subsist there in the first place. Thus, most of the displaced apes are bound to starve anyway.

Therefore the primary effect of logging a particular area of habitat causes the expulsion of some apes and a serious reduction of habitat – in extent as well as quality – for all. The secondary effects are, however, more serious: Logging which leads to the displacement of some apes can readily cause what may be called a 'shock-wave' of refugee impact on surrounding populations and on the carrying capacity of the invaded habitat. Such a situation soon leads to general starvation (see e.g. Hardin, 1968; Catton, 1980). This breakdown of the ape population in conjunction with the habitat continues until it meets equilibrium again at a remaining base level of carrying capacity. It is probable that for some areas this will readily lead to the extinction of the sensitive ape population.

It will be evident that a similar shock-wave of socio-ecological impact can also be caused by the introduction of (rehabilitant) orang-utans into a resident population (Rijksen, 1982). It may be assumed that an existing wild orang-utan population lives close to the carrying capacity of its range; the high incidence of commuters in all studied demes may well reflect the carrying capacity limit of a particular area of habitat.

Finally, field research elsewhere has revealed that 'fragmentation of the forest canopy (...) increases the vulnerability (...) to predation' (Skorupa, in Johns, 1985). It can also influence the susceptibility of animals to disease, due to a general decline in physical condition, and may facilitate a higher probability of infestation due to a greater incidence of parasite and disease vectors in the degraded and more intensively utilised habitat. Thus, in conclusion, logging, and in particular industrial timber extraction, generates such an impact that the deme will suffer a disproportional loss of members, even if orang-utans are not directly targeted for harassment.

> In principle many of the negative impacts of timber extraction can be (and to some extent are formally) regulated so as to minimise the damage and supposedly optimise regeneration for new timber production, and for the conservation of some biological diversity. Ecologically sustainable timber extraction from a wildland forest is not only feasible from the viewpoint of nutrient conservation (Brouwer, 1996), but may also allow a biological diversity structure to be conserved if (1) the logging blocks in the annually allowed allocation (RKT) are well planned according to ecological rather than economic dictates, (2) the gaps that result from extraction are smaller than some 700 m^2, (3) the number of trees or m^3 of wood extracted per ha is limited to less than 2%, (4) collateral damage is minimised and the extraction of the logs is carried out with appropriate care, and (5) no 'silvicultural treatment' is conducted.
>
> The Indonesian selective logging system provides a framework for the regulation of harvesting which can minimise logging impact in a forest area (Smits et al. 1992). That regulations are commonly flouted is primarily due to a lack of adequate control.

In conclusion, the orang-utan is an extraordinarily sensitive species with regard to habitat disturbance. It is specialised in feeding from particular communities of tree species occurring in patches which may take up no more than 50% of a larger forest expanse, and usually cover less than 35%. These patches are called ecotones, and are areas where one is likely to find the highest biological diversity in that particular expanse of forest.

As a consequence of its sensitivity, and its particular habitat requirements distributed over a large expanse of forest, the orang-utan can therefore be considered a major representative of the biodiversity of rainforest in western Malesia. In the field of nature conservation such a position is referred to as a 'focal species', or 'umbrella species', i.e. an organism whose existential requirements are believed to encapsulate the needs (or 'niches') of the largest number of other species in the particular ecosystem type under consideration (e.g. Murphey and Wilcox, 1986; Noss, 1990).

A HISTORY OF HUNTING AND POACHING

Prehistoric hunting

The caves of the Padang highlands in West Sumatra, at Niah in Sarawak and at Madai in Sabah retained evidence of prehistoric hunting in the great quantities of orang-utan remains (Hooyer, 1948; 1961) that had been left behind by tribal hunters (Medway, 1979; Bellwood, 1985). This, and more recent evidence of persistent persecution of the ape, suggests that the disappearance of orang-utans from several of the permanent Holocene land areas was due mainly to displacement and persecution by *Homo sapiens*. The ape became locally extinct in the largest part of Sarawak's coastal lowland forests north of the river Rejang, and including Brunei; it vanished from much of the flood-plain area of the Kapuas river, and from the fertile valleys of the Kayan Mentarang area in East Kalimantan; and it disappeared from virtually all the lowland areas east of the Barisan mountain chain in Sumatra. It completely vanished from the Thai-Malaysian mainland, as well as from Java. The reasons for displacement may have been manifold, including, for some people, a common predatory incentive, or competition for a fruit crop, and, for some specific tribes, a religious incentive akin to the head-hunting tradition (Rijksen, 1978).

Traditional hunting

Hunting of orang-utans for subsistence and/or religious purposes has continued in more recent history; most of the Bornean tribal peoples and those on Sumatra's east coast are known to have hunted the ape. Yet, the most persistent and effective hunting impact has been exerted by the archaic hunter-gatherers of central Sumatra and Borneo.

Seven major tribes of hunter-gatherers have been identified in Sumatra: the Abung, the Kubu, the Mamaq, the Sakai, the Akit, the Lubu and the Ulu (Loeb, 1972). All are considered to be of 'Proto Malay' stock, and they live(d) on the eastern side of the Bukit Barisan mountain range, occupying the banks and dry ridges of the extensive flood plains and peat swamps of eastern Sumatra south of lake Toba. Interestingly, their traditional belief system prevented them from hunting in the higher hills (*Pegunungan Tigapuluh, Pegunungan Duabelas*) and the higher foothills of the Barisan mountain complex to the west. Their traditions and culture had strong affinities with those of the traditional Batak communities (around lake Toba) before colonial domination, and with the Mentawaians. The early ethnographers noted that the favourite prey of all these tribes comprised 'monkeys and apes', which were hunted using blowpipes with poison-tipped darts, or with dogs and spears.

At the close of the nineteenth century the German explorer B. Hagen, in his booklet *Die Pflanzen- und Tierwelt von Deli* (1890), reported that the orang-utan was already rare in the forests of Deli and the northern Batak country (i.e. the present districts of Simalungun and Tanah Karo) because the Bataks frequently hunted it. He noted that the old staffs indicating the supreme status (*tunggal papaluan*) of Toba and Karo Batak chiefs (*raja*) as well as the magic wands (*tongkat malehat*) of lower ranking shamans (*datu* and *guru*) were often decorated with orang-utan hair[32]. Loeb (1935) described that in order to draw the spirit of the *pangulubalang* into the staffs, a human sacrifice had to be made involving a kidnapped child, who was fed beforehand with the best of foods, 'and especially the liver of apes.'

> At present only relics of these tribal entities survive; the majority of the people have been settled and hence are absorbed into the common Indonesian mainstream. According to Loeb (1935), the Abung were the original inhabitants of Lampong, along the river Abung; they were notorious head-hunters, at times organising themselves into bands of raiders. Sometime around the 14th century the political changes and conquest of Sumatran territories by outsiders may have caused bands of such people to move to Borneo, founding the Iban tribe.
>
> Traditionally the Kubu ranged in the swamp forests between the rivers Musi, Rawas, Tembesi and Batang Hari. They considered the hills and mountains *tabu* because they were the abode of the spirits (*halom dewa*) (Persoon, 1994).
>
> The Mamaq people were considered to be of Veddoid origin, inhabiting the swamp forests in the regency of Indragiri on the right side of the river Kuantan; they were separated from the Kubu by the (Malay) Sultanate of Djambi.
>
> The Sakai ranged north of the Mamaq in the alluvial plains of the regency of Siak, along the upper stretches of the river Mandau (a tributary of the river Siak), and the upper reaches of the river Rokan.
>
> The Akit were a 'mixed race' of coastal people, ranging close to the mouth of the rivers Mandau and Siak.
>
> The Lubu and Ulu were closely related, living in the foothill valleys of the mountainous Toba plateau rim, the current regencies of Padang Lawas, eastern Angkola and Mandailing.
>
> The archeologist Robert van Heine-Geldern (in Loeb, 1972) noted that around the first millenium the Batak people had for several centuries been under the direct colonial influence of tantric Vajrayāna Buddhism, combining Buddhist and Sivaitic elements. A number of ruins at and near Gunung Tuwa in the county Padang Lawas have retained evidence that a large Indian monastry (*biara*) must once have stood in the area. The Asian missionary monks were devoted to Lokanātha, a four-armed image of Bodhisatva Avalokiteśvara, Lord of the World, and forced their subjects to sacrifice to Heruka, a god who demanded flesh and blood, preferably human. Van Heine-Geldern notes that this gave an impulse to the notorious 'magic' cannibalism of the Bataks, which survived until the end of the nineteenth century. It will certainly have had a serious impact upon the orang-utan population southeast and east of Lake Toba.

In 1883, in his report on the customs of the peoples of Borneo, Mohnicke stated:

> (...) the [orang-utan] meat is either cooked or sun-dried. It is supposed to have a somewhat sweet, but nice taste. Europeans and Malay people never eat the meat. The Dayaks, and part of the Chinese population, however, seem to enjoy the consumption of this meat.

In his review the *Album der Natuur* Van Balen (1898) commented that orang-utans were hunted by *Oeloe Bejadjoes* from the river Kahayan water catchment area and

---

[32] In the book 'The Batak, Peoples of the Island of Sumatra' (A. Sibeth, 1991) several of the illustrations of old *tongkat malehat* provide evidence to support Hagen' observation, e.g. pp. 127, 129 and 131.

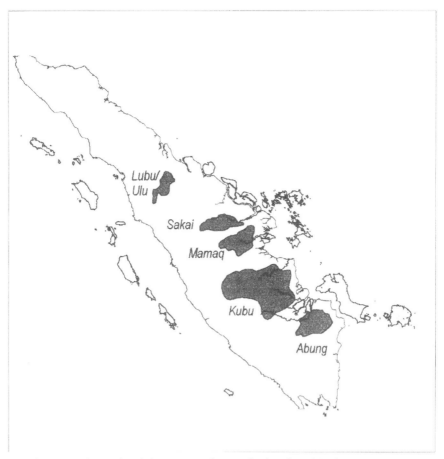

Map of Sumatra indicating the tribal ranges up to the time of Independence (1945).

from the river Kapuas, by Dayaks from *Sampiet, Kotaringin* and by 'others', and he noted that 'the continuing persecution has already driven away the orang-utan from many areas.' He echoed Mohnicke's finding, and adds that 'the adult apes are usually very heavy and fat', and that 'the whitish meat (...) is unpleasant if not repulsive of taste for those who are not used to it.'

During his expeditions in what is now West Kalimantan province, W.L. Abbott (in Lyon, 1911), also noted that the Dayaks were extremely fond of orang-utan meat. Echoing van Balen, he also stressed that it was useless to look for orang-utans anywhere in the neighbourhood of a 'Dyak settlement.' In 1938, Westermann observed that despite the improved Animal Protection Ordinance of 1931, the Dayaks still hunted orang-utans, for the meat, or for the illegal sale of the juveniles, which, in the upper Kahajan, sold for some US$ 3-6. According to Dammerman (1937) the Dayaks usually hunted the ape by means of a blowpipe (*soempitan*) charged with poisoned darts that had been tipped with a lethal mixture of the sap of the tree

Antiaris toxicaria and of the liana *Strychnos ignatii*. Ironically, the fruits of these lethal species are highly favoured by the ape.

Again, van Balen (1898) also mentions that the ape was frequently hunted by the Dayaks in order to catch an infant, which was then raised in the longhouse 'for the amusement of the clan.' He refers to the report of naturalist S. Müller who, in the early nineteenth century, had bought just such a tame ape for 40 *realen*, approximately US$ 400, which was a considerable amount of money at the time. Van Balen had already acknowledged that an infant or juvenile ape could be caught only by killing its mother.

Nevertheless, the gaps in the current ape distribution pattern suggest that the wide-ranging forest-dwelling gatherers of Borneo, the Punan (Bukitan, Bakitan, Ketan, *tau'ukit, tau toan*) and Penan, had the greatest impact. Unfortunately, accurate eye-witness records of the orang-utan-hunting activities by these efficient hunters are rare. This is undoubtedly because the ethnography of these peoples remained poorly documented until very recently (Kedit, 1982; Hoffman, 1986, Langub, 1989), after acculturation had exerted its impact over several decades, and the apes had vanished from their respective ranges. It is also unfortunate that their culture gives little opportunity for the conservation of artefacts (e.g. skulls, skin, teeth, etc.) as with the more sedentary Dayak shifting cultivators. Yet it may be inferred from their common hunting culture (Kedit, *op. cit*; Hoffman, *op. cit*.; Brosius, 1986, Langub, 1989; Tillema, 1990), and a preference for animals with a high body-fat content (e.g. bearded pig), that they must have had an effective exterminating impact on the orang-utan in the lowlands and alluvial valleys. After the fruiting season, orang-utans may be overweight, carrying reserves of fat in preparation for a subsequent lean season. Unfortunately, the more powerful and predatory human hunter-gatherers in the forest suffer such seasons as well.

Brosius (*op. cit*.) estimated the hunting success of the Bornean Penans at 90%, which is extremely high for traditional hunters. Many of the closely related Kenyah also openly admit that they consider all primates, including orang-utans, to be prized food (Chin, 1985), while during the present surveys in Central Kalimantan local

> In a discussion during the UNESCO Symposium on Ecological Research in Humid Tropics Vegetation (1965) between the Government Ethnologist and Curator of the Museum, T. Harrisson and the US botanist F.R. Fosberg, the latter asked: 'Was the disappearance of the large animals perhaps the result of the Dayaks' use of blowpipes with poisoned darts?'. Harrisson answered: 'Yes, (...) the development of a highly effective blowpipe in Borneo was extremely important in the development of fauna and related flora patterns. The Dayak blowpipe (...) is far superior to the related weapons in Malaya (bamboo) and elsewhere. It (...) can kill a rhinoceros. I think this [effectiveness] plus the aggressive, masculine values of the Dayaks themselves, contributed to a great reduction in game before the introduction of modern fire-arms which have, of course, greatly accelerated the process. There are 60,000 licenced shotguns in Sarawak alone, as of 1963!' (Harrisson, 1965).

people not only classified orang-utan meat as *paling enak* (or most delicious) but also as *obat kuat* (i.e. sexual potency-generating medicine).

Important circumstantial evidence for a religious incentive in the hunt for orang-utans comes from the many ethnic artefacts in which parts of the ape have been used. Skin and hair of the ape was used up to the present time for war-cloaks, jackets and caps, and to decorate the handles of the *mandaus* (swords) in order to confer *semangat* ('soul-power') onto the wearer (Schlegel and Müller, 1839-41; Brooke, 1841; Ling Roth, 1896; Nieuwenhuis, 1900; Harrisson, 1960; De Silva, 1971; Rodman, 1973; and Rijksen, 1978). The 19th century artefacts, in addition to the rich folklore involving the ape (Ling Roth, 1896), reveal that in connection with the head-hunting tradition the indigenous tribes attributed a special religious significance to all preserved body parts of an ape. Indeed, well into the twentieth century many Dayak hearths retained one or several traditionally prepared orang-utan skulls hanging amidst the human-skull trophies from ancient head-hunting raids (Hornaday, 1885; Beccari, 1904). Several museums have conserved characteristically prepared orang-utan skulls among their collections of head-hunting trophies (Rijksen, 1978). Such orang-utan remains also featured prominently in the major religious ceremonies until at least the end of the colonial era (MacKinnon, 1974; p.28).

Conversely, it is certainly known that the traditional religious system, i.e. *adat*, of some families or clans within tribal groups prohibited or restricted the hunting of the ape, i.e. a *totem*. Even among the Batak people in North Sumatra, who were once scorned by their (Muslim) neighbours for possessing a barbarous appetite, eating 'anything that moved,' one could find the occasional clan (*marga*) that did not touch a particular *totem* animal. Members of such a clan might traditionally retain a family taboo with respect to hunting, harassing or sharing the meat of, for instance, orang-utans. However, clan members were indifferent when their nearest neighbours hunted and ate the ape. The orang-utan as a *totem* has also been documented for Iban clans (e.g. Bennett, 1993).

> Some authors have reported such a taboo on the consumption of orang-utans (Hose and McDougall, 1912; Medway, 1976; Bennett, 1993), suggesting that the taboo is generally held by the entire tribe. However, the tribes or cultural units supposedly maintaining this taboo are at the same time known to have a hunting record with respect to the ape. Therefore the contention cannot be accepted without serious reservation, and may in fact apply only temporarily for some period before or after a festival, or exclusively to particular families (i.e. a totem) rather than to the whole tribe. In any case, many modern Indonesians, and in particular newly converted Muslims of tribal descent, when confronted with the question, will strongly deny in public that they ever hunted or ate the meat of an ape – for its supposed magical properties – although they may nevertheless have retained the habit.

A totem function of the ape is also indicated by the fact that the orang-utan has featured in several tribal myths. It was often acknowledged as the aboriginal king or lord of the forest (Brooke, 1866), and usually represented a temporary or permanent transformation of a legendary clan elder. In legends it was depicted as having a predilection for kidnapping – and siring offspring on – human women.

Among the Iban the *maias* was seen as the representation of the 'war god' (i.e. *Klieng*) (Hose, 1926). Thus, some quasi-sedentary farming tribes entertained a taboo with reference to the ape which may have safeguarded it against regular persecution, but in all tribes it could nevertheless be hunted for its head on special occasions, such as to initiate a war ceremony. For some ethnic groups the 'soul substance' (*semangat*) represented in the ape's head was valued even higher in this respect than that of a human (Brooke, 1841).

It is, however, hardly relevant to argue about whether the hunting of orang-utans is, or was, motivated by religious, magical or food-related incentives when the cultural context is ignored. Tribal societies in transition, when exploring new frontiers – like the Iban for the last two centuries – or when being subjected to acculturation and religious conversion, usually come to ignore traditional religious taboos and customs, and advance to opportunism with respect to their natural environment (see Chapter Poaching).

Great cultural changes in the tribal communities of Borneo were induced through external trade relations during the early nineteenth century, through missionary activity and subsequent colonial rule. These changes and the conquest of the Iban from Sumatra, moving from coastal West Kalimantan northwards into Sarawak during the nineteenth century (Freeman, 1960), caused mass migrations of indigenous tribes throughout the large island. It not only threw virtually all tribal entities into transition and a constant fear of warfare, but also caused the original

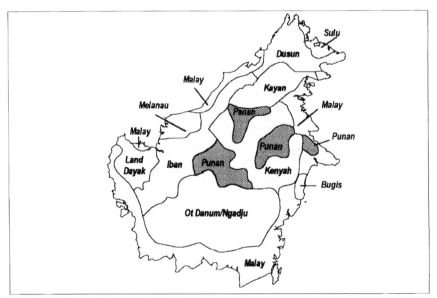

Locations of major indigenous tribal entities in Borneo; all the so-called Dayak, or Dyak, peoples have hunted orang-utans.

mythical image of the ape in Borneo to become diluted or to fade away in the shadows of modern consumerism.

> In his discussion of the survival prospects of the Sumatran rhino in Borneo, Banks (1931) inferred that the surge in head hunting which took place during the late nineteenth century gave the wildlife a temporary chance of recovering, because large areas of forest were avoided by hunters fearful of roving bands of (Iban) head-hunters. He noted that immediately after the colonial ban on head-hunting was enforced, many areas were 'opened up' again for hunting parties, leading to the rapid demise of rhinos in Borneo.

The more recent mass migrations into Borneo and Sumatra by 'transmigrants' and pioneer entrepreneurs have spurred the cultural development process. Be that as it may, much of the traditional regulatory system or *adat*, and in particular its *tabu* structure, has now vanished[33] in the age of communication and education. Communities which would traditionally be prevented from hunting the ape except on special ceremonial occasions have lost the socio-religious mechanisms to prevent their members from violating the outdated *adat*, especially when it concerns the *tabu* structure that refers to land use and the animistic relation with wildlife. The belief in magic powers often lingers on, however.

Tribal youths trying to sell forest produce of any kind, here a slow loris. (Sumatra, 1996).

---

[33] Generally the supposed land-use 'ights'and customs which came with *adat* have been retained expediently in people's mind, but invariably the powerfully religious *tabu* structure concerning **the constraints** of land use and the relation with wildlife was abolished during the colonial period through missionary activity, and, since independence, through modern education; any traditional restrictions remaining refer solely to socio-sexual mores.

Despite the common illusion of tribal societies living in harmony with nature, a notion falsely perpetuated in tourist brochures and in richly illustrated coffee-table books, the indigenous tribes of Sumatra and Borneo at the close of the twentieth century are scarcely different from other cultural entities in transition. Following careful anthropological scrutiny, the majority of subsistence farmers, forestry labourers and forest scroungers of indigenous descent can only be classified as modern Indonesians or Malaysians, no matter how remote the area in which they live. In dress, language, personal perspectives and desires they are virtually identical to the average person in the streets of Jakarta, Kuala Lumpur, Medan, Kuching or Kota Kinabalu (see e.g. Vayda *et al.*, 1980). In the view of a Sumatran or Bornean peasant, whether indigenous or transmigrant, the orang-utan has either an economic, a 'medicinal', a nutritious or a nuisance value, any one of which warrants its persecution.

Scientific collecting

The collecting of specimens by western scientists has added to the impact of indigenous people on the local distribution of the ape, perhaps not so much in terms of absolute numbers as in terms of amplifying the false image of the ape as valuable big game (see Brooke, 1866; p. 64). Nevertheless, collecting and hunting-for-sport expeditions in the nineteenth and twentieth century cost the lives of many hundreds of orang-utans. Mohnicke (1883), for instance, shot fifty of them between 1852 and 1854, mainly in what is now West Kalimantan. The famous explorer Beccari (1865) shot twenty-four apes, and another forty-eight were shot by Abbott so that they could be preserved (Lyon, 1911). A further twenty were bagged by te Wechel (Hooijer, 1948). Alfred R. Wallace (1869), the famous professional specimen collector and discoverer of the principle of natural selection, also collected seventeen orang-utan specimens, not for his research, but to sell to the highest bidder among European museums and private collections. In his conquest of Sarawak, the adventurer who became the first white *Radja*, James Brooke (1866), shot five orang-utans, and the renowned self-proclaimed conservationist Hornaday (1885) bagged another forty-one. However, Selenka (1896) must have the world record, with a total of two-hundred and seventeen orang-utans shot and collected in an amazingly small area in West Kalimantan.

It should not go unmentioned that all the collectors killed an even greater number of apes than they took home and recorded; many of the prospective specimens were simply left to die among the canopy branches if they could not be readily taken. Notably the smaller specimens, after being shot and becoming stuck in the canopy of a tree, were rarely retrieved, because they were reportedly 'of comparatively little interest' anyway (Wallace, 1869; p. 40). One seriously wonders what the scientific point of such carnage could have been.

Amateur collector in colonial times; a note written on the back of the old photograph says: 'Greeting from the land of the head-hunters; East coast of Borneo.'

Commercial collecting

After 1924, the hunting of orang-utans became illegal. Yet, as has been mentioned, the persecution did not cease. Particularly notorious was the Dutch professional animal collector van Goens, who specialised in capturing the more spectacular adults rather than trading juvenile apes which had been caught by local people.

In 1927, van Goens returned to Amsterdam with a shipment of 25 orang-utans, which he sold for 25,000 German Marks a pair. Four months later he imported a new group of 33 orang-utans, which he sold to the American circus belonging to John Ringling, and in 1928 he returned to Amsterdam again with 44 orang-utans, all caught in Sumatra. Zoo records show that at least 218 orang-utans were exported from the Netherlands East Indies between 1924 and 1943 (Jones, 1980). In 1946 van Goens was the first to export adult apes again, immediately after the end of the Pacific war.

Sumatran adult male orang-utan caught to be shipped to a European zoo (1928).

The Zoo records of the western world reveal that between 1946 and 1978, when the international ban on the ape trade took effect, a total of at least 809 individuals were taken from the wild to be traded and exhibited in zoos. Most of these were juveniles. Since not all animal collectors were as professional as van Goens, who captured only adults, it is worth bearing in mind that each juvenile found in illegal custody reflected at the least one dead mother and possibly two or more other orang-utan casualties which remained undetected (Rijksen, 1982).

Direct references to poaching of orang-utans in the post-1930 period are extremely rare, but when one looks through photo albums and browses through army patrol reports, the impression is that poaching remained a major threat to the existence of the ape throughout its range, and continued to be a threat after Independence. In 1964, for instance, 136 orang-utans, which had been caught in the Gunung Leuser wildlife reserve in northern Sumatra, were sold in the Alas valley (Rijksen and Rijksen-Graatsma, 1975). However, one is unlikely to find references to poaching in Indonesian records.

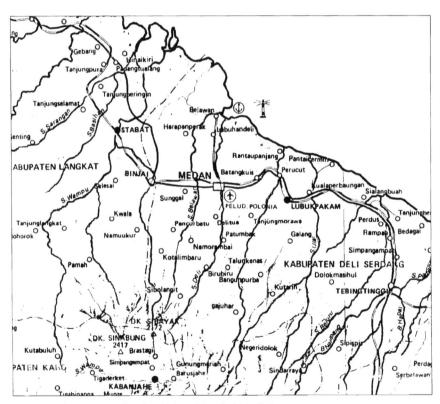

Geography of the environs of Medan (Deli and Langkat) North Sumatra.

In the late 1960s, first the International Zoo Association and then the IUCN recommended that international trade on orang-utans should be banned. In 1973, 21 countries signed the Convention on International Trade in Endangered Species (CITES) and the orang-utan was listed so that 'trade in it or its products are subject to strict regulations, and trade for primarily commercial purposes is banned.' Since the international ban, only a local or regional market has remained wherever law enforcement was lax or absent.

Orang-utans have been, and are, persecuted for a variety of reasons, but a commercial consumer-related incentive for body parts of the ape seems to be fairly recent. Notably army personnel have been reported to go on organised poaching expeditions targeting orang-utans for the acquisition of meat, as well as fat (oil or *minyak*) with supposedly magical (aphrodisiac) properties to be marketed. It is probable that a misguided belief in these magical properties also fuels the voracious consumer market for smuggled orang-utans in Asian nations, notably Taiwan, Hongkong, South Korea and China.

However, the economic marketing principles of supply and demand obviously determine the high degree of illegal persecution of the ape in the absence of law enforcement. The massive increase in tourism to Kalimantan during the late 1980s facilitated the sale of ape skulls as a (fake) traditional head-hunter artefact. Ignored by authorities, an international demand through the timber, plywood and shrimp trade meant that the smuggling of incredible numbers of orang-utan juveniles could re-emerge. It may be estimated that less than 0.5% of the illegally exported apes have been detected, invariably at transit points some distance from Indonesia and Malaysia.

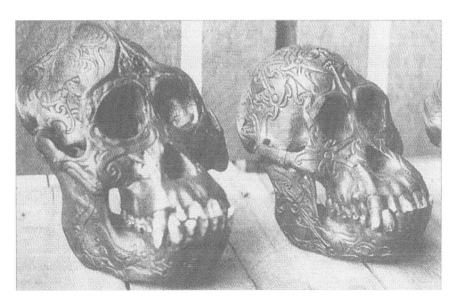

Hundreds of orang-utan skulls have been offered for sale during the 1990s.

With regard to the supply, it can be said that since the 1980s the increase in transport means and infrastructure within the remotest forest regions, due to timber exploitation and forest conversion, have tremendously increased the chances of human-ape contacts. If specialised animal catchers formerly set out to supply a demand, the post-1990 record of confiscated apes indicates that the majority of illegally captured orang-utans are a by-product of forest exploitation and conversion. Virtually all confiscated orang-utans originate from logging concessions, the immediate surroundings of new industrial plantations in converted forests and in particular from forests near new transmigration sites. Transmigrants are often settled in very remote regions and are usually forced to eke out additional income by looting the surrounding forest environment.

The popularity of orang-utans as pets is not a recent phenomenon. During the age of traditional slave-keeping, less than a century ago, the acquisition of an infant orang-utan may have been a major incentive for the hunt in the Dayak community, as was reported by Van Balen (1898), or it was the spoil of a hunt for the skull and meat of its mother. Yet Bock (1882) noted:

> '(...)[he] had a couple of large orang-utans, male and female, which had been caught in the interior. The big male was in a consumption, and lay most of the day wrapped up in a blanket, his great frame shaken incessantly by a terrible cough, which soon carried him off. His mate, (...), was apparently in good health, but on my return three months afterwards, she too had gone the way of all orangs. The Malays of Samarinda catch the orangs near the small creeks and streams falling into the Mahakam near the town. They told me that the animals only come to the banks early in the morning, returning during the day to the jungle. When they catch one alive they sell it for three dollars to the Chinese, who feed the animals first on fruit and afterwards on rice, but never succeed in inducing them to live long in confinement.'

These early accounts indicate that for more than a century, and probably for much longer, not only infant and juvenile orang-utans but also those of at least sub-adult size were caught to be kept as pets. Indeed, many of the earlier reports and diaries of naturalists in Borneo and Sumatra were enlivened with illustrations of immature – and usually miserable – captive orang-utans. MacKinnon (1977) believed that in Sabah young orang-utans were kept in Dayak longhouses for sexual entertainment well into the 1970s. In any case, the ape's human-like appearance, extraordinary sexual drive and keen intelligence has certainly contributed to its popularity in such malpractice.

In the 1980s, movies and TV soap operas created a new demand for pet orang-utans, especially in the Far East. Thus the popular movie *Any Which Way You Can*, in which a trained sub-adult male orang-utan featured as a main character next to the Hollywood star Clint Eastwood, and a Taiwanese television sequel entitled *The Naughty Family*, in which a young adolescent orang-utan named Hsiao Li featured prominently, were said to have initiated in Taiwan a booming demand for pet orang-utans (Lee, Phipps and Chen, 1991). The media abuse of orang-utans has not stopped; in 1996 another Hollywood production called *Dunstan Checks in Again* features a trained red ape as a leading character next to the movie star Faye Dunaway.

While it may have been difficult to obtain such a pet in the United States[34], in Taiwan it was very easy, because this country was not a signatory of CITES until 1990. In the period 1987-1990 the demand was so great that more than a thousand (and possibly as many as two thousand) orang-utan youngsters were smuggled out of Indonesia to supply a private market which was reportedly prepared to pay anything between US$ 11,000 and 20,000 for an individual ape (C. and E. Martin, 1991).

Quite distinct from the illegally kept orang-utans, hidden in private homes or shops, a considerable number of red apes are being kept in zoos and other institutes (excluding the rehabilitation centres). In particular in the United States of America several private institutions and commercial animal trainers keep orang-utans for movie performances and dubious amusement sessions (e.g. Berosini at Las Vegas, Nevada, Dunn at Sylmar, Wings at El Toro, and Michael Jackson at Neverland in California).

In the late 1960s, the Yerkes Regional Primate Research Centre instigated the registration of captive orang-utans, and in 1970 published a first studbook to facilitate breeding of the ape in captivity (Bourne, 1971). The research centre kept one of the largest collections of apes in the world, including 30 orang-utans, initially meant to be used for biomedical experimentation with relevance to biological warfare. In 1976, the Yerkes RPRC was subjected to financial constraints (Perry, 1976) and the authority for the studbook was transferred to the Zoological Society of San Diego, where it was administered by Marvin L. Jones (Jones, 1980). Whereas in 1969 some 460 orang-utans were reportedly held in captivity throughout the world, this number had increased to 625 apes in 1974. Of these, some two-thirds had been born in captivity (first generation), and an estimated 337 had been acquired in the wild, while some 179 wild-caught specimens had died (Perry, 1976). In 1980, a total of 778 orang-utans were held in captivity in over 160 collections. At the time, the Indonesian zoos of Surabaya (29) and Jakarta (27) held the second and third largest collections (after Yerkes RPRC).

In 1982, when segregation became an issue, the three captive populations of Sumatran, Bornean, and hybrid orang-utans comprised approximately 45%, 32% and 23%, respectively (Jones, 1982).

In 1989, the authority was transferred again, to the Atlanta/Fulton County Zoo. The current keeper of the international captive orang-utan database is Ms Lory Perkins.

The 1995 International Studbook has registered a total of 913 living captive orang-utans in 216 different collections; 377 of Bornean, 319 of Sumatran and 189 of hybrid (and 28 of unknown) origin (Perkins, 1995).

The largest numbers are to be found in western Europe, where 303 orang-utans are kept in 68 zoos; the second-largest numbers are in the USA, where 268 individuals are held captive in 64 different locations; in Indonesia some 80 are kept in five zoos, in Japan 64 in 29 locations, in Australia 27 in four zoos, in Singapore 27 in one zoo and in Malaysia 11 in two zoos (Perkins, *op cit.*).

---

[34] Owing to the persistent vigilance of the International Primate Protection League, a notorious international animal trader, M. Block, was sentenced to 13 months in prison, with three years probation, and a US$ 30,000 fine for conspiring to violate the US Endangered Species Act and CITES, by arranging an illegal shipment of six orang-utans which were discovered at Bangkok Airport in 1990.

Markham (1991) has evaluated the situation, and has concluded that 'the captive orang-utan is being demographically mismanaged on a large scale due to a lack of properly constructed species-management plans (...) [which] should be concerned with adapting the environment to suit the individual species, not forcing the species into preconceived niches in our zoos.' She found that 'over half the collections have unacceptably high mortality rates', due to stress- and obesity-related ailments, such as heart and kidney diseases, and to maternal mortality. 'Over 80% of captive orang-utans die before they are 25 years old, and the vast majority of orang-utans surviving infancy die before they have produced more than two offspring.'

More important, Markham also noted that 'captive births have begun to decline, and wild-caught imports have begun to rise. Between 1985 and 1990, the number of officially reported wild-caught imports accounted entirely for the increased population in those five years.' Some 63% of the wild-caught imports concerned Bornean apes. How this can happen under the scrutiny of CITES should be a matter of great concern.

Interestingly, the captive-breeding record of the (smaller) Sumatran population in zoos is better than that of the Bornean orang-utans; both the percentage of successful breeders and the average number of surviving offspring is higher. The breeding record for the captive Bornean population is well below replacement level. In view of such alarming facts, it is sad to note that some 22% of all captive orang-utans are considered 'impure' hybrids and are effectively banned from propagation, and that even 16% of the pure captive population has no chance to breed. Several zoos have only one specimen, or keep their individuals so separated as to preclude the proper establishment of relationship and mating possibilities. Most groups (76%) consist of three or fewer individuals, assuring inbreeding and a serious loss of genetic potential with each generation. At present only 25% of the males and some 30% of the females in captivity produce 75% of the offspring (Markham, 1991).

In conclusion, towards the end of the twentieth century, the media, the rehabilitation centres and to some extent *malafide* zoos have revived a demand for illicit trade in infant and adolescent apes, both within Indonesia and abroad, in particular in Eastern European, Asian and South American countries. Due to the implementation of CITES regulations the international trade is limited, but it certainly has not dried up; the largest leaks in the system have been mainly Chinese and local markets. In the 1990s, the major dealers in Indonesia were found in Jakarta, Sampit, Kumai, Pontianak, Banjarmasin, Samarinda and Bali. It is unsettling that the Agency for Nature Conservation seems ignorant of these facts.

Poaching

The hunting of orang-utans has been prohibited in Indonesia since 1924, and therefore it is to be formally designated as poaching, irrespective of whether people hunt for 'traditional', food-related, competitive or recreational reasons. Nevertheless, in spite of its protected status the orang-utan has been persecuted right up to the present by all manner of poachers.

In 1969, Schenkel and Schenkel-Hülliger reported that police, army and forestry officials in the Alas valley had readily shot an orang-utan when local people complained that it was raiding what they alleged were 'their' fruit trees in the adjacent forest. During the 1970s reports from the west coast of Aceh indicated that during field exercises, army patrols frequently practised their shooting skills on orang-utans and other wildlife, so that most ape juveniles on the illegal pet market were the spoils of such training activities. Even in the late 1990s it was reported that the industrial and army elite of North Sumatra and Kalimantan regularly organised weekend 'hunting safaris' into timber concessions, during which the poaching of orang-utans and other protected wildlife – notably tiger and elephant – was said to be common.

Orang-utan heads, hung on a fence to dry before being further prepared and turned into a tourist commodity; transmigration village in East Kalimantan (1994).

Orang-utans are still butchered nowadays and eaten by numbers of people, irrespective of traditional background. Blouch (1997) summarised evidence of orang-utan poaching in Sarawak from the reports of the Forestry Service, noting that although hunters vehemently denied killing orang utans, the evidence of butchered ape remains in poaching camps attests to the contrary. At the close of the twentieth century it is perhaps quite shocking for relatively enlightened people to learn that orang-utans are still slaughtered for food, even by supposedly modern urban elites in Southeast Asia. Indeed, the meat of poached orang-utans in Kalimantan and Sumatra has sometimes been offered for sale openly in local markets (e.g. Bontang, 1993). It is also rumoured that at least two restaurants in Jakarta, and possibly similar specialised 'wildlife restaurants' in other Southeast Asian cities, may offer, on special order and for exorbitant prices, brains (as well as liver and meat) from orang-utans and lesser apes. It is said that in particular wealthy customers of East Asian origin place advance orders and indulge in such primitive substitution cannibalism on special occasions.

That a tradition of killing and eating orang-utans has lingered on, or could readily be resumed (Medway, 1976), is hardly surprising, when law enforcement is lax, absent or dependent on authorities who are accessories in the illicit trade. At the same time, the situation is worsened by a modern consumerist perspective paired with the most primitive of attitudes, one lacking any socially restrictive context.

When the opportunity arises, numerous modern Southeast Asians seem to be possessed of the outdated 'big-game' hunter spirit, and do not exempt the orang-utan from their hit list. Indeed, a powerful incentive for the killing of orang-utans has been, and still appears to be, 'sport'[35] or 'recreation', while another is superstition. One wonders to what level the ethical and moral standards of Southeast Asian nations have advanced if large numbers of educated members of the upper class can still revel in and openly boast of the illicit slaying of such a human-like creature.

The impact of persecution

There must be a compelling evolutionary reason to be the largest and heaviest arboreal animal. For the red ape, arboreality is a dangerous lifestyle, but apparently it must be either much more efficient or somehow less dangerous than dwelling on the ground (Rijksen, 1978), given the prevailing circumstances over the last centuries. Although the orang-utan has become characterised as an exclusively arboreal ape, and indeed it displays advanced anatomical adaptations for its arboreal way of life, a growing number of observations indicate that in some areas orang-utans tend to travel along the ground frequently (e.g. in East Kalimantan and in central Sumatra). In such areas orang-utans seem to have an activity pattern reminiscent of the chimpanzee, climbing trees for feeding and nest building only (see Ghiglieri, 1985). Interestingly, the presence of tigers in Sumatra does not appear to influence this pattern.

---

[35] In particular local 'Big game' hunting clubs organise special safaris, in which any (protected) 'big game' including the orang-utan is bagged, and the 'Trophy' stuffed. It is noteworthy that as late as December 1978 the US Federal Register received an application requesting the formal import of hunting trophies of US hunters belonging to the international Safari Club. These trophies included five gorillas and five orang-utans (IPPL News 24 [1], April 1997: 25).

In October 1993, at Medan (Soemarna et al., 1995), an international workshop of the Captive Breeding Specialist Group (CBSG) – of the Species Survival Commission (SSC) of the World Conservation Union (IUCN) – on Population and Habitat Viability Analysis, demonstrated that existing populations of orang-utans are extraordinarily affected by hunting and poaching. The survival of the orang-utan is largely in question due to even a minor increase in adult mortality. This is mainly because the reproductive success per female is small even under the best of natural conditions. An adult female will not conceive before her 15th year (range 13-17). With an average interval of eight years between births, a single infant from each pregnancy and an estimated maximum longevity of some 45 years in the wild, she will, under optimal conditions, have no more than four offspring.

It is estimated that the adult age-specific mortality in a population is 'normally' around 2%. If this is correct then the Vortex computer model for analysing population viability demonstrates that an increase in adult mortality of 1% would lead to an inevitable decline in the population, and to extinction within some three to five decades. In practical terms, a mortality increase of 1% can be transcribed as the removal of only five adults per year from a population of one thousand apes, a number which is currently well exceeded by the poaching pressure in many areas of Kalimantan. If the primary target were to be females, as is the case when people set out to catch ape infants for the pet trade, then the increase in annual mortality would be twice as high. Under continuous poaching pressure this would cause a collapse of the local population within two decades (Working group CBSG 1993; Ellis and Seal, 1995).

Thus, poaching can readily result in the ape's local extermination, because:
- the orang-utan is by far the slowest and easiest prey in the forest,
- its particular habitat preference increases the risks of detection and fatal confrontations with human forest dwellers,
- its extremely slow and limited reproductive output means that persistent persecution by only a few hunters readily drains its reproductive potential.

In conclusion, it may be inferred that the extraordinary impact of hunting on orang-utan populations can have two major consequences, dependent on the incidence of impact: First, under the typically persistent hunting pressure for subsistence by nomadic hunter-gatherers, the ape has little chance of survival; indeed, the orang-utan has been effectively hunted to extinction in Borneo and in Sumatra, wherever these efficient human nomads scoured the forest.

Second, where hunting was restricted to magico-religious occasions or competitive events involving shifting cultivators, it has selected for elusive behaviour adaptations in the surviving orang-utans.

In forest, elusiveness can be expressed in different adaptive behaviour patterns dependent on different habitat types. It can be expressed in (solitary) dispersal of

Section I: the orang-utan

> Whereas the nomadic Penan and Punan in Borneo subsist almost entirely on what the forest yields in terms of game, molluscs, snails, fruits and palm- and tuber-starch, ethnological studies have demonstrated that even shifting cultivators rely for anything between 47 and 65% of all additional food items (next to rice) in their diet on products foraged from the surrounding forest (e.g. the Kenyah; Chin, 1985). In both the gathering (Penan/Punan etc.) and shifting cultivator ('Dayak') tribes the majority of adolescent and adult males are engaged almost full-time in scouring the forests for items to eat and to trade (Langub, 1989).

individuals of a deme, in a shift from diurnal to nocturnal activity patterns, and in either total arboreality, or a predominantly ground-dwelling existence, where trees are ascended only for food gathering and sleeping, as is the case with chimpanzees. Whereas an arboreal ape has higher survival chances in high-rising dryland forest with sparse ground vegetation, in swamp forest and alluvial forests with dense ground cover the survival chances are presumably highest for ground-dwelling apes[36]. Indeed, orang-utans may, when the undergrowth is dense and the observer not persistently attentive, silently descend and sneak away along the ground, especially in Borneo (Rijksen, unpubl.).

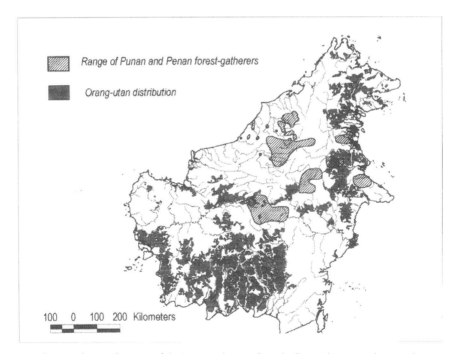

Map of Borneo showing the ranges of the Punan and Penan forest-dwelling gatherers, in relation to the present distribution pattern of the orang-utan.

---

[36] It is significant in this respect that when encountering humans, pig-tailed macaques invariably flee along the ground, jumping down from the canopy to disappear in the undergrowth.

In general, the orang-utan has adapted so as to live in a social structure of limited gregariousness, while in dryland forest areas (where virtually all studies have taken place) it lives an almost entirely arboreal life. How the ape lives in swamp forests with dense undergrowth is still largely unknown. In some areas (e.g. Aceh's west coast) its lifestyle does not seem to differ from the traditionally known pattern of predominant tree dwelling. According to local reports, however, in other areas (swamps in Central Sumatra and Central Kalimantan) it may have adapted (or retained) a predominantly ground-dwelling habit for its ranging. In the Rimbo Panti area of Central Sumatra it even seems to have adopted as its survival strategy a largely crepuscular pattern of activity.

History of orang-utan conservation

The administrative structure of protection

Protection of a species implies that a legal framework exists which prohibits people from disturbing, catching, and killing that species or from keeping and trading it (or parts of its body). Protection can be effective only if the public is aware of and acknowledges the legal framework, and if law enforcement is in place and active.

Nature conservation in Southeast Asia has been incidental and poorly organised from the beginning. It was, and still is, a political side-issue. This in turn is conducted by a bureaucracy typically embedded in a government structure whose primary aim is the exploitation of natural resources and the use of land for cash-crop production. Even in colonial Indonesia, conservation measures would probably not have been on the agenda had it not been for the vigorous lobbying and active booster activities of the Netherlands-Indies Society for Nature Protection, as well as the commitment of a few colonial staff members of *'s Lands Plantentuin* (i.e. the Botanical Gardens).

> The colonial organisation of the archipelago was concerned mainly with trade interests, and when nature conservation became a political issue in the first decade of the twentieth century, the authorities were either extremely reluctant or claimed to be unable to simply reserve areas in Kalimantan and Sumatra for nature conservation. It is true that the colonial organisation had usurped only direct jurisdiction over the island of Java and some smaller regions in the outer islands; most of the *Buitengewesten* territory was under guided self-rule. However, where the establishment of enormous colonial plantations was desired, e.g. in Deli, North Sumatra, the land-allocation procedure seemed to go smoothly within this dualistic system. Eventually, after the 1920s, the Netherlands Indies Society for Nature Protection could propose establishment of a reserve, and the proposal was then presented to the local indigenous government (*Zelfbestuurders*) for approval.

In spite of the legal protection framework which came into effect in 1924, the orang-utan has been subjected to waves of what may well be called persecution since the beginning of the twentieth century. The records show that the establishment of legal protection had little effect on the ongoing capture of orang-utans as long as law enforcement was neglected or lax. Indeed, it is evident that the orang-utan has survived largely by default, owing to its own shyness and the hitherto poor accessibility of its habitat, rather than to the existence of a steadily growing conservation organisation.

Shortly after the Animal Protection Ordinance came into effect in 1927, shipments of freshly captured wild orang-utans were still being formally cleared by customs in the port of Medan (Belawan), to be sent off to European zoos. Coomans de Ruiter (1932) reported shipments of 26, 60 and 46, while other sources mention 25, 33 and 44 apes shipped abroad by one dealer alone[37] (Jones, 1980). It was only after the East Indies Nature Protection Society had spurred the government to

---

[37] The apes fetched prices between some 7000 Dutch guilders and 25,000 German Marks in Europe (i.e. between US$ 5,000-10,000 at the time).

organise an information campaign among the authorities in northern Sumatra and Borneo that the persecution and trade temporarily declined. Nevertheless, in 1929 another 30 orang-utans were smuggled through Penang to Europe (Coomans de Ruiter, 1932).

After Independence (1945), the Indonesian government established a nature conservation service (*Dinas Perlindungan dan Pelestarian Alam* – PPA) within its Directorate of Forestry of the Ministry of Agriculture. It had close links with the few zoos in the country and with hunting clubs, but its presence in the field was barely noticeable. During the late 1960s, however, the service's leadership, notably Gen. Walman Sinaga, invited international support for better conservation of its endangered animals.

In the 1960s, the socio-political situation in Sumatra was such that local people again freely captured and traded orang-utans. An average of 20 orang-utans were shipped out of Medan every year; in 1964 two teams of poachers managed to catch 136 juvenile orang-utans in the Alas valley alone. This must have cost the lives of a similar number of reproductive ape mothers at least.

In 1965, at the IUCN Conference on Conservation of Nature and Natural Resources in Tropical South East Asia, convened in Thailand, Barbara Harrisson summarised the 'conservation needs of the orang-utan' by emphasising two integrated actions: First the protection of the species, and second the establishment, management and protection of reserves or sanctuaries covering a significant sector of its habitat range[38]. She noted that orang-utans need a large area for survival, quoting Schaller (1961) who estimated that one ape needed at least 1.5 square miles of forest for its subsistence. In order to stop illegal trade, Harrisson and her husband Tom, the curator of the Sarawak Museum, had established the Orang-utan Recovery Service (OURS), 'to help all governments of the region in enforcing existing legislation, and to advise and help in the care of requisitioned infants' (...) 'under the auspices of the Survival Service Commission of IUCN' (Harrisson, 1965). The recovery services seemed not to have been in great demand, however, as OURS silently disappeared, concurrently with some 70% of the forest habitat in Borneo and Sumatra.

When it became known that orang-utans were subjected to increasing hunting pressure, especially in North Sumatra, the international conservation community sought to support the small powerless Indonesian Nature Conservation Service (*Dinas* PPA). The pioneer among the organisations was the van Tienhoven Foundation, the driving force behind the Netherlands Committee for International Nature Protection and the major initiator of the Netherlands Appeal of the World Wildlife Fund. All three organisations were headed by Jan H. Westermann, who coordinated several initiatives to support the small dormant Service for Nature Conservation in Indonesia.

---

[38] She specifically recommended improved protection for the 'Loeser Reserve' in Sumatra, and the establishment of a large orang-utan sanctuary in the 'Ininhabited and inaccessible area of the Ulu Segama'in eastern Sabah, for which the Harrissons had issued a formal proposal in 1963 (Harrisson, pers. comm. and 1965).

Initially, the financial and advisory assistance of the van Tienhoven Foundation was directed exclusively at boosting the performance of the Indonesian conservation service in western Indonesia. Soon afterwards (during the early 1970s) Westermann arranged, through his government connections, for the establishment of a special (expatriate) conservation advisor in Indonesia under the *aegis* of the UN Food and Agricultural Organisation[39]. The advisor's function specifically concerned technical assistance for improved management of conservation areas and for protection, and also included the coordination of financial support from the Netherlands Appeal of the World Wildlife Fund and other organisations. The advisorship focused on the policy level and presented a considerable number of reports and management plans for existing conservation areas, while offering proposals for the establishment of new protected areas. Unfortunately it also initiated a major change in the legislative basis of conservation, which not only caused nature protection to operate in a legal vacuum for almost a decade, but also yielded, in 1990, a deficient and strongly 'resource-oriented' new regulation.

After a vain attempt to initiate an Indonesian WWF national appeal by depositing a starter fund in the newly established Indonesian Wildlife Fund (IWF) in 1972, the Netherlands support for the FAO representation gradually diminished, until in the early 1980s the International World Wildlife Fund, in collaboration with IUCN, took over the functions, reviving the major support of the WWF Netherlands Appeal (WNF).

The ultimate effects of this sequence of well-intentioned support activities were hardly noticeable in terms of active conservation in the field. In the early 1980s, however, the Service for Nature Conservation (PPA) was upgraded to Directorate-General Forest Protection and Nature Conservation (PHPA) and considerably expanded.

Towards the end of the 1980s, representations of many more international conservation organisations established a foothold in Indonesia. The growing complex of vested interests ironically caused the Indonesian conservation structure to wean itself from effective international assistance and assume a position of managing the flows of support. Presently the international organisations, rather than coordinating their actions and seeking to guide more effective conservation, began to develop their own independent missions, mainly in the eastern frontier of the archipelago where central authority was still deficient. For orang-utan conservation it meant that effective international attention evaporated and that whatever protective effect had locally prevailed for the ape or its habitat now readily disappeared. The new missions had no policy for the protection of species, and instead installed the dogma of

---

[39] Westermann, in his capacity of chairman of all major conservation organisations and secretary of the Organisation for the Advancement of Tropical Research (WOTRO) in the Netherlands, was instrumental in arranging strong governmental support for FAO to establish a special conservation advisory project in Indonesia, headed by John Blower, a former game warden in Africa.

'sustainable utilisation of resources' to be achieved by 'people's participation' in 'collaborative management' of protected areas (see Hails, 1996; Golder, 1996). The deplorable findings of the present report reflect to a great extent this sweeping conservation policy change in Indonesia and Malaysia.

> The conservation movement gained momentum during the 1970s, and the political arena, spurred by public opinion, began to allocate substantial financial resources in order to further arrest the perishing of unique species (e.g. whales, seals, rhinos, the tiger, etc.). Then, mysteriously, the major international conservation NGOs underwent a radical change. Their identity became that of a commercial corporation, the traditionally committed volunteer service was replaced by 'professional efficiency' and the policies were transformed into the euphemistic jargon of bureaucracy. A sequence of authoritative publications accompanied this upheaval, tendentiously denouncing all former efforts in nature conservation as failures, due to 'outdated protectionist approaches to conservation' especially where it concerned 'species conservation' (see e.g. McNeely, 1989). Instead it was advocated to consider 'higher levels of organisation such as ecosystems and landscapes', to promote 'sustainable utilisation' and to pay primary attention to 'traditional rights' and the alleviation of the poverty of rural 'stakeholders'. Even the concept of species was transformed into something of a higher order, namely 'biodiversity'.

The new approach was described in a joint publication of IUCN, WWF and UNEP, called the *World Conservation Strategy*. For the diplomatic and political bureaucracies of the world it may have been a welcome ideological policy document, but in the field the new intellectual concepts resulted in paralysing confusion, as the strategy could not be translated into any practical programme approach with relevance to nature protection. Applied to the real world, it facilitated a *laissez-faire* approach involving uncontrollable exploitation of protected areas and wildlife. A conservative focus on a high media-profile umbrella species, like the orang-utan, would, with the current availability of resources for 'sustainable development', never have led to the progressive decline in rainforest, as is being witnessed today under the banner of the revolutionary approach ushered in by the international conservation corporations.

Reserves

It was mentioned earlier that formal protection of the orang-utan dates back to the mid-1920s, when protection was to have been offered to the ape and implemented through effective law enforcement. However, this status fails to accommodate the conservation of the ape's living conditions, i.e. the quality of its habitat, and is barely effective where poaching is concerned. It was also mentioned that the ape has survived mainly because of its shyness and the protection offered by its relatively inaccessible habitat. Hence, in order to give the ape a chance of survival, there is an overriding need to safeguard significant areas of its habitat, in addition to effectively protect its existential form. The traditional approach has been the establishment –

and subsequent protection – of wildlife reserves (*Suaka Margasatwa*) or other kinds of conservation areas (such as the nature sanctuary *Cagar Alam*, and since 1980 the national park *Taman Nasional*). This was, of course, recognised at an early stage, and during the 1930s the Netherlands-Indies Society for Nature Protection succeeded in having the colonial government establish some of the currently most significant reserves in Borneo and Sumatra,e.g. Tanjung Puting, Gunung Palung and Kutai in Kalimantan, and the Gunung Leuser/Langkat complex in northern Sumatra.

Although 21 conservation areas are known to harbour orang-utans in Borneo (14) and Sumatra (7), the design and establishment of these reserves has rarely taken into account the ecological requirements of the structure of biological diversity it was intended to protect (see also Dinerstein *et al.* 1995). Even during the colonial era, a reserve could usually be established only if its location were remote, inaccessible and demonstrably in no-one's immediate exploitative interest. Hence, most of the protected areas are leftovers from a land-allocation process which exclusively focused on short-term economic gain. Usually the smallest possible perimeter 'for protection' was drawn around a site in which a particular species was known to occur.

The Netherlands-Indies Society for Nature Protection proposed several reserves. Some of the proposals indeed materialised, others were rejected, usually because a claim for mining was anticipated for the area or the local government (*Zelfbestuur*) simply declined to accept. However, all officially sanctioned reserves were significantly smaller than had been proposed. In Sumatra the proposed area for the western Gunung Leuser area (the Gajoe and Alaslanden reserve) was more than twice as extensive (i.e. 9,000 km^2) as what was eventually gazetted in 1936 (i.e. 4,165 km^2). The uninhabited lowland areas of the county Trumon, the Singkil swamp, the coastal swamps near Meulaboh and the water catchment of the river Tripa were excluded. The proposal to turn the biologically and hydrologically important Protection Forests of Dolok Sembelin and Simbolon into wildlife reserves was rejected. In Kalimantan the 195 ha Mandor nature reserve was what the government gazetted following the proposal for a 4,000 ha reserve; of the Gunung Palung area some 70% of the proposed area was gazetted (300 km^2); the proposal for a reserve including the 'sacred' Bukit Raya was rejected; and the Kutai wildlife reserve (3,066 km^2) was only 20% of what had been proposed (see 10e and 11e *Verslag van de Nederlandsch Indische Vereeniging tot Natuurbescherming*, 1935 and 1936-38).

It is interesting that at least one major nature reserve, in which orang-utans occur, was established at the instigation of the local communities by a self-rule decree (*Zelfbestuurs besluit*), namely Gunung Leuser in Aceh (Rijksen and Griffiths, 1995). For this unique mountain reserve around the Gunung Leuser mountain complex the incentive was in a sense anti-colonial. The mountain complex was once a sacred forest area, and by establishing it as an officially protected nature reserve the local Gayo communities hoped to safeguard it against desecration by the colonial powers for conversion into plantations or for mining (Ir F.C. van Heurn, pers. comm.). However, the area is patently unsuitable for cultivation, and of value for wildlife conservation only along its very fringes. Nevertheless, the local *Zelfbestuurders* could never have foreseen that one day immigration as well as the population explosion of their own people would encroach into the forested flanks of the mountain complexes, and hence develop into the greatest threat to their traditionally sacred grounds.

The mountain Bukit Raya in Central Kalimantan was also once considered sacred, and hunting or harvesting was strictly *tabu* for the indigenous tribal communities (Ngaju and Ot Danum). Like the Leuser mountain complex in Aceh, it was 'the mountain of the deities' and the major source of *karohei*, i.e. spirit power (Schärer, 1963). In the 1930s the area was proposed for reservation but the colonial government rejected the proposal in 1938. Eventually an area of 1100 km² was established as a wildlife reserve in 1978 by the Indonesian government.

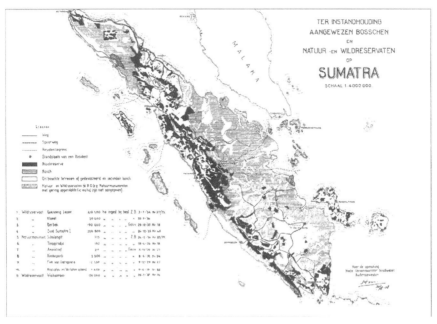

Map of Sumatra showing the forest cover and Protection Forests (in black) in the 1930s.

Up to the present, conservation has never been effectively integrated within the framework of regional land-allocation planning. In general it must operate in isolation from local government affairs, which are typically aimed at consumer-related development, the exploitation of natural resources and an increase in productivity. Development aid or cooperation, even when it specifically concerned the integration of conservation (of natural resources) and development, never broke this pattern of isolation, despite statements suggesting the contrary (Brandon, 1997). Some main reasons for this will be discussed shortly. In the framework of international cooperation, integration has meant that the resources for conservation came to be deployed for rural socio-economic development – usually at the expense of the integrity of a conservation area. Until the late 1990s much of the technical assistance in the regions has been offered by international conservation organisations which are openly ignorant of, and uninterested in, the simple conservation of an ape, even if it concerns an 'umbrella species.' Instead, they have been fully engaged in ways and means to alleviate poverty and strengthen the autochthonous awareness of the local people, measures which unfortunately often degrade the habitat of the 'flagship species' that served to generate the project funds.

In 1990, J.R. MacKinnon noted that orang-utans occurred in 40 protected areas in Borneo, apparently including large Protection Forests and proposed reserves[40]. The present survey found, however, that some of these areas appear to have no populations of orang-utans (e.g. Muara Kaman Sedulang, Kayan Mentarang), while Protection Forest (or forest reserves) and proposed reserves do not warrant inclusion in a list of protected areas indicating prospects of survival of the ape, due to the absence of any protective measures. Indeed, in the field such areas are in no way distinguishable from regular Production Forests, and commonly suffer the impact of scheduled timber exploitation, encroachment, timber theft and poaching.

Orang-utans are now known to occur in 14 formally established conservation areas in Borneo[41], covering a total of 21,559 km^2. We will shortly elaborate upon the fact that the forest in these reserves is composed of anything up to 35% of habitat (see chapter 2), while several of the reserves have been – or still are – subject to disturbance and logging impact, thus reducing the habitat quality.

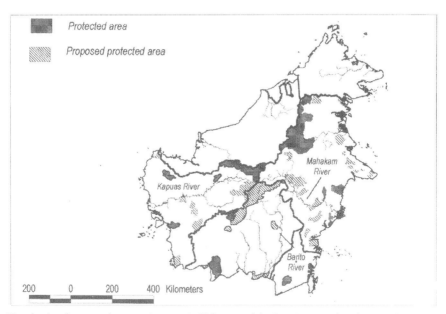

Map showing the proposed conservation areas in Kalimantan (after MacKinnon and Artha, 1981).

---

[40] The extent of this supposedly protected area network is unclear, ranging from 62,018 (p. 214) to 70,572 km^2 (p. 215) (MacKinnon, 1992).

[41] The latest official map of the DG-PHPA Planning Section (July 1996) has omitted the Kendawangan and Danau Sentarum conservation areas. A proposal to establish the Kinabatangan Wildlife Sanctuary in the logged-over flood plain of the river Kinabatangan, and the Malinau Wildlife Reserve in the Southern Forest reserve in Sabah, is in progress (Payne, pers. comm.).

TABLE VI

Overview of the conservation areas for Bornean orang utans, and their potential habitat content if all the area were to be covered by forest and well protected; the Bukit Baka – Bukit Raya conservation areas are usually considered as one; the Danau Sentarum area (*) is in the process of being gazetted (1996).

Conservation areas in Borneo	Total area in km^2	Potential habitat
Gunung Palung (W. Kal.) – 1936	900	270
Gunung Niut Penrisen (W. Kal.) – 1982	1,800	450
Danau Sentarum (W. Kal.)* – 1982	1,320	280
Gn. Bentuang & Karimun (W.Kal) – 1992	6,000	1,400
Bukit Baka (W. Kal.) – 1981	705	142
Bukit Raya (C. Kal.) – 1978	1,106	165
Tanjung Puting (C. Kal.) – 1936	3,050	900
Kutai (E. Kal.) – 1936	1,986	495
L. Entimau/B. Aie (Sarawak) – 1984	1,988	217
East Tabin (Sabah) – 1984	556	160
Crocker Range (Sabah) – 1984	1,269	50
Kinabalu (Sabah) – 1964	754	51
Kulamba (Sabah) 1984	207	72
Danum Valley (Sabah) – 1995	438	131

Of the 21 conservation areas in Borneo and Sumatra, most would have less than 300 km^2 of habitat if one considers that the ape favours forest below an altitude of 500 m. It is remarkable that the East Malaysian State of Sabah has established reserves with the smallest extent of potential habitat (i.e. an average below 100 km^2), while Sarawak has only one reserve of any relevance for orang-utans.

In terms of size, most orang-utan conservation areas in Borneo are either small, or isolated and mountainous, or of low general quality, while all have been more or less badly degraded since the 1970s.

In Sumatra the originally 9,000 km^2 complex of Leuser and Langkat-Sikundur reserves was expanded in the 1990s to some 25,000 km^2 by including all surrounding State forest land between the mouth of the river Alas in the south to the environs of Lake Tawar in the north. This was done in order to redesign a feasible National Park within what is called the larger Leuser Ecosystem, and render it fit to conserve the essential ecological functions of the area, including viable populations of orang-utan, as well as elephant, Sumatran rhino, tiger and numerous other endangered forest species. This Leuser Ecosystem currently offers anything up to some 4,500 km^2 of habitat for the ape. In 1994, it was granted a special 'conservation concession' status by Ministerial Decree, which was enforced in 1998 by Presidential decree (No 33/89). One hopes that the new structure of protective organisation, as outlined in the master plan of the Leuser Development Programme (Rijksen and Griffiths, 1995), will afford the orang-utan a better chance of survival.

> In 1992, the Indonesian Ministry of Forestry invited the Institute for Forestry and Nature Research (IBN) to draft a proposal for an integrated conservation and development programme for the designated Gunung Leuser national park. The inception phase of the Leuser Development Programme was sponsored by the European Commission. It resulted in a partnership programme which is designed to evolve step-by-step, seeking integration of nature conservation in development from the planning level down to day-to-day existence. The habitat and ranging requirements of the orang-utan were of primary concern in its design. Initially considering the designated national park[42] as the focus for strict conservation, it soon became apparent that the 9,000 km² conservation area was far too small and unsuitable for the protection of the habitat of the larger mammals and birds; virtually all of the ranges of these major faunal elements were on the very fringes of, or outside the boundaries of the park. The much expanded design of the Leuser Ecosystem, within the confines of State forest land (*Hutan Negara*), was determined by ecological requirements such as water catchment, and the ranges and minimum population size of at least three major faunal representatives, including the orang-utan.
> The Programme operates with a conservation-centred objective, and was designed to deploy a *quid pro quo* approach: it may provide support for locally desired rural development in exchange for hard commitments to conservation by the local communities living around the Ecosystem. Implementation of the programme started in 1996 with a budget, covered by EU support of some 32 million ECU, in addition to 18 million ECU in Indonesian contribution. It can be considered a first major investment for the protection of the 25,000 km² Leuser Ecosystem concession, which is in the custody of the Leuser International Foundation (YLI). The Foundation is an NGO composed of influential Indonesian citizens having a strong affinity with northern Sumatra.

Towards the brink of extinction?

At least three times during the twentieth century the alarm has been sounded over the anticipated fate of the orang-utan. On each occasion action was initiated because of international concern regarding the detection of large numbers of wild-caught apes being exported from, or smuggled out of, the archipelago.

The first alarm was in the late 1920s, when European zoos and circuses received large shipments of apes, despite their formal protected status since 1924. Usually less than 40% of the apes survived the transport, and virtually all died within a year of arrival in Europe. In May 1928, an article in the English newspaper *The Times*[43] accused the Netherlands East Indies colonial government of an appalling 'slave-trade' in orang-utans. The article was translated and accompanied by an editorial in several

---

[42] The 'National Parks' in Indonesia are in fact designated as such (*calon Taman Nasional*) as long as detailed regulations, boundary delineation and zoning have not been formally established; such designated National Parks have a very weak legal basis, and in practice offer fewer opportunities than a timber concession for protection against encroachment (Rijksen and Griffiths, 1995).

[43] The article was written by Sir Hesketh Bell and Dr Charles Mitchell; in September 1929 eight orang-utans arrived at Rotterdam Harbour, the survivors of a shipment of 32 individuals which had been smuggled out of Sumatra.

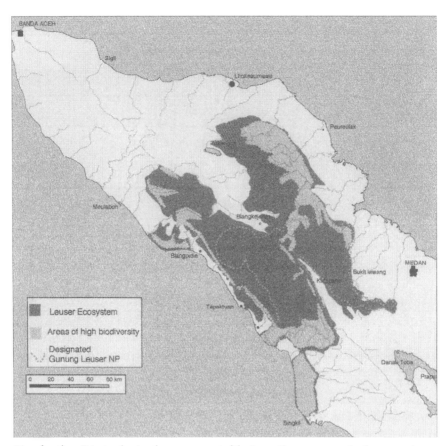

Map of northern Sumatra showing the current extent of the Leuser Ecosystem concession for conservation; the dotted line indicates the boundary of the former designated Leuser National Park.

leading Dutch newspapers (Coomans de Ruiter, 1932). The Netherlands Commission for International Nature Protection reacted with a call for a formal ban on imports, and passed on the alert to the Netherlands Indies Society for Nature Protection, which then issued an awareness campaign for authorities.

Following the commotion, the Netherlands-Indies authority tried to involve neighbouring nations in solving the problem, and sought to persuade the British colonial authorities to help tackle the market by issuing a ban on the import and transit of orang-utans, especially through Singapore. They also requested prohibition of the export of apes from Sarawak and British North Borneo. In 1929, the North Bornean authority issued an export ban for orang-utans, while the Sarawak government established an export levy of some £ 300 per ape. The Singapore authorities were adamant, however (K.W.L. Bezemer, in *Mededeeling* No 7, Netherl. Comm. for Intern. Nature Protection, 1929).

Under the auspices of the Netherlands Commission and the Netherlands-Indies

Society, a first attempt was made as well in the 1930s to assess the whereabouts of orang-utans. As elaborated earlier, notably Zondag (1931), Witkamp (1932) and Westermann (1936-38) had collected all available data from contemporary records and inquiries for the Kalimantan region, while Heynsius-Viruli and van Heurn (1935) as well as Carpenter (1938) had done the same for North Sumatra. However, if these publications yielded a fair picture of the distribution range of the ape, very little relevant information was provided about the orang-utan's numbers and the threats to its survival, and the material had a minimal effect concerning the ongoing decline of its numbers and habitat extent.

During the 1960s, the orang-utan came once more to the attention of the conservation community in the western world, mainly at the instigation of Ms. Barbara Harrisson (1960, 1961, 1965). It had been noticed that American and European institutions had created a demand for large quantities of wild apes, including orang-utans, to be used in biomedical research (Jarvis, 1968; see also Bourne, 1971). Conservationists as well as the management staff of some major zoos, faced with a paucity of data on the situation relating to the orang-utan's condition in the wild, rightly feared that the ape was on the very brink of extinction. Short surveys by naturalists in Borneo (Schaller, 1961; Davenport, 1967) and Sumatra (Milton, 1964; Schenkel and Schenkel-Hülliger, 1969; Kurt, 1970) seemed to corroborate these fears. Hence, in 1964, the International Zoo Association finally banned the purchase of wild-caught orang-utans which were not accompanied by the appropriate documents.

As a matter of fact, it was believed that the total number of orang-utans in the world at the time did not exceed four thousand individuals (Harrisson, 1961; Simon, 1966). This forced the World Conservation Union (IUCN) decision, in 1966, to place the red ape on its Red List of Endangered Species, Category One, which implied that the ape 'is facing a very high probability of extinction in the wild in the near future, if the causal factors affecting its decline in numbers continue to operate'. However, no-one could really know even approximately how many orang-utans were still surviving in Sumatra and Borneo. Such ignorance was initially the result of delayed political stability in those countries, which in turn prevented any biological interest in wildlife before the late 1960s[44]. It was undoubtedly also due to the remoteness and extremely poor accessibility of the vast stretches of forest in Sumatra and Borneo, and the exceptional elusiveness of Indonesian wildlife in general, and the orang-utan in particular.

In 1972, many of the locations mentioned in the early literature for Sumatra (Heynsius-Viruli and van Heurn, 1935) were checked, as an essential component of a research project regarding the conservation situation of the orang-utan (Rijksen, 1978). Unfortunate as it may seem in hindsight, the early records of orang-utans

---

[44] The first post-war wildlife surveys were cover-ups for US intelligence operations aimed at acquiring information on the political situation in Indonesia's revolutionary hot spots, namely, Aceh and West Java.

south of Lake Toba were hardly taken into serious consideration. It was found that although the supposed range of the ape north of Sibolga bay was almost identical of that of the 1930s, the habitat was coming under severe pressure due to timber extraction and slash-and-burn (*ladang*) encroachment. Some of the findings were corroborated by the results of a series of field surveys to determine the distribution of the Sumatran rhinoceros, covering some 30% of the known orang-utan range (Borner, 1976). In particular the extensive lowland forests along the east coast were found to be badly degraded and demolished through timber extraction, and through conversion for the development of rubber plantations. And although the estimates of surviving orang-utans were at least ten times higher than the hitherto supposed figures, it was concluded that the ongoing and planned habitat destruction justified the ape's formally declared endangered status.

> In 1844, an excellent monograph on the orang-utan was published in The Netherlands. It was based on the field observations of the Dutch naturalist Salomon Müller, and it noted that the ape was a rare creature in the vastness of the Bornean jungle. And, notwithstanding the fact that nineteenth century naturalists were able to bag appalling numbers of specimens in any one location within a few weeks' time, the image of a rare, solitary creature was set. Well into the 20th century it was believed that the red ape was scattered over enormous areas of forest, with few possibilities of finding a mate (MacKinnon, 1974). Indeed, the first field studies corroborated this image. After two months of searching in the swamp forests of Sarawak, Schaller (1961) reportedly felt that the arboreal ape was either so rare, or so exceptionally solitary, that it was virtually impossible to study its behaviour. He estimated that no more than 700 orang-utans existed in the whole of Sarawak, and noted that the species had declined drastically since the preceding century. In any case, his short publication strengthened the feeling that the ape was on the very brink of extinction.

The collateral long-term studies on wild orang-utans in the surroundings of rehabilitation centres during the early 1970s – Ketambe in Sumatra (Rijksen, 1974) and Camp Leakey in Central Kalimantan (Galdikas-Brindamour, 1975) – revealed that orang-utans have a much more complex ecology and social life than had previously been supposed. These findings did initiate a growing interest in the study of orang-utans, also in Indonesia, but ironically they had scant effect on the conservation situation and contributed little to the knowledge of the ape's whereabouts. After the first well-intentioned attempts at rehabilitation in the early 1970s had initially drawn massive public attention, followed by justified criticism (MacKinnon, 1977; Rijksen, 1978), the conservation community quickly lost interest in the plight of the orang-utan.

Even without special surveys it must have been evident to the international conservation community of the early 1990s that the orang-utan population had declined steadily since the late 1960s. It was well known that the rainforests in Borneo and Sumatra were under mounting pressure of timber extraction, human settlement and conversion (see Chapter entitled Direct causes of the decline). When one considers the relative distribution of different types of forest (MacKinnon and Artha, 1981) on the islands, and the well-documented exponential impact of logging, transmigration, plantation development and population expansion since the early 1970s, it can readily be surmised that of the extent of forest cover in Borneo, which was estimated by FAO in 1980 at some 350,000 km², more than 50% has been demolished. The situation in Sumatra has been no different.

> It may be surmised that in about 1900 the Bornean orang-utan population was distributed over a still very extensive, more or less contiguous, area of gently undulating lowland and swamp-forest habitat covering approximately 495,000 km², i.e. 67% of the 733,000 km² total land surface of Borneo[45]. The relict Sumatran population must have occurred in at least 120,000 km² of forest in the lowlands, coastal swamps and the valleys of the more mountainous terrain of western Sumatra.

Since hardly anyone has looked seriously into the issue of orang-utan distribution and the deficient protection measures, it should come as no surprise that since the colonial era the ape has steadily but silently disappeared, along with the forest cover, from all of the more accessible regions of its former distribution range. It was displaced when human settlement and forest conversion usurped the best sectors of its habitat, and it has been pushed back into ever remoter regions on Sumatra and Borneo.

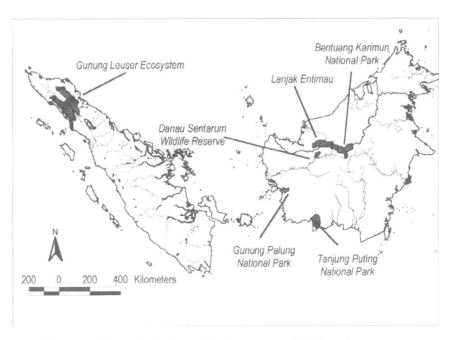

Map of Sumatra and Borneo showing the established nature reserves of significance for orang-utans.

---

[45] cf. K. MacKinnon (1986), who estimated 415,000 km² in Borneo.

Conservation awareness?

A more detailed analysis of the waxing and waning of interest in orang-utan conservation is in order, because it is a cautionary tale of how varying priorities of the large conservation organisations are hardly conducive to effective wildlife conservation, despite ongoing fund-raising campaigns suggesting the contrary.

Several times during the twentieth century field biologists have attempted to investigate the status of the orang-utan in the wild. In the early 1930s the Netherlands Commission for International Nature Protection played a major role in facilitating these investigations. In the late 1960s this role was taken up by the World Wildlife Fund in general and its Netherlands Appeal in particular. However, the surveys were typically limited in scope, of short duration and generally of low quality, due to a poor understanding of ape biology, a lack of coordination, logistic constraints and insufficient financial support.

In 1987, WWF Malaysia commissioned J. Payne to investigate the status of orang-utans in Sabah. He published his findings in 1988. The report revealed that the ape was in fact under incredible pressure from development and forest conversion in North Borneo, but also gave a much higher total estimate of the number of surviving apes than had been previously assumed. In fact the report concluded that all was well; it was apparently not meant to alert anyone.

Signpost in the centre of Kutai National Park, East Kalimantan, along the newly cleared Trans-Borneo 'highway'; it announces that it is 'forbidden to reside, cultivate fields, fell trees, construct buildings, and hunt', but, in the absence of law enforcement, it seems to have attracted, rather than thwarted, encroachment.

Towards the end of the 1980s, tourists in Kalimantan began to report the sale of so-called traditionally decorated orang-utan skulls[46] procured by the 'head-hunters' of Borneo. It was also reported by an expatriate WWF Indonesian Programme staff member that orang-utans were persecuted as vermin in a new oil-palm plantation in East Kalimantan, established by means of a World Bank loan (K. MacKinnon, pers. comm.).

> Since the mid 1980s, forest areas have increasingly been converted for the establishment of large industrial oil-palm plantations; many of these forest areas were the home of orang-utans. With the development of these plantations the displaced apes began to be considered a pest species despite their protected status. In their desperate search for food, these apes went after the edible palm-heart and destroyed the young oil-palms. It has been documented that plantation managers in the period 1986-1990 issued orders to exterminate orang-utans in the surrounding forests and provided financial rewards between US$ 5-10 for evidence of a kill. As a result, thousands of transmigrants and local farmers must have scoured the surrounding forests for orang-utans to kill, and in some transmigration villages one could find the severed heads of apes drying on stakes in front of the houses (Djuwantoko, pers. comm.).

In 1988 and 1989, the first alarming signs of increased orang-utan poaching and habitat decline inspired a few expatriate conservationists in Indonesia to try to boost the protection of the ape at the Directorate-General PHPA. The (verbal) reports, addressing the massive drain on the wild orang-utan population since the mid 1980s, caused some embarrassment within a hitherto uninformed Directorate-General, and it was readily decided to appeal for international help to rescue the ape. The few informed expatriates were asked to help raise the international support which Indonesia expected. This proved to be virtually impossible. The first pleas fell on deaf ears – even at the offices of the international organisation representatives within Indonesia – because 'species conservation is an outdated concept' (R. Betts, pers. comm.).

Then, in the early 1990s, the disconcerting problems burst into the open. In December 1990, the US magazine *Wildlife Conservation* contained an article entitled 'In Indonesian Borneo Orangutans Are For Sale' (E.B. and C. Martin, 1990). At the same time the regional Asian press revealed that large numbers of orang-utans were being smuggled out of Indonesia into Asian countries in spite of the CITES regulations. Several shipments of ape juveniles were seized at international airports (Singapore, Bangkok). On November 8, 1991, the respectable Indonesian daily *Kompas* reported that 'hundreds of orang-utans are still in the hands of people in East Kalimantan.' On December 14th, 1991, the same newspaper revealed that, since the mid 1980s, more than two thousand five hundred orang-utans had been illegally

---

[46] In 1985 the coffee-table book *Indonesia – Paradise on the Equator* (Muller and Zach) carried an illustration (p. 176-77) showing an orang-utan skull being used in an enactment of a traditional 'head hunting' raid.

exported to Taiwan, smuggled in the cargo of fishing and timber vessels[47]. The English-language *Jakarta Post* (Nov. 8, 1991) was more conservative and carried the headline 'At least one thousand Indonesian orangutans still in Taiwan'. In 1991, even the widely read East African magazine *Swara* (14, 3: p.19) reported on the deplorable conservation situation of the orang-utan in 1991, as tourists had noted that ape skulls and live juveniles were freely offered for sale in Indonesian cities (C. and E. B. Martin, 1991).

> In Indonesia, the press articles and alarming reports of the field experts caused anxiety at the Ministry of Forestry, but the hopeful waiting for international support seemed to preempt any serious national initiative. Nevertheless, the National Bank BNI deposited US$ 100,000 for the conservation of the rhino and the orang-utan at the Indonesian Wildlife Fund[48].
> In 1993, the Director General PHPA invited the Institute for Forestry and Nature Research (IBN) to organise a partnership involving an Indonesian and a European surveyor to chart the orang-utan's conservation situation. The Ministry of National Development Planning and the Ministry of Forestry issued the official Biodiversity Action Plan for Indonesia in the same year. And while the Action Plan carried a US$ 1 million proposal for the conservation of the orang-utan in Kalimantan on its priority list (Proj. 6), and the Directorate-General PHPA could draw upon the allocated BNI fund at the IWF, it expected the Institute to finance the partnership unilaterally. Finally, in the spring of 1994, the Netherlands' *Wereld Natuur Fonds* (WNF) declared itself prepared to support in part the present survey[49].

The media hype concerning the mounting local hunting pressure in Kalimantan[50] exposed the embarrassing lack of law enforcement for protection of the ape. If it did raise any concern within Indonesia, the rest of the world seemed to ignore it entirely (Rijksen, 1995), although in the early stages, in 1990, the 18th General Assembly of IUCN adopted a resolution, prepared by its SSC Primate Specialist Group, calling for 'strict law enforcement in Asian countries to end the illegal traffic in Orangutans.'

---

[47] During the present survey it was learned that until 1994, when regulations were gradually becoming better enforced, some ships departing from Sampit (Central Kalimantan) during the night reportedly carried as many as eighty orang-utan juveniles with false permits (acquired at a price of some $ 25/ape) to Singapore, Taiwan and Japan; in Banjarmasin orang-utan juveniles were auctioned among ship crews of foreign cargo vessels.

[48] The IWF is an endowment Fund for nature conservation under the *aegis* of the Ministry of Forestry; its board consists of (former) top functionaries of the Ministry. Since communication beyond the private sphere at the Ministry is poorly developed, there is no way of knowing how or whether this money has been effectively allocated.

[49] Support for the present survey was outside the framework of the WWF Indonesia Pogramme. A partner team of the DG PHPA and IBN was formed comprising Mr. A. Muin and Mr. E. Meijaard. When the DG was succeeded in 1994, Mr. Muin was transferred to another assignment, and the active interest of the agency waned. Mr. Meijaard was subsequently accepted within the framework of the International MoF-Tropenbos Programme, under the guidance of the Agency for Forestry Research and Development (AFRD), and found a new (PHPA) partner in Mr. Bachtiar.

[50] In addition to newspaper and magazine articles, some TV and radio producers immediately recognised the enormity of the problem and did their best to inform a larger public of the ape's predicament; foremost among these were Gerard Baars Productions (Tros TV – Holland), Sarah Cunliffe TV Productions (BBC/NBC), Mike Searle (Storyteller Productions, Discovery Channel, Australia); and Marijke van der Meer (Wereldomroep the Netherlands).

In Indonesia, members of the Indonesian Primatological Society wanted to make a modest beginning at charting the distribution range of the orang-utan (Sugardjito and Van Schaik, 1992; Sugardjito, pers. comm.). However, much of the initiative suffocated under the logistic problems, while the BNI funds were apparently too limited to guarantee results. The Red Alert programme of the Flora and Fauna Preservation Society also sponsored a preliminary survey, the organisation and results of which, however, continue to remain unknown.

At the Great Apes Conference, held in Jakarta and Tanjung Puting in December 1991, the scope of illegal exports to Taiwan was brought to light by members of the Orangutan Foundation (Lee *et al.*, 1991), and the alarm regarding the plight of the ape due to the developments in Borneo and Sumatra was publicly raised (Rijksen, 1991). In addition, the frightening decline of the ape's distribution range in Sarawak was discussed (J. Bennett, 1991).

The excellent advertisement which drew attention to the plight of the orang-utan in the Netherlands during the early 1990s; the text reads: young man, desperately seeks shelter, preferably in forested environs; Care about the future. WWF

In that conference's public opening address, H.E. the President of the Republic, Suharto, reiterated the official appeal for international support for orang-utan conservation. Since the World Wide Fund for Nature already had a long-standing Indonesia Programme, established under its former image of World Wildlife Fund, it was now formally approached for help as well. However, even Indonesia's formal appeal went unheeded. The urgent warning that *Homo sapien's* Asian relative was vanishing was lost in the quagmire of international bureaucracy and a modified policy which turned a blind eye to the very wildlife that, ironically, featured so prominently in global fund-raising campaigns. The major argument used to turn down the request for support was that 'species conservation is an outdated policy'; another was that 'too scanty information on the status of the ape is available.' Most ironically, however, it was falsely argued that 'sufficient resources were being spent in the conservation of the ecosystem of the ape.'

> As a matter of fact, the WWF Indonesia Programme had just terminated its long-standing interest in the Bohorok rehabilitation project at the fringe of the designated Leuser National Park in Sumatra. The programme was active in Kalimantan only in the Kayan Mentarang wildlife reserve where orang-utans no longer occurred.
>
> In the late 1980s the World Wildlife Fund had changed its name and logo. Reflecting the changed policy which considered species protection outdated, it removed the word 'wildlife' from its name. Henceforth called the World Wide Fund for nature, its policy came to focus on 'ecosystem conservation' in the form of 'sustainable utilisation' through 'participatory management' in accordance with the message of the *World Conservation Strategy*.

Remarkably, only the Captive Breeding Specialist Group (CBSG) of the IUCN SSC reacted. In its justification to convene a gathering of experts for a Population Viability Analysis, it stated, 'suitable orangutan habitat has declined by more than 80% in the last 20 years and orangutan numbers have declined by 30-50% over the last 10 years' (*Problem Statement of the Orangutan PHVA Workshop*). Apparently the group possessed information not available to the major conservation corporations.

In 1993, the CBSG organised a workshop in Medan, North Sumatra, in order to somehow clarify the conservation situation of the ape. How this was to be achieved with no recent survey data had apparently not been considered. The Great Ape Conference in 1991 had called for detailed surveys in order to determine the distribution of orang-utans as a priority in its recommendations, but no-one had provided the resources to implement such surveys (or was able to do so).

Despite the paucity of accurate data, however, the workshop at Medan did produce vitally important predictions concerning the population dynamics of the orang-utan under pressure. Up-to-date data on the life-history characteristics of the ape, yielded by the international accumulation of orang-utan expertise, were entered into the special Vortex Population and Habitat Analysis computer model, in order to simulate scenarios of population dynamics under various pressures. The results demonstrated that the slightest impact due to hunting pressure could drive the orang-utan to extinction within a time span of two generations at most (Ellis and Seal, 1995; Leighton *et al*. 1995).

Thus, the third, and latest, alarm, sounded in the early 1990s, failed to receive the ready attention which had been given the somewhat exaggerated alarm of the 1960s.

> In its crude re-evaluation of the surviving populations, the workshop was obliged to consider in particular the situation in protected areas and those areas apparently proposed for protection, because data on the distribution outside the reserves were lacking altogether. Acknowledging the assessment's many uncertainties, the meeting adopted a rational, and most prudent, correction factor of 0.6 for the Bornean and 0.75 for the Sumatran situation on the basis of the estimated population size.
>
> The crude assessment came up with a total of 22,360 km^2 for the supposedly protected Bornean forest range and 11,700 km^2 for the Sumatran range as it had been – erroneously – estimated at the time. Applying a standard average density estimate of apes for this range, while supposing that all forest represents habitat, this yielded an estimated Bornean population between 10,828 and 15,546, and an estimated Sumatran population between 9,200 and 12,275 apes. However, it was evident – at least to the few ape biologists – that the scarcity of accurate geographical information and the problem of realistic relative density valuations rendered the exercise of limited value only.

The impact of numbers?

International attention and support for effective protection of the orang-utan ground almost to a halt after the early 1970s. It is illustrative that when a sequence of alarms was sounded in the late 1980s and early 1990s, each of them was more or less waved aside, the main argument being that insufficient information was available. This may seem odd considering that several international organisations had representations in Indonesia, and should have been the first to know. If independent press articles and the reports given by concerned, acknowledged experts were not deemed sufficient to raise a general alarm, and to spur international concern sufficient to at least verify the information, then what was required? What prompted the abysmal lack of concern? Is the international conservation community truly on the alert for real and accurate information? Was it confused by the contradictory reports emanating from various supposed experts concerning allegedly surviving numbers of apes? And if so, why did it fail to investigate the situation immediately? What precisely is required to expediently mobilise available funds to react in an emergency?

In both the national and the international context, the issue of orang-utan conservation has been plagued by a total absence of guidance and direction, insufficient cooperation and fruitless debates among ape biologists competing for scanty resources and public exposure. To this predicament was added a deteriorating image of wildlife protection, as well as a growing misunderstanding of ape rehabilitation, a scarcity of field data and a sequence of flawed desk-top assessments (owing to a lack of reliable information) of the ape's conservation status. After the crude surveys of the 1970s had revealed that the estimated number of 3,800 apes – which had been published by IUCN in 1966 – was unrealistic, confusion about numbers became rampant during the 1980s. At the same time the so-called rehabilitation of orang-utans began to usurp the image of ape conservation and resulted in feelings of disgust among purist ecologists.

In 1982 the IUCN Conservation Monitoring Centre published a Conservation

Strategy for the Great Apes, in which all contemporary information about the distribution and status of the orang-utan was compiled. No new facts underpinned the publication, however; it simply reiterated the numbers provided earlier by Rijksen (1982) at the 1979 Workshop on the Conservation of the Orang-utan in Rotterdam. The publication seemed only to falsely reassure the international conservation community that the orang-utan's status was relatively secure. Information relating to endangered species apparently may be filed as it filters in, but no special policy has been implemented to actively collect accurate data.

In 1985, at the International Primatological Conference in San Diego, an estimate of 37,000 orang-utans for Borneo was presented (Rijksen, 1986) almost simultaneously with another presentation estimating as many as 156,000 surviving orang-utans on the island of Borneo (K. MacKinnon, 1986). Unfortunately no-one seriously discussed this discrepancy, or those following, which soon reached an inconceivable magnitude.

> In 1990, J. MacKinnon presented a *Species Survival Plan for the Orangutan* at the International Conference on Forest Biology and Conservation in Borneo (Sabah)[51]. The presentation tables indicate the assumption that the Bornean population lived in approximately 352,000 km^2 of habitat at that time, purportedly implying a range between a minimum of 10,000 and a maximum of 116,000 apes. The following year, MacKinnon's wife (K) reviewed this sweeping statement to note that the total Bornean population would be around 61,000 apes, of which 27% (i.e. some 16,000 apes) were said to live within protected areas and in proposed reserves (MacKinnon, 1992).
>
> At the Great Apes Conference (Jakarta, Pangkalanbun; December 1991) later in the same year, J. MacKinnon issued another presentation entitled *Orangutans as Flagship Species for Conservation* (MacKinnon and Ramono, 1992) noting now that 'the wild population in Borneo alone may number 40,000 animals.' That the MacKinnons' estimates had plummeted – from 156,000 in 1986 to some 60,000 in 1990 and 40,000 in 1991 – would, one might suppose, either alert the interest, or challenge the credibility of any scientific discipline. It is difficult to find such extremely dramatic reductions in a population in a very short time span without biologists becoming very upset, in particular since the last estimates closed in fast along with other, more informed, assessments. After all, during the same conference the presentation entitled *Orangutans: Current Population Status, Threats and Conservation Measures*, (Sugardjito and van Schaik, 1992) stated that 'the total population in protected areas is between 19,000 and 30,000.' It cautioned that total numbers are meaningless and that few protected areas contained populations considered to be viable in the long run.

The discrepancies in published numbers caused no concern in the international conservation community. Is it possible that the estimates of surviving apes may still have been considered too high to arouse misgivings within the international conservation establishment?

The theoretical estimates of minimum viable population sizes produced by the newly fledged discipline of Conservation Biology (Soule and Wilcox, 1980) may have led international conservation organisations to believe that any number above

---

[51] The title may suggest that the article was a reaction to the expanding emergency concerning massive poaching of the apes and habitat destruction, but no mention is made of this, apart from the comment: 'Last year it was even discovered that farmers were being paid bounties for killing orangutans as agricultural pests in E. Kalimantan from a project set up under a World Bank loan' (MacKinnon, 1992; p. 213).

an effective meta-population size of 500 is not scientific ground for professional concern regarding the survival of a species. As a matter of fact, empirical evidence for human populations suggests that for apes the viability threshold may well be above 5,000 individuals. Could it, however, be possible that the orang-utan has been evaluated against this hard-nosed, unsubstantiated minimum standard, and hence was disregarded for action?

Irrespective of whether the dearth of information was a valid reason for persistent neglect by the conservation community, it is certain, however, that the orang-utan suddenly began to gain academic – as well as media – attention[52] in the early 1990s. That this had little effect on the conservation of the ape indicates how much the international conservation community is divorced from the reality of field biology (see also Wavell, 1996). It meant that for at least a decade considerable time and money was spent on convening international meetings to exchange and reprocess largely outdated information, while the international conservation organisations were usually not even represented at these gatherings.

Since the 1980s, no significant international support has been made available for the realistic, down-to earth conservation of the ape or its habitat. And, despite the heightened awareness in the western world regarding environmental issues (thanks to a series of TV documentaries on the beauty of the rainforest and on the orang-utan) few conservation organisations stirred when the first danger signals were evident in the early 1990s. A resolution calling for 'strict law enforcement in Asian countries to end the illegal traffic in Orangutans' (IUCN) at a gathering of conservation experts was as far as the world community would go to support conservation of the ape.

Fortunately, since 1995 international attention has slowly re-emerged. After the International MoF-Tropenbos Programme for Applied Research in Sustainable Forestry established a new model rehabilitation centre at Wanariset Samboja (East Kalimantan), the International Primate Protection League was the first to provide encouragement with some modest financial support. The *Wereld Natuur Fonds* could also be persuaded (in 1994) to provide funding covering the basic costs of the present survey. Especially the animal welfare organisations soon began to show a more substantial interest. In particular the World Society for the Protection of Animals (WSPA) and the International Federation for Animal Welfare (IFAW) have now acknowledged the emergency and have decided to provide structural support for some major issues in the Orang-utan Survival programme.

Recently, the *Wereld Natuur Fonds* (i.e. WWF Netherlands), following the lead of the WSPA and IFAW, has shown rekindled interest in the conservation of orang-utans. It not only bolstered its fund-raising campaign but also manifests genuine attention

---

[52] The establishment in 1994 of the Golden Ark Foundation was an attempt to raise funds for the active protection of endangered species like the orang-utan.

for active support. Still, all campaigns typically rely on the media appeal of the new Wanariset Samboja reintroduction project for confiscated orang-utans, thus enforcing a dangerously false public image that orang-utan conservation simply means handling apes.

If the new rehabilitation project generated any precursory interest in ape conservation, the massive forest fires of autumn 1997, which attracted worldwide attention and yielded shocking media coverage of scorched and butchered orang-utans, did finally reinstate 'wildlife conservation' as a matter of World Wide Fund for Nature policy. However, it remains to be proved whether this organisation, will be able to effectively deploy its resources for the protection of the ape.

The rehabilitation of orang-utans has turned into a circus, attracting hundreds of visitors per day.

REHABILITATION OF ORANG-UTANS

Introduction

Rehabilitation is an attempt to enable a captive animal to re-adjust to living independently under (more or less) natural conditions. It is prescribed where confiscated animals of an advanced mental capacity are to be set free, and involves providing individuals with the experiences and/or training which is believed to be necessary to survive and reproduce successfully under feral conditions. Rehabilitation is therefore a management tool in the field of nature conservation, and although it may attract disproportionate public attention, it is no more than a minor tool, often deployed with dubious results. It should by no means be confused with 'species' conservation, which is a complex of conservation actions focused on one particular wild-living 'umbrella species', or endangered species.

The rehabilitation of apes in general, and orang-utans in particular, is a difficult, time-consuming and expensive process. Orang-utans are intelligent apes, and do not have an innate ability to survive in the complex patchy wilderness of the tropical rainforest. Indeed, they must learn to survive under wild conditions. From birth, orang-utans, like all Hominids, are mentally shaped by a learning process in which the mother plays a crucial role for at least six years. A new-born wild orang-utan is fully dependent on its mother for care and training during these first years of its life. The youngster learns about the survival possibilities within its mother's home range, and so acquires a deep-seated knowledge of the whereabouts and seasonal availability of edible plants, as well as of inherent dangers and how to evade them. In this period, when the infant is physically developing into an adolescent, the mother is also indispensable for guiding it through social and potentially dangerous inter-specific interactions. Hence, it would be irresponsible to release ex-captive apes into the wild without such early, guided experience, because without the necessary basic knowledge an ex-captive ape's chances of survival are extremely small.

It is therefore understandable that orang-utan youngsters – between 3 and some 8 years of age – with some prolonged experience in the wild have a greater chance of being rehabilitated successfully, whereas infants, i.e. those younger than one year, have the least chance. It must be realised, however, that whatever the juvenile ape has experienced in terms of normal orang-utan culture and coping in the wild, this experience is quickly superceded by captive conditions: After all, the human standards of submissive dependency are forcibly imprinted on the captive ape.

Consequently, the return of an an ex-captive orang-utan to the wild state implies a profound mental reassembling on the part of the ape. Having been raised to full dependency in a captive situation, it now has do away with everything it has learned. It must become both independent and either indifferent to, or outright afraid of, humans, and should actively try to avoid any interaction. The point is that once the mental pattern of the ape is 'set', at around the onset of sub-adulthood, i.e. approximately 10 years of age, it is impossible to erase this knowledge and to adjust the animal's perspective significantly in terms of accepting a more insecure and strenuous lifestyle. While it may be relatively easy to corrupt wildness in an ape by means of habituation, or even to 'tame' it, e.g. through the regular provision of food, it is virtually impossible to reverse the process in a positive, unharmful way. Instilling in an ape a sense of awe and wariness concerning humans cannot be accomplished by any ethical means. The ape, after all, has become well aware of its comfortable dependency on, and physical dominance over, its human caretakers. Perhaps it should be realised that, where choice is involved, in the mind of a Hominoid the conditions of regular provisioning and care in captivity are usually more favourable than 'freedom' and the prospect of independency in a largely unknown and possibly hostile forest.

Thus, it is perhaps understandable that the time in captivity, the age and the mental disposition of the individual ape determines the success or failure of its rehabilitation. In fact, these factors should determine the protocol during the rehabilitation process. Conversely, the outcome of the rehabilitation process will be different for each individual, dependent on its age at capture, the duration and conditions of captivity and its age on reintroduction into the wild. Empirical findings indicate that both the age classes of (1) infants taken from the mother under one year of age, and (2) sub-adults with a long history of captivity have little chance of ever becoming independent of and indifferent to humans. Some will be permanently dependent on human ministering and provisioning, others will become a dangerous nuisance both for humans and for fellow apes. Even of those apes captured and confiscated during the most expedient juvenile and adolescent age classes, no more than some 80% can successfully be rehabilitated. In other words, some orang-utans can never be enabled to survive under wild conditions; and all of them will be in some manner damaged mentally.

The value and pitfalls of orang-utan rehabilitation

Orang-utan rehabilitation is a consequence of law enforcement. It is held that consistent law enforcement to remove all illegally captured and captive apes from the private sector would significantly reduce poaching. Hannah and McGrew (1991) reviewed ape-rehabilitation and concluded that it is a valuable instrument for conservation.

In the case of orang-utan protection, law enforcement is almost exclusively focused on the confiscation of orang-utan youngsters illegally kept or traded. Once the confiscated ape is in the custody of the authority, the issue is: what should be done with it? The legal framework prescribes that a confiscated protected species specimen must be returned to its natural habitat, unless this is not feasible for one reason or another. For orang-utans, return is impossible without rehabilitation, that is, a special training to adapt to wild conditions.

In order to rehabilitate an orang-utan successfully, and to reintroduce it into a suitable habitat, it would ideally require that an infant or juvenile is reared in a manner that closely approximates the normal situation. Although much discussion has been devoted to this subject, and some ideas have been published (e.g. Yeager, 1997), the essential problem is that there is no way a human caretaker can ever substitute for an arboreal ape mother. The result of the regular care and attention that a caretaker must apply in order to keep the infant alive and somehow sane is that a human model is imprinted onto the ape. Even a perfect ape disguise would not change that result.

Confiscation is meant to stop poaching, but what is to be done with the victim?
D. Sinaga of the Forestry Department played a major role in law enforcement

Evidently, an ape's attitude towards people is shaped in the same way that it would be among its kin, and is determined by whether the ape has been exposed to many or just a few different people, and by the kinds of interactions it has experienced with humans. In particular during late adolescence and early sub-adulthood, orang utans actively engage in assertive interactions in order to establish their social status. Captive apes will attempt to include humans in such interactions, and since they have become physically stronger and tend to gnaw-wrestle (or manhandle) for status in a fairly unpleasant way, they soon come to recognise their physical dominance[53]. Such a position may readily lead to dangerous situations in potential conflict situations, for instance in triadic interactions where the opposite sex is involved, and when food is at stake. In other words, the kind of attending and the number of different attendants is probably crucial for the mental development of the ape during its rehabilitation process.

---

[53] Keepers and trainers of orang-utans, in order to be able to maintain face-to-face interactions with the adult apes in their custody, therefore usually assert their dominance aggressively, which commonly involves physical abuse by means of implements (e.g. sticks or electric devices); note, however, that G. Brandes maintained friendly face-to-face relations with the *wild-caught* adult orang-utans kept at Dresden zoo (in the 1930s).

Therefore, it may well be that for rehabilitation the least harmful approach is to minimise and carefully orchestrate human contact, to deploy a few suitable sub-adult orang-utan individuals as role models and to maximise peer interactions. That in this way the youngsters develop into more social apes, in comparison to their wild peers, should not be considered of any consequence.

It is in the context of rehabilitation that people fall readily into the trap of either the potential commercial approach to ape-viewing or the sentimental approach of ape-care, or both. Neither approach is beneficial for the conservation of the orang-utan as a wild, independent relative, and, perhaps most significant, both approaches are in fact illegal.

The historical development of the rehabilitation concept

The rehabilitation of orang-utans is prescribed to facilitate law enforcement in order to curb the illegal hunting of, and trade in, apes. As a consequence, it should be seen as a temporary emergency measure to remedy a weakness in the legal framework concerning the conservation of protected species.

During the late 1960s and early 1970s it was originally believed that where and whenever possible all confiscated apes should be used to replenish a steadily declining meta-population in the wild. It was consequently held that rehabilitant apes should be returned to the wild populations. The possible risks were insufficiently – if at all – contemplated.

It is not realistic to expect that after several years of dependency in a cage the mentally abused ape can be rehabilitated and transformed into a normal orang-utan.

> Young orphaned orang-utans suffer a high level of stress and are susceptible to diseases. Ape poachers take into account that they must sell the juvenile or adolescent ape within some three months or it will die. The mortality among captive apes under the usually stressful conditions in Indonesian rural villages is very high. Infants under one year of age cannot usually survive in any case, even with special care. Since the incidence of tuberculosis, meningitis and hepatitis, and infestation with intestinal parasites and scabies among the rural human population is high, most of the captive apes readily contract these diseases, to which they soon succumb if their general condition is not upgraded and the appropriate medication given. Thus, ultimately, any illegally kept ape which is eventually detected and confiscated is believed to represent a loss of anything between a minimum of two and as many as ten apes, extracted from the wild population: first its mother, and next any number of peers (and their mothers) that died before being detected.

The first attempts to rehabilitate confiscated orang-utans were carried out at Bako Park by Barbara Harrisson in Sarawak during the 1960s (Harrisson, 1962; 1963). British North Borneo (i.e. Sabah) soon followed suit, and in 1964 a rehabilitation centre was established at the tiny (43 km^2) Sepilok Forest Reserve by the Wildlife Department (De Silva, 1971). Although the extent and quality of the area for reintroducing the apes is grossly inadequate, it must be mentioned that the Sepilok station from the beginning has maintained excellent quarantine facilities. When the Harrissons left Sarawak shortly after its independence, they transferred the first rehabilitants to this Sabah station.

In 1971, the first Indonesian rehabilitation station was established by the Indonesian Nature Conservation Service (PPA), with modest support from the World Wildlife Fund Netherlands Appeal (WNF), at Ketambe (Aceh, northern Sumatra)[54](Rijksen, 1972). It was set up and managed in conjunction with a research project concerned with the ecology and conservation of wild orang-utans, financed by the Netherlands Foundation for the Advancement of Tropical Research (WOTRO) and sponsored by PPA (Rijksen, 1974).

Shortly afterwards, a rehabilitation centre for Bornean orang-utans was established in the Tanjung Puting wildlife reserve in Central Kalimantan by Birute Brindamour-Galdikas with financial support from the National Geographic Society of the USA (Galdikas-Brindamour, 1975).

Subsequently, a few months later, a second rehabilitation project for Sumatran orang-utans was established by the Frankfurt Zoological Society (FZS) at Bohorok on the eastern side of the Gunung Leuser wildlife reserve (Frey, 1976; 1978; Borner, 1976; Aveling, 1982). In 1980 the FZS handed the administration of the station over to the Directorate-General PHPA.

In 1977, a new rehabilitation centre was established in Sarawak, in a tiny 6,4 km^2 forest plot adjacent to the Semengok botanical research facility and tree nursery (Aveling and Mitchell, 1982; Bennett, 1992; Lardeu-Gilloux, 1995).

---

[54] The total WWF support in 1971 for the establishment of quarantine cages at the Ketambe rehabilitation station amounted to US$ 8,794; the annual WWF contribution to the station in the period 1972 – 1978 was on average about US$ 5,000 (Rijksen, unpubl.; see also WWF Yearbooks).

Finally, in 1992 another station was established at Wanariset Samboja in East Kalimantan, according to a new conceptual approach (Rijksen, unpubl.; Smits, et al. 1993) which had been designed at the instigation of the Directorate-General PHPA and was formally approved by the Minister.

> In 1990, the management of the designated Kutai national park took the initiative to use confiscated orang-utans as a means of attracting visitors to its Teluk Kaba recreation area, under the guise of rehabilitation. The establishment of primitive facilities was supported by the Indonesian Wildlife Fund (IWF) and local industries. Due to substandard medical conditions this facility was closed in 1994, but despite the official closure summons, in early 1996 the Park management acquired three new juvenile orang-utans to keep in cages as a visitor attraction.
>
> In 1991, the Indonesian Directorate-General PHPA set up an additional 'alternate' rehabilitation facility (Tanjung Harapan), at the border of the Tanjung Puting reserve. This was to supplant the original Camp Leakey set-up, which had been ordered to discontinue its rehabilitation activities, because it was recognised that the facility 'had serious shortcomings in its prime function to rehabilitate the apes in its custody' (Anonymous PHPA report, 1991). The rehabilitation programme at Bohorok in North Sumatra was also terminated in 1995 for the same reason. The establishment of new facilities, according to the Ministerial guidelines for rehabilitation (i.e. SK Menhut 280/Kpts-II/95), is in progress – quarantine facilities will be established in Medan (North Sumatra) and Palangkaraya (Central Kalimantan), well separated from the reintroduction sites. Public access to all facilities will be prohibited.

During the first five years of operation the two Sumatran stations did demonstrate that a rehabilitation programme is an effective means to facilitate law enforcement and that it can curb the illegal catching of, and trade in, orang-utans. Conditions for this success were that (1) confiscation of apes was consistent, (2) the objective of rehabilitation was focused solely on the application of law enforcement and its follow-up, and (3) the rehabilitation process was financially supported and hence fully independent of any need to generate its own funds. It is probable that only continuous international supervision and support can guarantee such crucial conditions for the time being.

Three years after its establishment in the mid 1970s, it became evident at the Ketambe station that some essential premises of the rehabilitation technique were wrong. The major premise initially was that confiscated apes should be put back into an existing wild population, in order to compensate for some supposed loss. It was insufficiently realised that the loss referred to the meta-population but hardly applied to local populations in shrinking habitat conditions. This supposition also failed to take into account that a local wild population is usually composed of individuals who are familiar with each other, and its numbers are already up to the carrying capacity of the habitat. Any foreign addition will boost competition for scarce resources and may cause considerable social unrest as well.

Another important finding was that if rehabilitation allows for regular close contact between apes and visitors, the risk of serious disease transfer from humans via the rehabilitants to the wild population becomes dangerously high.

It was, moreover, felt that if a rehabilitation centre were increasingly subject to visitor attention and eventually became a tourist attraction, it would readily corrupt its major objective and create problems for conservation rather than solve them. It was realised that an economic profit incentive would creep in with tourist development, which would erode the need for full government subsidy for the rehabilitation process. It was foreseen that in such a terminal state a station would become a permanent, self-perpetuating facility, which could in fact inhibit protective law enforcement, so that an ample supply of apes would need to be secured in order to attract a regular flow of tourists. The resulting commercial dependency would then even be likely to create an illicit demand for apes, just like any ordinary zoo or circus.

Initially, the risks of disease introduction, the possible dangers of over-stocking areas of habitat, or the problem of causing serious unrest in an established deme were never thoroughly considered. When these issues were finally raised, all existing stations, except Ketambe in Aceh, simply ignored the problem, although token quarantine had become a standard issue in the stations' public relations programme. Consistent with the findings and fears, the Ketambe rehabilitation station was closed in 1978, under protest from the local authorities (Schürmann, pers. comm.), who indeed had hoped to eventually turn the essentially temporary rehabilitation project into a sustainable tourist attraction.

When the negative findings were publicised (MacKinnon, 1977; Rijksen and Rijksen-Graatsma, 1978, Rijksen, 1978), the international conservation community quickly lost interest in any further support for orang-utan conservation. If conservation of the endangered ape had so far mainly concerned law enforcement and rehabilitation, the necessary further steps to secure the ape's survival were scarcely considered, and after the late 1970s orang-utan conservation was entirely neglected (Rijksen, 1995). Indeed, after rehabilitation and orang-utan conservation had somehow become identical in the minds of the international conservation community and the public, the fate of the ape almost became a taboo subject.

Rehabilitation and the King-Kong archetype

The image of the 'weak' human somehow dominating the 'mighty' beast was the archetypal stuff of popular myths and legends that made the King-Kong and other 'Beauty and the Beast' stories so appealing. It was not coincidental that the orang-utan was originally named the Indian Satyr, while its supposed lewdness with respect to indigenous women was especially mentioned (Tulp, 1641).

Human nature seems to be particularly sensitive to the psychological notion of the implicitly sexual meeting of compassionate feminine care and the potential threat of bestial violence in masculine form, which is so powerfully represented in the male ape. In a given situation, it is also the predilection of human nature to respond very quickly in a parental, or caring, manner. These two basic instincts can create a powerful psychological impediment to the proper implementing of rehabilitation.

In the 1960s, biologists began to study apes in the wild. Men like Adriaan Kortlandt, Vernon Reynolds, Junichero Itani, Helmut Albrecht and George Schaller spent months in equatorial rainforests in the hope of gaining some insight into our own hominoid prehistory, collecting the very first field data on the great apes. Nevertheless, it was women who ushered the great apes into the international limelight. The pioneering attempts of the male biologists soon faded into the background when an attractive young woman, sponsored and publicised by the National Geographic Society of the USA, was sent out into the wilds of Tanzania by the paleo-anthropologist Louis Leakey.

Due to her outstanding work, Jane Goodall will forever be remembered as the foremost scholar of wild chimpanzee behaviour, although a score of field biologists subsequently added to her work and findings. Rather than submitting to the role which photographers and film-makers attempted to foist upon her, Goodall soon learned to utilise the media, in particular to draw attention to the plight of the chimpanzee, and she never ceased striving to turn publicity into a vigorous force for ape conservation.

Then, almost a decade after George Schaller had pioneered the study of wild mountain gorillas, it was another woman, directed by Leakey[55] and sponsored by the National Geographic Society, who drew the world's attention to the gorillas, albeit in a most unfortunate way. Soon after she had published the results of her excellent work, and had earned her Ph.D, Dian Fossey made world headlines when she was murdered in Rwanda, as a result of her dedication to the apes and the ensuing strained human relations. Although her commitment and work under arduous circumstances were in themselves outstanding enough to elevate her to world fame, it was in fact the publicity relating to her death, plus a movie cashing in on the King-Kong archetype, which really captivated world audiences: The movie *Gorillas in the Mist* highlighted all of the possible sensational aspects of her work, culminating in her violent end. Yet despite this false image, the movie served to lend support to organisations which, right up to the present, have effectively conserved the mountain gorillas living on the Virunga volcanoes.

Why did public fame fall so readily to women? Clearly it was primarily due to their perseverance and the excellent quality of their work. But there was something else. The women studying the African apes could hardly avoid the false glamour which the sensationalistic media – and even *National Geographic* – attempted to project onto them. One can not deny that for the general public the pictures suggesting a 'King-Kong's maiden' image, showing close physical contact between woman and ape[56], were apparently considered to be immensely appealing, and may have been a determinant for shaping the public image of ape biology. The answer to the above

---

[55] As a matter of fact, two other young women, also sent by Leakey, had tried to study the mountain gorillas before Dian Fossey; they were Rosalie Osborn and Jill Donisthorpe.

[56] See Fossey, 1970: pp. 53, 60, 61; 1971: pp. 578, 579; 1981: pp. 504, 505.

question may therefore well be that western culture, steeped in complex myths featuring classical satyrs and fantastic apes, has long been prepared to provide a hall of fame especially for women actually demonstrating their mental and physical 'control' of the supposedly mysterious, dangerous and potent Satyr.

Illustration (by Arthur Rackham) from Edgar Allan Poe's late nineteenth' century bestseller The Murders in the Rue Morgue.

In order to succeed, supporters of ape conservation need to heighten public awareness and to create an appeal through the media, but it is questionable whether the false sentimental image which has so far been generated by the media is required. One would hope that the public would support the conservation of apes because it acknowledges that our wild, closest relatives are entitled to their own independent place in the world and must be allowed to continue existing in the wild. After all, the essence of conservation is respect for wildness. It is also the seeking of constraints on inherently expansionist and assertive human behaviour, such that other organisms are allowed and enabled to survive in self-sufficient freedom.

However, the sensational aspects of ape-human contact unfortunately tend to blur the issue, and readily present the conservation of apes as simply meaning that individual humanoids are to be cared for by attendants. Not only the commercial media, but also serious organisations like the International Primate Protection League exploit the strong appeal of the King-Kong archetype, by depicting attractive, caring women making a point of demonstrating, for their constituency, their 'love' or 'motherhood', while controlling the survival of the humanoid beast. Few people realise that this is *not* what conservation is about, and under such circumstances the public will never be made properly aware. But certainly such exposure does generate funds and cause the constituency to grow. It is difficult for the underpaid and undervalued conservationist striving to restore the independence, dignity and wildness of apes, to resist the temptation of riding a false but lucrative wave of sensationalist publicity.

Contrary to the situation regarding the African apes, international interest in the orang-utan centred around care and rehabilitation from the very beginning. Barbara Harrisson initially drew attention to the fate of the orang-utan, and then established the first rehabilitation project, but the publicity machine in Southeast Asia in the early 1960s was not yet ready to elevate her to international fame.

Yet, with the growing impetus of the international media and the ready availability of ex-captive orang-utans, the King-Kong archetype soon appeared to be a psychological trap, even more difficult to avoid than in the African situation. All the western women associated with the red ape played major roles in the handling and dedicated care of confiscated orang-utans. Perhaps this is not so surprising when one considers that the wild orang-utan is elusive, exceedingly difficult to study and far less spectacular in its daily affairs than the African apes. Indeed, it is not easy to picture a ground-dwelling person and the wild arboreal ape together. Nevertheless, the orang-utan's orphaned, captive infants and juveniles are readily at hand, and they are the most attractive, gentle and docile of all the apes, appealing to any maternal or parental instinct.

After the pioneering work of Barbara Harrisson, Birute Galdikas swept on to the stage in 1971. Also directed by Louis Leakey, and sponsored by the National Geographic Society, while being assisted by her husband Rod Brindamour, she sought the role of champion of the red ape, closing ranks with her two famous African colleagues, in order to fulfill Leakey's dream of beholding the behaviour of all three wild relatives of early humans.

Galdikas also intended to study primarily wild orang-utans (Galdikas, 1995), but the Indonesian situation almost immediately obliged her to become involved in the rehabilitation of ex-captive apes. In its pictures, the *National Geographic* emphasized this 'mothering' aspects of her work (Galdikas-Brindamour, 1975) even stronger than it had done with her colleagues[57], and she readily adopted the irresistibly appealing role. One of her subsequent articles even appeared under the title: *Living with the Great Orange Apes* (Galdikas, 1980).

Be that as it may, the global publicity, in particular involving women caring for apes, meant that for the general public ape conservation has become synonymous with cuddling, care and custodianship, or what in the case of orang-utans has become rehabilitation. When the idea of a need for 'awareness' in conservation is coupled to such a false image, it is not surprising that the concept of rehabilitation is entirely misunderstood as the public display of feral apes under human custodianship.

---

[57] See Galdikas-Brindamour (1975) *National Geographic Magazine* cover, and pp. 444, 448, 451, 452, 453, 456, 462, 463, 465, 472.

> A few months after Birute Galdikas and her husband Rod Brindamour had settled in southern Kalimantan, two young Swiss women, Monica Borner and Regina Frey, were sent out by the Frankfurt Zoological Society (Germany) to Sumatra to help the Indonesian PPA agency with orang-utan rehabilitation in Langkat (Bohorok, North Sumatra province), although the rehabilitation station at Ketambe, some 80 km to the west (in Aceh province), had just been established by PPA with the support of the World Wildlife Fund Netherlands Appeal and WOTRO. Rather than being guided by ambition to carry out a field study on the wild ape, the two Swiss scientists dedicated themselves fully to orang-utan conservation in a modest yet effective way, trying to boost the performance of the PPA. However, they never achieved – nor did they seek – worldwide public recognition comparable to that of their Canadian colleague in Borneo.

For a person in a Southeast Asian culture, to be associated with orang-utan research or rehabilitation does not confer the same status as is accorded in the west, possibly because in Southeast Asian countries the notion of evolutionary kinship is less well accepted and there are few classical Satyr myths and romantic stories of the 'Beauty and the Beast type'. Nevertheless, though oriental people do not seek status by means of caring for apes, a common human characteristic is revealed by the fact that in a social context many ape attendants scarcely differ in their approach to apes; rather than unmitigated concern for the dignity and welfare of the creature in their custody, an element of domination and posturing is equally discernible.

Nevertheless, in the wider context of orang-utan conservation marked differences are revealed when one compares the approach of an oriental conservation worker to that of his/her western counterpart. The former seems barely influenced by any inherent instinctive appeal on the part of the ape, and operates in constant awareness of a socio-economic and political sphere, while under obligation to function according to the regulations concerning a protected animal. The latter is often readily – and almost blindly – committed to protecting the orang-utan individual, albeit in a largely paternalistic way.

> At a confiscation site, the oriental employee goes about his/her task in the most careful and opportunistic manner, merely considering the ape an object. The western colleague, however, while seeking to take the ape into custody, is usually captivated immediately by the creature's personality and feels compelled to establish a relationship with it.
> Interestingly the orang-utan itself may play an active role here. It is quick to recognize compassion and will seize the opportunity to make things a little easier for itself by facilitating such a relationship.

The point is that a psychological trap lies at the basis of the ape conservation issue. It appears that for most people the ape is either considered to be a human caricature in need of help and care, or an inherently fearsome object which can be manipulated to facilitate a public display of human domination. In their interactions with orang-utans, both western and oriental people often intuitively act as though the creature were a domestic pet rather than a self-contained sentient being who deserves to be free and independent. Few people – even those active in the field of orang-utan conservation – can bring themselves to respect the apes unselfishly, and consequently to stay as much as possible out of their personal sphere, while granting them their own cultural mores and private dignity.

It is undoubtedly a superb sensation to establish friendly physical contact with an ape. Such interaction has even been described as a 'spiritual experience' (Galdikas, 1995), and is often interpreted as an expression of the 'love' needed for ecological awareness. However, if love is characterised by an essential element of unselfish, loyal dedication to the personal integrity of another, then the context and behaviour of most people confronting (rehabilitant) orang-utans represents anything but love. In reality, much of the desire for contact with an ape seems instead to be rooted in consumerist sensationalism laced with a clear streak of human self-aggrandizement[58], at the expense of the ape's independency. In commenting on the visitor situation at the Tanjung Puting rehabilitation centre, Russell (1995) noted, 'interactions were characterised by a desire for physical contact with the ex-captives, satisfaction in feeling needed and thus helpful, a tendency to evaluate orangutan behavior by human standards and an emphasis on the orangutan's physical attractiveness and entertainment value.' In any event, the false 'love' for apes is far removed from the required respect for the integrity of a wild relative, which is at the core of the conservation concept.

It may be argued that it is often the rehabilitant ape who initiates contact. This may be so, but one must realise that an ex-captive orang-utan is in a peculiarly enslaved situation. Humans were the cause of its predicament, and although it might now enjoy some freedom of movement, it cannot survive without human intervention. Devoid of normal ontogenetic experience the orang-utan is to a considerable extent dependent upon human care, protection and the provision of food. Being intelligent, the ape probably understands fully its precarious position, and will readily adjust its way of thinking and behaviour in a most expedient manner. Indeed, it is conceivable that under such bizarre conditions the ape has actually come to need and to desire human attention and affection.

The essence of rehabilitation, however, is to encourage or restore in the ape as much mental independency as possible under the particular ontogenetic conditions; it is not intended to cater to the imposed, unviable needs and desires of an individual ape. And if, within the context of a personal relationship being necessary for the mental health of the ape, a personal caretaker is attuned to provide as much care as is required, then he/she should be the sole attendant. No other person should be allowed to accept or seek such contact, irrespective of whether it concerns the non-assigned managerial staff of a rehabilitation station or (paying) visitors. It may be hard to resist the charm of a clever orang-utan orphan begging for attention and affection, but yielding effectively 'spoils' *in toto* everything of the creature's potentially wild self.

---

[58] Unfortunately some managers of ape rehabilitation stations, despite general rules laid down to prohibit human-ape contact, often behave as though they were exempt from such rules, and take for themselves the privilege of showing off 'heir' apes to larger audiences or cameras. Such behaviour undermines the required discipline, renders regulations useless and degrades rehabilitation to a cheap publicity act.

Many people feel an irresistible urge to establish physical contact with apes, especially in front of an audience or for a photograph; at Singapore Zoo this yearning is exploited in the well paid for ritual of daily 'breakfast with an orang-utan'.

For sound emotional development, infant orang-utans, like many mammals, and in particular primates, need a mother (Harlow and Harlow, 1965). For orphaned orang-utan infants a human caretaker may adopt the temporary role of a 'mother' in order to guarantee the desired emotional development, but the growing ape eventually comes to perceive itself in a human (domestic) rather than a wild context. This is contrary to the objective of rehabilitation. The experimental set-up of the new approach to rehabilitation suggests that – in keeping with the findings of Harlow and Harlow (op. cit.) – a peer-group can, to some considerable extent, substitute for the 'mother' function (Russon, pers. comm.) and lead to a more appropriate self-image in the growing ape.

Nevertheless, in the rehabilitation process the human caretakers do provide food, and at times seek contact, and hence the developing ape reciprocates by seeking contact as well, when need be. The major problem of rehabilitation is therefore that the ape must be trained to respect humans, and should come to consider them as potentially dangerous. After all, seeking contact with a human in an indiscriminate manner and under feral conditions could lead to life-threatening situations for the ape, as well as for the person if the response is not in accordance with the ape's expectation. In all rehabilitation programmes, western managers, as well as the local attendants and even visitors, have occasionally been attacked and injured by the rehabilitant orang-utans. No doubt in such instances the ape's reasons for seeking contact were essentially different from, and much less exalted than, those of the person approached.

Rehabilitation and the commercial trap

Considering the cultural bias, it is perhaps not surprising that orang-utan rehabilitation was initiated by people from western countries, and that the sponsoring of rehabilitation attempts was mainly from western NGO sources. Whether the non-indigenous initiatives, expatriate technical assistance and extraneous support have fostered sufficient local commitment is doubtful, however. Perhaps the lack of local commitment is partly because in the world media much credit for ape conservation has been given to the zealous western caretakers and western sponsors, whereas local contributions were usually downplayed or ignored.

Most important, however, is that after the extraneous sponsorship for the rehabilitation process was eventually phased out, the stations set off on a course that appeared to be oriented more towards economic self-support and exploitation of the potential for awareness rather than towards a primary conservation objective. Since then, station management has come to see visitor service and ape-viewing possibilities primarily as a gambit to please the public. The idea that rehabilitation – along with law enforcement – was initially to serve the conservation and protection of the ape has receded into the background.

> Neglect on the part of the international conservation organisations and the biased view of local authorities were undoubtedly the main reasons that rehabilitation veered away from its primary function. The numbers of visitors attracted to the stations must have sparked some ideas involving 'self-reliance' in terms of income, and indeed in the early 1990s the Bohorok and Tanjung Puting 'Camp Leakey' station in Indonesia and the Sepilok Station in Sabah embarked upon a new course, inviting visitors and levying fees in order to become at least partly self-supporting. The international sponsors, notably the World Wide Fund for Nature, the Frankfurt Zoological Society and the Leakey Foundation apparently encouraged such a quasi-commercially oriented approach[59], although its dangers for orang-utan conservation had been outlined clearly at an early stage (Rijksen and Rijksen-Graatsma, 1975; Frey, 1976; Rijksen, 1978). These stations must now be considered simply as two more commercial ventures exploiting wildlife. Rather than serving conservation, they have become a serious liability to the ape's survival.

Hence, rehabilitation according to the intent of the law disappeared. Confiscated apes were no longer trained to live an independent wild existence, but were essentially kept for display.

It is realistic to note that rehabilitation stations, due to their new visitor-oriented function, in fact became largely dependent upon exploiting the display of rehabilitant orang-utans. Hence, one can imagine that it is hardly an incentive to force the apes to leave the feeding site and return to the wild. And if apes nevertheless become feral and independent of food provisioning, the commitment to paying visitors will dictate that those apes which have returned to the wild be replaced. As a consequence, the sustainable need for new rehabilitants will certainly not help prevent poaching, but facilitate it instead.

---

[59] At the Bohorok station in North Sumatra an attempt was made to have the best of both worlds. The head of the PHPA and the international technical experts were aware of the possible dangers but felt that the station management would be able to control the process, allowing no more than 50 persons to attend a feeding session. An effort was made to adhere to the rule that visitors be kept well separated from the apes, but in practice this could not be enforced consistently, because the numbers of attending visitors soon spilled over the limit. The lure of profit from paying visitors led perhaps to extraordinary leniency and to corruption of the best intentions.

As long as so many people fail to perceive such treatment as abuse, it will scarcely be possible to build up a significant constituency in support of the conservation of orang-utans.

The fact is that after 1980 the stations failed to have any influence on orang-utan poaching. Indeed, the illegal capture of and trade in apes revived in massive proportions during the late 1980s, especially in Kalimantan, despite the existing rehabilitation projects that were attracting ever more visitors. Whereas a rehabilitation project should in fact be a temporary emergency measure, phasing out the illegal acquisition of wild apes, and, as a consequence, its own functional existence, in reality the supply of orang-utans on the illegal trade circuits increased. The projects apparently had neither a lasting effect on the inhibition of poaching, nor any role even in the recognition of the resurgent problems. If there was an interest in illegally kept and traded orang-utans, it was for the acquisition of new rehabilitant apes, rather than for purposes of law enforcement. Indeed, the smuggling of orang-utans in the period 1986-95 was at its worst in Sampit, Central Kalimantan, but no information was forthcoming from the nearby rehabilitation centre.

It is ironic therefore that visitor attendance at rehabilitation projects was commonly justified as being a major contribution to environmental awareness and orang-utan conservation (see e.g. Galdikas 1991; 1995). In reality, all the current rehabilitation projects for orang-utans developed into open-air circuses for tourists (Russell, 1995), paying only lip-service to orang-utan conservation while pursuing a commercial or public relations interest.

In the outdated rehabilitation centres orang-utans are misused as an attraction for visitors; contacts between humans and apes are unavoidable, and 'rehabilitation' is a fraud (Sepilok, 1996).

It is virtually impossible to scrutinise the administrative records – if there are any – of the rehabilitation stations in order to obtain any reasonable idea of the income from tourism. However, it is possible to estimate a baseline income, as has been done below. Established in 1964, the Sepilok orang-utan rehabilitation centre started as a scientific operation. Admission to the centre was at first restricted, but gradually became more relaxed. It was free of charge until the early 1990s, when annual visitor numbers surpassed 30,000. By about 1980, the Sepilok centre was attracting 17,000 visitors annually (Aveling and Mitchell, 1982), and Davies (1986) reported 'more than 25,000 visitors each year' some four years later; in the 1990s the numbers have increased to an estimated 40,000 at least. Foreign tourists are currently charged M$ 10, local visitors M$ 2. If foreign tourists comprise no more than 30% of the total, the annual income exceeds US$ 100,000.

In the late 1980s, the Bukit Lawang (Bohorok) region in North Sumatra attracted an estimated 16,000 visitors annually; in 1996 this had increased to over 500,000 visitors, of which some 40,000 were foreign tourists (R. Frey, pers. comm.). In general the foreign tourists come to visit the rehabilitation project, while most of the domestic visitors come to amuse themselves along the river banks. The official figures for (foreign) visitor attendance at the feeding site in 1994-96 amount to some 18,000, with a peak of 21,577 in 1995. In terms of impact on the feral orang-utans at Bukit Lawang this implies that a daily average of anything between 49 and 110 people attend the feeding sessions, with peaks of up to 300 spectators per session. Since the early 1990s, foreign tourists have paid Rp 4,500 or approximately US$ 2 per day (for a maximum of two visits). Indonesian visitors commonly do not pay an entrance fee, yet in terms of economy there is still a gross average annual income of anything between some US$ 43,000 and US$80,000.

In 1991 Galdikas was formally refused permission to go further into the park and manage rehabilitation within its boundaries, but the Sekonyer (Camp Leakey) Station (Tanjung Puting) in Central Kalimantan is nevertheless used for eco-tourism arrangements by the Orangutan Foundation International, beyond the authority of the park organisation. Until 1997 it had been the focus of a number of eco-tourism operators (Earthwatch, Wolftrail, Animal Watch), from June to November, attracting paying guest-students or 'volunteers' who carried out tasks or participated in simple short-term research assignments for a price of some US$ 2,000 per person for a 10-day period. Some 80% of the payment (or US$ 1,600) was said to go to the Foundation, hence supporting Galdikas. If the trips were fully booked during the regular season, some 15 groups of an average of 10 persons per group would add up to an annual income of at least US$ 240,000.

It is significant that, except in the Orangutan Foundation International brochure (1991), hardly any relevant interpretation of the reasons for and principles underlying rehabilitation has been provided at any of the stations. One rarely finds hard facts on the need for rehabilitation as an extension of law enforcement with its flaws and dangers, or even any indications concerning the problems of orang-utan conservation due to poor management of the forest resource.

> The official Sepilok Station leaflet provided by the Ministry of Culture and Tourism notes: 'Sabah has one of the largest numbers of Orang Utan (...) they range from a minimum of 2,000 to a maximum of 10,000 to 20,000 in the State. The loss of the Orang Utan's forest habitat due to agricultural activities is seriously affecting its survival. Logging and poaching are, however, trivial problems. Young Orang Utan, abandoned by their mothers when forests are depleted, are brought to Sepilok.'

Except for one leaflet issued by incidental WWF support for the Bohorok Visitor Centre (1990), brochures commonly carry neither rules and regulations for visitor behaviour with reference to the rehabilitant apes, nor any warning that orang-utans can be extremely dangerous in close-contact situations[60]. On the contrary, notably the brochure of the Orangutan Foundation dealing with the camp Leakey facility at Tanjung Puting (Central Kalimantan) actually invites close contact between humans and apes, to generate a supposedly mystical experience which would ultimately lead to sympathy for and support of conservation (Galdikas, 1995). It flatly ignores – if not outright denies – the potential dangers for both humans and apes[61], which could result from such contacts.

It seems that the spirit of commercialism has also made it increasingly difficult for an average visitor to gain an accurate picture of the numbers and the fate of apes taken into the stations after the early 1980s[62], when international supervision terminated. The only information which is invariably offered rests in a general exhibition depicting the natural history of the rainforest, and a few neutral facts about ape behaviour.

---

[60] It is perhaps not surprising, albeit unacceptable, that all brochures and leaflets designed and issued by national organisations contain biased propaganda rather than adequate information, usually avoiding the issue of regulations and a warning note on the dangers of ape-human contacts. It is noteworthy in this respect that an official PHPA leaflet for the Bohorok station (1989) carries a section entitled 'Controversy over rehabilitation'.

[61] At the Bohorok and the Camp Leakey Stations, people have frequently been attacked and seriously bitten, in particular by sub-adult males and females with infants (see e.g. Yeager, 1997).

[62] Since the 1980s the Bohorok, Samunsam and Sepilok rehabilitation stations have been secretive about the numbers of apes taken in and returned to the wild; all typically claim to have rehabilitated 'more than one-hundred.' In Bohorok the formal station record lists a total of 176 apes in 1996, although the 1991 brochure noted that in the period 1973-1988 a total of 166 had been brought to the station, of which at least 35 had died. In reality a total of at least 190 orang-utans were taken into the station between 1973 and 1991, and in 1991-96 another 14 were added, three of whom died.

Table VII

Rehabilitation at Tanjung Puting 1971-1996
Total numbers of orang-utans taken in for rehabilitation 1971-1990 (Galdikas): 108
Total numbers taken in during 1991-1996 (PHPA): 54
Total recorded deaths 1971-1996: 32
Born from rehabilitants: 25
Still regularly provisioned at the stations (T. Harapan, Pondok Tandui, Camp Leakey): 23
Apes said to be in the illegal, covert custody of Galdikas in 1997: 68-89 (MoF information)

The attendants rarely control visitor behaviour and often display inappropriate conduct in their dealings with, and handling of, the apes: The rehabilitant ape is an object of show, and it is hardly likely that any visitor would be filled with a sense of wonder and respect for this wild-living, close relative who in fact requires a dignified, non-interfering approach in order to survive in the wild. Indeed, the apes are being conditioned to associate the presence of groups of visitors with the provision of food.

> Considering this official example, it is not surprising that several enterprising local guides at Bukit Lawang (Bohorok) have privatised the ape-show during jungle-treks with tourists. They trained some of the rehabilitant apes to look for bananas at a predetermined spot, some distance away from the regular feeding site. They then lead their tourists to the spot, and pretend that they have come upon a 'real wild' orang-utan, which can subsequently be fed by the tourists, and possibly even enticed to join in a photograph. This has led to extremely dangerous situations in several cases, where the ape became so bold as to attack guides and tourists along the forest trail, apparently in order to go through the people's bags in search of food. Not only have people been bitten during these skirmishes, but at least four orang-utans have been killed for their aggressive behaviour, sometimes in direct conflict by means of a jungle knife, but, in at least two cases, by means of poisoning.

Be that as it may, it is clearly a violation of the intent of the law regarding protected species if the rehabilitation process is in any way hindered or set back. Visitor attendance not only exposes the animals to health hazards, but also effectively stalls or reverses the learning process through which a domesticated ape can become free-living or semi-wild again. Hence, in reality the traditional rehabilitation stations for orang-utans, which allow visitor attendance, function illegally. Consequently, on April 23, 1991, the Director General of PHPA issued an instruction to close the rehabilitation centres at Bohorok (North Sumatra), Camp Leakey (Central Kalimantan) and Teluk Kaba (East Kalimantan).

The cost-benefit economy of the rehabilitation process has sometimes been challenged (e.g. Brambell, 1977; MacKinnon, 1977; MacKinnon et al. 1986). Considering that the cost of confiscating, transporting, quarantining, socialising, transporting again, and finally reintroducing one ape is currently anything between

US$ 1,500 and $ 5,000 (Sajuthi *et al.*, 1991; Smits, pers. comm.), there is reason for concern. It is certainly a high price to pay in order to attempt arresting the poaching and illegal trade, and it is not surprising that when such amounts are mentioned, uninformed people of limited vision readily contend that the money might better be spent on the conservation of the ape's habitat.

> Because rehabilitation is meant to prepare an ex-captive ape mentally and physically for living under wild conditions, independent of human care, the regular exposure to visitors and the manual provisioning of food makes this impossible. Indeed, allowing visitors to view the orang-utans during the rehabilitation process tends to increase their dependency, making them tamer rather than more feral, and therefore violates the intent of the law.
>
> During adolescence and sub-adulthood the orang-utan becomes increasingly assertive, and, if given the chance, it will increase its efforts to gain superiority over individual people (especially women) in its surroundings. The extremely consistent – and at times forceful – treatment which is required to force ex-captive orang-utans to become indifferent to – as well as mentally independent of – humans is absolutely incompatible with visitor attendance. Visitors are an attraction for the rehabilitant ape and, if provoked, they react inconsistently and usually such that the ape quickly learns how to manipulate rather than to fear them. In the wild state, it would cost the ape its life to display assertive inquisitiveness or intimidation in a confrontation with an armed local person.

It has also been noted that spectacular manoeuvres involving endangered species, like 'rescue actions', captive breeding, rehabilitation, translocation and reintroduction, draw public attention away from the real issue, namely, how to combat habitat destruction (MacKinnon *et al.* 1986; Galdikas, 1991). Indeed, such manipulations can even undermine habitat protection as the prime focus of conservation (Povilitis, 1990), while providing ineffective conservation agencies with a false mask of functionality.

It is telling, however, that since the mid-1980s when this academic debate over cost-allocations and supposed negative effects of rehabilitation for conservation began, no support whatsoever for either the protection of orang-utan habitat or protective law-enforcement and rehabilitation of the ape itself has been forthcoming from international donors, although the media campaigns may suggest otherwise. This demonstrates that no conservation authority, whether national or international, has had any vision or any clear sense of direction concerning the ape or its habitat. Whatever kind of orang-utan rehabilitation as a follow up of law enforcement has been carried out since the 1980s, it has been facilitated by meagre government allocations and fund-raising by individual people trying to somehow couple their own welfare with that of the apes in their custody. Their only option has been to delve into the pockets of visitors.

The issue of course is not a matter of choosing between protection of habitat or protection of the ape. The wild orang-utan can be saved from extinction only if large areas of wildland forest containing sufficient habitat are effectively conserved and the ape is safeguarded against further persecution. Both issues are intricately linked, and cannot be weighed or separated, but require leadership to be dealt with appropriately.

Attempts to stop poaching and illegal trade must be followed by a proper procedure to deal with the ape within the context of the legal framework. It is a matter of ethics as well as of primary allegiance to the legal system. Indeed, in order to give the wild orang utan and its habitat a chance of survival, the illegal waste of orang-utans must first be stopped by legal action, which prescribes reintroduction, whatever the price. After all, strict law enforcement would instill in people a necessary respect for the ape, so that the notion becomes readily accepted that such a well-protected creature must, as a matter of course, also be granted a place to live within the regional area.

> Moreover, it must be realised that when a population is under the kind of pressure that the orang-utan is currently facing, every female specimen which can be saved and rehabilitated to become a reproductive (free-living) adult counts disproportionally for the survival of the species. Hence a survival programme should not just stop the illegal hunting, but also pay close attention to rehabilitation of the surviving confiscated result of the hunt.

The shift in rehabilitation, from serving law enforcement to accommodating a flourishing tourist industry, has been neither regulated nor properly guided. An unscheduled swell of visitors and tourists incited by possibly well-meant media coverage of rehabilitation has simply overturned the original function of rehabilitation, while the educational potential has remained undeveloped.

In any case, it is unacceptable to consider rehabilitation an economic asset, because the economic incentives entirely displace the original conservation objective, violating the intent of the law. Hence, the traditional rehabilitation projects have become a liability to conservation in general and are a threat to the survival of the ape. Obviously the methods of traditional rehabilitation needed revision, because the development of an appropriate tourism model with reference to the orang-utan is of high priority (see chapter Orang-utan Survival Programme).

> Since rehabilitation is a formal extension of law enforcement, one would expect rehabilitation stations to operate under a special permit, outlining terms of reference for functioning. One would also expect such stations to be subject to regular government control. The Sepilok station in Sabah has an official government licence to handle orang-utans. However, contrary to expectations, until 1994 none of the stations in Indonesia had a formal mandate or were subject to such supervision; the only officially licenced station is the new Wanariset Samboja station in East Kalimantan, SK No 132/Kpts/DJ-VI/94 (08.08.1994). None of the stations had ever been formally evaluated until 1997, hence there has been no control whatsoever of their practices.

The developments in orang-utan rehabilitation have demonstrated that if orang-utan protection is forced to operate without significant government or other commercially independent funding, and is allowed to run a quasi self-supporting course, the primary conservation objective can not be met. As with many other issues of nature conservation, rehabilitation is incompatible with commercial exploitation, and is not in keeping with misplaced capitalistic ideas of self-support through 'sustainable' utilisation. It must be supported by government funding.

Even before the economic crash of 1998 it was already obvious that Southeast Asian governments could not generate enough political commitment, management ability and funds to properly run the rehabilitation centres for confiscated protected orang-utans in their custody. It bodes ill for the conservation of biological diversity if even the red ape is unable to engender sufficient commitment in a significant proportion of the people with whom it must share its environment. No doubt this was at least partly due to the misguided management of the centres and a lost opportunity to raise public awareness with respect to the ape's dilemma. In any event, the manner in which rehabilitation of the ape is supported and organised may well be considered a reliable measure of the level of conservation interest in a country. It is evident that without international support of conservation in the future, the outcome will be bleak for the ape and other wildlife.

It is an encouraging sign that in Indonesia the undesirable developments in rehabilitation were acknowledged in the early 1990s, and the primary focus on law enforcement for rehabilitation projects was restored under the guidance of the Minister of Forestry, Ir Jamaludin Suryohadikusumo. Whether the initiative can be sustained consistently under the changed conditions of political turmoil and a weakened economy is doubtful however.

Modern rehabilitation: reintroduction

Rehabilitation programmes have been in operation in Indonesia since 1971. Despite the initial success of the Ketambe, the Bohorok and the Sekonyer (Camp Leakey) stations, however, an evaluation of the conceptual premises revealed that the original design of the programmes had serious flaws. Indeed, the original methods of rehabilitation added to the latent risk of extinction rather than serving conservation of the species. As a consequence, a new rehabilitation programme was drawn up.

After the Orangutan Foundation International (based in Los Angeles, and supporting the work of Galdikas-Bohap) had repatriated ten orang-utans from Taiwan with the intention of rehabilitating them at the Camp Leakey station in Tanjung Puting wildlife reserve, the Ministry of Forestry refused to issue permits for the local transport of the apes. An earlier shipment of six confiscated apes from Bangkok had died at the Camp Leakey station and because the government had been bypassed by an international network aimed at acquiring orang-utans for rehabilitation from outside Indonesia, the confidence of the Minister in the direction of ape rehabilitation at Tanjung Puting had evaporated. New working methods including a major role for the PHPA conservation agency had to be developed in order to make rehabilitation serve ape conservation again.

A newly adjusted procedure, called reintroduction, was designed at the instigation of the DG-PHPA by IBN-DLO (the Netherlands) and the International MoF Tropenbos Programme (Rijksen and Ramono unpubl.; Smits et al. 1995). It was based on the experience gained at the Ketambe station and on the international consensus concerning reintroduction

Section I: the orang-utan

> procedures (Brambell, 1977; Konstant and Mittermeier, 1982; Caldecott and Kavanagh, 1983; Kleiman, 1989; Sajuhti et al. 1991; IUCN SSC Guidelines for Reintroduction, 1992).
>
> A sequence of draft concepts involving a new approach to rehabilitation were discussed at all major international and national conferences, symposia and workshops dealing with the ape. As the result of a final working session of selected experts at the Ministry of Forestry in 1994, the Minister, at the recommendation of his Director-General of PHPA, declared the new rehabilitation design (i.e. reintroduction) a matter of policy in Indonesia (*Surat Keputusan MenHut* No 280/Kpts-II/95).

There was an acute need to accommodate the ten orang-utans illicitly imported from Taiwan, which had caused considerable commotion in Jakarta, and large numbers of illegally kept apes were to be confiscated in Kalimantan. This emergency situation hastened the establishment in mid-1991 of a new station in Kalimantan built according to the latest design.

> In June 1991, the Ministry of Forestry appealed to the team leader of its International Ministry of Forestry – Tropenbos Project at Wanariset Samboja (East Kalimantan)[63] to implement the new method of rehabilitation, called reintroduction (Smits et al., 1995). The team-leader, in collaboration with the head of the Balikpapan International School[64], then created the Balikpapan Orang-utan Society (BOS) to raise funds for the establishment of the station. The Society raised such considerable goodwill and resources by means of an 'adoption programme' and an appeal for donations among regional industries, that a proper quarantine and rehabilitation facility were able to be established. The initial investments were less than US$ 100,000. Since 1994, the Ministry of Forestry has been partially carrying the running costs of the project.

The Wanariset station in East Kalimantan was accorded official quarantine station status in 1992, and acquired the first official licence in Indonesia for keeping and rehabilitating orang-utans in 1994. The regional head of quarantine in East Kalimantan was appointed head of the medical staff of the Wanariset quarantine operations. A confiscation campaign was boosted in East Kalimantan through the DG PHPA with the support of the Armed Forces. Law-enforcement campaigns in the other provinces and Sumatra were slow to start, however, and were still sub-optimal in 1997.

The Sungei Wain protection forest was selected as the most suitable area for an initial experimental reintroduction. It is a vital watercatchment for the oil industry and for citizens of the city of Balikpapan, and is effectively protected against encroachment (but not against arson during a long dry season, as became apparent in 1998).

The Protection Forest consists of a mixed swamp and upland forest in which no wild orang-utans occur. It has not been affected significantly by the occasional timber poaching, and has a forested connection with adjacent and considerably extensive

---

[63] The International MoF Tropenbos programme is implemented with technical assistance from the Institute for Forestry and Nature Research (IBN-DLO) and the Agency for Forest Research and Development (AFRD) of the Indonesian Ministry of Forestry.

[64] I.e. Dr Willie T.M. Smits, IBN-DLO researcher and team leader of the International MoF Tropenbos project, and Dr Jonathan Cuthbertson, head-master of the Balikpapan International School.

permanent forest blocks (i.e. the logged-over Production Forests Batu Ampar and Mentawir). No wild orang-utans occur in these adjacent forest areas or the near surroundings.

Visitor attendance to the rehabilitation process is not permitted, although it is increasingly difficult to prevent high-ranking officials and their guests from visiting, especially since the international media coverage of these visits draws such a swell of public attention.

TABLE VIII

Comparison of outdated 'rehabilitation' and modern 'reintroduction' of confiscated orang-utans.

Original design: rehabilitation	New design: reintroduction
• Confiscated apes are collected and primitively quarantined, then set free and provisioned in an area where wild orang-utans occur.	> Confiscated apes are collected in a separate quarantine facility to stay for no more than 6 months of rehabilitation, including socialization, after quarantine.
• The rehabilitant apes come into contact with wild orang-utans, causing unrest and assertive competition in the wild community, and an 'overshoot' of the carrying capacity.	> The rehabilitant apes are released in carefully selected groups, in suitable habitat in which no wild orang-utans occur; after release; the group is provisioned daily for as long as is required.
• The rehabilitant apes are in daily contact with human attendants and visitors, increasing the risk of serious infectious disease transfer; a diseased rehabilitant ape can easily pass its infection on to the wild population, which is beyond cure.	> During the reintroduction process no visitors are allowed close to the apes or in the forest; contact with attendants during provisioning is kept to an absolute minimum.
• Rehabilitant apes can remain near the feeding site, or go walkabout to become wild; their competitive advantage may force wild residents to leave and search for a new home range, with considerable risks.	> When provisioning is no longer required the group is left to its own devices; new, subsequent groups are reintroduced in a similar manner at another location; diseased or unwilling and dangerous individuals are removed from the feral group.

In the period May 1992 – April 1996, when a last group of 21 orang-utans was released, a total of five batches of well-quarantined, healthy apes were reintroduced into the Sungai Wain area, bringing the total of reintroduced apes to eighty-two. No more apes are to be released at the site[65]; a new, much larger area of forest has been selected for subsequent reintroductions in the Meratus mountain complex. Monitoring in the Sungai Wain area during the first four years suggests that reintroduction is successful for anything between 20 and 45% of the confiscated apes (Frederiksson, unpubl. 1995; Peters, unpubl. 1995). It seems that a considerable number of young apes take up wandering and spread all over the area and beyond – into the logged-over forests adjacent to the area – and sometimes straggle into home-gardens where they may be captured again or killed and eaten. An adaptation of the procedure involving new reintroduction sites is therefore being considered.

> It is essential that the confiscated apes are reintroduced into a suitable habitat as soon as possible; that is, after the obligatory quarantine and checks indicating their state of health, all apes are to be released into forest surroundings with a varying regime of attendance, dependent on how quickly they can become independent of provisioning and human care. Provisioning and attendance must be carried out by small teams of trained attendants attached to one group of apes. Interference with the ape group should be restricted to the provisioning and directed in order to allow all group members a fair share of food. Diseased and socially unaccepted individuals are returned to quarantine to be treated or reallocated to another group. For chronically unsociable- or dangerous nuisance-individuals an alternative solution must be sought. Only the smallest infants, up to one year of age, may remain in quarantine for a longer period of time, until they are judged to be fit for release into the forest within their infant group with, and with full, long-term attendance, but no longer than one year after their arrival. Apes in quarantine are not to be allowed to breed. Birth in the quarantine facilities must be regarded as a form of mismanagement.

Table IX

Reintroduction at Wanariset Samboja – Sungei Wain 1992-1996	
Total number of orang utans accepted at the Wanariset quarantine station: 208	
Arrival	in 1991: 29 (incl. 7 from Taiwan)
	in 1992: 38
	in 1993: 20
	in 1994: 77 (incl. 22 from Taiwan)
	in 1995: 32
	in 1996: 12 (until July)
Total confiscated in East Kalimantan: 126	
Total confiscated in West Kalimantan: 34 (restricted campaign)*	
Total confiscated in Central Kalimantan: 10 (no campaign)	
Total from outside Borneo: 10 (+ 29 from Taiwan)	
Total recorded (known) deaths: 49 (23%)	
Total reintroduced at Sungei Wain Conservation Area: 82	
*) in West and Central Kalimantan the confiscation campaign has so far been sub-optimal.	

---

[65] During the drought of 1998, some 65% of the Sungei Wain forest went up in flames, and although no ape casualties have been recorded, it is evident that the carrying capacity of the area has been decimated. which may necessitate a return to prolonged provisioning of the apes. That some 35% of the forest has been saved was due to a small group of up to 20 fire-fighters battling fires continuously for some two months under the command of Gabriella Frederiksson.

As has been stated, rehabilitation and reintroduction are designed primarily to boost law enforcement, and to enable the confiscated apes to lead a feral existence independent of human attendance, in accordance with the legal framework for protected species. The scope of poaching and the large numbers of rehabilitant apes mean that this concerns a very significant potential breeding stock of at least three hundred and possibly as many as a thousand orang-utans in Kalimantan, before further poaching and habitat destruction can be arrested.

It is probable that reintroduction – without rehabilitation – will also soon be required for considerable numbers of displaced wild orang-utans, especially from Central and East Kalimantan. Suitable areas for this purpose may be the Barabai area in the Meratus range in South Kalimantan and the Berau (Inhutani I) forest concession, as well as the Kayan Mentarang wildlife reserve in East Kalimantan, if the protection and management of these areas can be extensively revised and improved, and further hunting by indigenous tribes can somehow be prevented.

Supposing that on average at least 10,000 km^2 of natural forest is required to harbour an existing minimum viable meta-population of some 5,000 apes, one would need for the rehabilitation of one-hundred apes at least 200 km^2 of such forest to justify the sustainability of the investment. Considering the extent of prime quality forest area required for reintroduction, it will be clear that, conversely, reintroduced apes could be deployed as token symbols indicating that a particular area of natural rainforest must be protected. Such deployment will psychologically boost the efficaciousness of protection, because everybody can understand that if a powerful organisation has invested hard cash in the area it is likely to retaliate if that investment is being wasted. After all, the orang-utan has extraordinary attention-attracting properties in the media, and the considerable, clearly visible, investments required for reintroduction place an acknowledged value on the area. This may, with a minimum of protective effort, inhibit potential encroachers. Thus, the reintroduction of apes can also be an effective means to safeguard rainforest areas of considerable size.

In ideal instances, when a sufficient area of forest is available for reintroduction, the forest should be (re-) stocked with no more than approximately half its supposed carrying capacity, i.e. for 100 apes approximately 500 km^2 of forest would be required, in order to give the reintroduced population an opportunity to grow and adapt. Unfortunately, there are no longer many safe forest fragments of the required (double) size available in the regions where orang-utans are absent. In any case, it is prescribed to conduct regular monitoring of the reintroduction process. Monitoring can be a way of establishing effective 'local participation', as has been demonstrated in Africa by T. Kano for bonobos and by J. Goodall for chimpanzees. The monitoring activity can serve as a protective patrol measure, while the monitoring results are meant to direct the management of the ape stock as well as their habitat; that is, for instance, to conduct additional temporary provisioning of food when fruit availability is too low during a certain season. Or, where required, it may involve increasing the incidence and diversity of fruit trees in gaps and/or particular sectors of the forest.

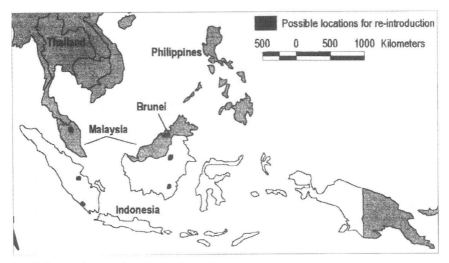

Possible locations for reintroduction in Southeast Asia.

In conclusion, appropriate rehabilitation and reintroduction, in addition to their primary function, i.e. serving law enforcement for protection, have the following important psychological functions for nature conservation in general and orang-utan conservation in particular:
- it is a continuous public reminder of effective law enforcement for species protection, when properly presented in the media;
- it instills public respect for the government in general and the protecting agency in particular, because consistent law enforcement and an ethically appropriate follow-up is the basis of a harmonious society;
- it can instill public respect for the ape, a creature deemed worthy of a protected status by the government, which hence must be allowed – like citizens who enjoy the same protection – a place to live within the State;
- it facilitates protection of the forest areas in which the ape is reintroduced because people will more readily appreciate the financial investments in the area and hence may fear retaliation for any infringement;
- it saves apes that would otherwise be lost as free-living orang-utans, even if they entered a captive facility, demonstrating the government's appreciation of the wild state;
- it can help to expand the Protected Forest area network beyond the established scatter of conservation areas, and may provide models for more effective protection of conservation areas;
- it can provide habituated feral ape populations in natural surroundings for well-regulated eco-tourism and education, thus preventing the exposure of originally wild populations to disturbing and dangerous contacts with humans.

Section II
Orang-utan Distribution

The prime habitat of orang-utans is to be found in alluvial valleys and in peat swamps.

Section II: orang-utan distribution

If detecting the elusive orang-utan in a rainforest is not easy, the traces of its presence in the form of a nest or sleeping platform are readily found; a male about to leave his nest at dawn, lifting the cover that protected him from the rain during the night.

SURVEY METHODS

Introduction

The orang-utan is the largest arboreal animal on earth. Its long waving body hair varies in colour from orange-red to a very dark maroon-black, and it is therefore remarkable how difficult it is to detect an orang-utan in its natural environment, the rain forest. Even trained naturalists, relying on sight alone[66], have great difficulty finding orang-utans, and unless the apes are habituated one can hope to surprise only an occasional individual in unfamiliar, and usually poorly accessible, terrain.

> When sensing the approach of a human, unhabituated wild orang-utans will usually either sit immobile in the canopy, or very cautiously go into hiding (Rijksen, unpubl.). Indeed, the average unhabituated orang-utan will begin an intimidating display only when it knows it has been detected (see Radermacher, 1780; MacKinnon, 1972).
>
> Experiments involving tourists, who came to see orang-utans at the Ketambe research field station in the early 1970s, revealed that both individual persons and small groups, despite extensive wildlife experience elsewhere, invariably failed to notice immobile orang-utans even when they passed under the canopy at a distance of less than five metres. In 1995, near the Danum valley field station in Sabah, this finding was corroborated when two experienced field ecologists were observed to pass entirely unaware underneath a group of seven immobile orang-utans (and past their human observer). After one hour of observation the apes were apparently accustomed to the observer but they froze when they noticed new people approaching (Rijksen, pers. observ.). Again, the nearest distance between human and one of the apes was less than 6 metres, which is well within the average distance at which a human can detect the ape's odour. This obvious absence in many people of a 'searching image' for orang-utans may also explain why even experienced birdwatchers and other naturalists usually failed to notice the presence of the ape in several areas of Sumatra and Borneo.

For spending the night (or for playful fun), however, like all great apes, the orang-utan builds a nest of broken branches, lined with twigs, in the crown of a tree. Because such typical nest platforms remain visible for a number of weeks, their presence betrays the occurrence of the ape in the area.

The land area of Borneo is some 740,000 km^2, of which about 500,000 km^2 (69%) was said to be covered with natural forest during the early 1970s (Collins *et al.* 1991), excluding mangroves. The extent of forest land on Sumatra in the supposed range of the orang-utan was believed to be anything between 100,000 – 150,000 km^2, dependent upon whether one includes the sporadic reports of orang-utan sightings outside the northern section of the Barisan mountain range. Only a fraction of this forest is inhabited by orang-utans, as at least 50% of it is located at an altitude of well over 1,000 m. Considering the difficulties in finding orang-utans, it is virtually impossible to simply look for apes and chart the distribution range in this enormous extent of poorly accessible forest area without the deployment of a well trained, technically equipped organisation, coupled with sufficient resources. Therefore it

---

[66] Payne (1988) reported that it took his staff of trained forest rangers an average of 16.5 km of survey to spot one ape.

will be clear that there is an apparent trade-off involving time, financial input, and the level of survey data detail.

Fortunately, the crude extent of the distribution range had already been described in the 1930s (see Chapter Historic Distribution). Hence, for the present survey it was initially believed that orang-utans were absent from the whole of South Kalimantan province, from Brunei Darussalam and the upper part of Sarawak, although it soon transpired that some recent discoveries in Sumatra and South Kalimantan made this belief untenable. Be that as it may, the elusive ape is too erratically or sparsely distributed, and the area still too extensive and in many places too remote or inaccessible, to acquire an accurate picture on the basis of field checks by a few persons alone.

In order to overcome these practical constraints, the method applied in this study comprised the following issues:
– acquisition and interpretation of the official forest maps to delineate the extent of possible habitat;
– interpretation of satellite imagery to assess the real extent of forest habitat;
– application of the current knowledge of orang-utan ecology with respect to habitat preference;
– distribution of written enquiries among forest concessionaires and field biologists to collect actual sighting data;
- review of Environmental Impact Assessments of timber concessions;
– field visits in preselected locations and interviews with local people to verify and cross-check information;
– brief forest surveys to fill in gaps in the collected information;
– research into the origin of museum specimens;
– research of the literature.

Inquiries

A questionnaire was distributed to 152 logging concessions in Kalimantan in order to inquire about the presence and status of orang-utans inside each of the concession areas. The questionnaire was sent through the Association of Timber Concession Holders (APHI) and accompanied by a an official letter stating that a response was obligatory. In addition to the questionnaires for concessionaires, inquiries were sent to experienced field workers at a number of research and industrial projects in Kalimantan.

For practical reasons the concessionaires in only three of the four provinces of Kalimantan (West Kalimantan, Central Kalimantan and East Kalimantan) were approached. South Kalimantan province was largely excluded from the survey because it was believed since the beginning of this century (Lyon, 1911) that orang-utans were absent east of the river Barito. This belief was rectified somewhat by the reports of orang-utan sightings near Tanjung by Witkamp (1932) and Westermann (1937), so that the eastern boundary of the distribution range in reality is the river Negara, a tributary of the Barito.

Unfortunately, no more than 20% of the questionnaires were returned.

All individual presence/absence reports were recorded in a geographical database. Each record contains the source of information, the latitude and longitude of the location, the local name of the location, the date (month or year) of reported sighting, status of the area and numbers of orang-utans sighted. In the rare instances where research was at the base level of statements about abundance or density, this was also mentioned. In some instances it was impossible to provide the exact geographical location of a sighting, for instance, when a large region was mentioned. In the logging concession reports, the approximate central point of the area was used as the geographical location of the record. For the present report this database has been summarised so as to reveal presence/absence only.

In addition to the presence/absence data, information on poaching and on illegal trading in orang-utans was recorded separately.

Literature, document and thematic map research

All available literature on orang-utan distribution was studied. This included not only formal publications, but also a large number of 'gray' reports and documents[67]. A review of the obligatory Environmental Impact Assessment (*AMDAL*) reports of logging concessionaires and transmigration projects were meticulously screened for references to orang-utan presence.

> It might be argued that the information which was provided by concessionaires in questionnaires and obligatory Environmental Impact Assessments is of dubious accuracy. The management of a timber concession is undoubtedly aware that timber extraction has a considerable impact on the ape's chances of survival. It may therefore be inclined to deny the presence of orang-utans in its concession. Indeed, in a few instances the responses indicate that this has been the case. And although most negative responses concur quite accurately with the known gaps in ape distribution, some responses precisely overlap with a recently – and reliably – confirmed occurrence of the ape. The positive responses frequently carried statements on the relative abundance of orang-utans (e.g. 'many', or 'only one'). It is doubtful what value can be attributed to such statements. In fact the statements 'only one' or 'rarely seen' may reflect the concern of the management about revealing their impact on the forest and the hope that their answers do not invite controlling actions. Consequently, the information was interpreted as probable presence or absence only.

In order to obtain an image of the distribution range of the ape with reference to areas of the size of Borneo and northern Sumatra (above the equator some 200,000 km^2), it is not sufficient to collect a few presence/absence data and extrapolate these

---

[67] At the instigation of Dr J. Sugardjito, surveys were conducted between 1991 and 1993 as part of a research assignment carried out by A. Yanuar for the DG PHPA, supported by the Red Alert Programme of FFPS and IWF.

over entire geographical units of forest areas. This method, used by the early twentieth-century naturalists was perhaps acceptable under the circumstances, but at the close of the twentieth century, with detailed satellite imagery now available, a more detailed analysis is due. Currently virtually all forest areas have been allocated for timber exploitation, while large sectors are destined for conversion into oil-palm plantations as well as for transmigration and settlement. All forest areas have also been explored for gas and oil, as well as for gold, diamond, and coal mining, and are nowadays crisscrossed by tracks and roads. Hitherto inaccessible forest areas have now been brought within the reach of many people, large numbers of whom may nowadays know where, within their own radius of activity, orang-utans occur. Such knowledge, however, is rarely obtainable without considerable effort.

The method applied in the present study was to collect information and organise it in a Geographic Information System, using PC ARC/INFO and ARCVIEW software. The GIS application is a database of stored layers of information and references. The following layers are relevant for the present report:
– accurate topographic base information, (rivers, lakes, 500 and 1000 m contour, etc.)
– thematic map information (land allocation; towns, villages, roads, concession outlines, etc.)
– satellite imagery for recent fragmentation of forest cover
– orang-utan presence / absence data.

Accurate geographical base information on a scale required for the relatively fine-grained purpose of this study is not commonly available in Indonesia; detailed topographic maps have been produced only recently but are classified. Despite two decades of international assistance in the geographic delineation and satellite monitoring of the extent of forest cover in Indonesia, accurate information on this issue appears to be jealously guarded[68].

In 1977, the Indonesia National Institute of Aeronautics and Space (LAPAN) was commissioned to implement a 5-year remote sensing project to accomplish integrated resource mapping of the entire country for some US$ 60 million. The National Coordination Agency for Surveys and Mapping (Bakosurtanal) also launched a multi-lateral development project to provide thematic maps of the country, including land use and land capability. In addition the Ministry of Forestry, with the support of the UN Food and Agriculture Organisation (FAO), embarked upon a resource inventory which should have been completed towards the 1980s. It was held that by the mid 1980s Indonesia would have an accurate Geographic Information System for rational land allocation (Myers, 1980). The report of the National Forest Inventory for Indonesia (FAO), based on satellite imagery, was finally published in 1996. Nevertheless, the information concerning forest resource,

---

68 Maps and aerial photographs of all timber concession areas, down to the detail of individual trees of commercial value, are produced by the HPH concessionaires by contractual obligation. Interpretation regarding forest types on composite satellite images showing the entire extent of State forest land has been carried out for Kalimantan and Sumatra, and is present at the regional forestry level (Kanwil). However, the planning for timber harvesting and conversion rarely seems to take into account and implement the accurate information contained in these images. It is deplorable that such detailed information is not commonly available.

> and the plans for land-allocation, does not seem to be consistently up-to-date and on maps is interpreted on a scale of 1:250,000 and 1:5 million. Most important however, is that the geographic information is not yet widely and consistently available at the regional and departmental levels. Therefore, twenty years after the first statement of intention, proper integrated land-use planning and land allocation on an ecological basis for sustainable development is still virtually impossible.

Hence, the information used in this report has unfortunately had to be superimposed upon a somewhat crude topographic basis, notably for Borneo. Although an accurate topographic map of Sumatra (produced by the national topographic service Bakosurtanal, 1986) was used, such an official map for the whole of Kalimantan is as yet unavailable. Even the commercially available Digital Atlas of the World provides very incomplete coverage for Borneo, and topographic maps based on aerial photographs or satellite imagery have not yet been produced for the entire island. As a consequence, for Borneo the relevant contour lines (500 m and 1,000 m altitude) have been traced and digitized from the 1 : 2,500,000 Times Atlas of the World (500 m) and were drawn from the mountain forest interpretation outline of the WCMC digital map (1000 m). An interpretation of known terrain features in a few locations, and a comparison with other maps of Borneo, suggests that the accuracy may be reasonable.

All known thematic maps of Sumatra and Kalimantan were scrutinised. The soil distribution, land allocation (TGHK) and timber concession maps were especially of relevance for the present study. Unfortunately, land-allocation maps in particular are usually outdated or inaccurate, and invariably also extremely difficult to acquire; for instance, the forest land-allocation maps which can be admired hanging on the wall in any major regional office refer to intergovernmental agreements made in the mid-1980s.

> Composite satellite images and aerial photographs of all forest land and an interpretation of their allocation are to be found at the Ministry of Forestry (INTAG) as well as at the Association for Forest Concession Holders (APHI) and a commercial mapping outlet (MAPINDO), yet are classified such as to be virtually unavailable to anyone lacking the most thorough bureaucratic clearance. Only recently has access to these vital sources become possible for the staff of the Directorate-General PHPA, but it is still rarely applied for.

Thus, the basic geographical and thematic information was obtained by re-evaluating the results of many different sources, including commercially available maps and digital data sets originating from the World Conservation Monitoring Centre (WCMC), the Directorate-General PHPA, the MoF/FAO National Forest Inventory Project, the Tropical Ecosystem Environment Observation by Satellites (TREES) Project and in particular the regional offices of the Forestry Department (Kanwil). In this way it is hoped that a reliable picture of the potential orang-utan distribution range has been achieved. Comparison with the MoF timber concession maps (*Peta Saran RTRWP*), however, indicates that in Indonesia all remaining habitat has been, or will soon be, affected by logging, and consequently is, or will be, of reduced quality. Considering Payne's report (1988) it is evident that the same applies to Sabah.

Checking presence

As noted, orang-utans are difficult to detect even by experienced students of orang-utan biology, especially in the more inaccessible areas where they may be presumed to survive. The apes may be found by implementing every conceivable bush skill, deploying in particular one's sense of smell (to detect their typical odour), hearing (to perceive the rustling sound of their movement through the foliage and the dropping of fruits and their shells from a dense canopy) and vision (especially the movement of branches and trees). The ape's presence may also be revealed by its rare, loud vocalisations, notably the typical 'long call' of adult males of high status (carrying up to a kilometre in a forest), the 'lork call' and the mating squeals of adult females and the distress cries of sub-adult females being raped or of infants unwilling to follow their mothers. Furthermore, the ape's presence can be inferred from the finding of typical traces of feeding, notably the discarded 'wadges' of chewed-up fibrous vegetation (especially of the climbing Araceae, e.g. *Pothos, Rhaphidophora, Scindapsus*, and the palms, such as *Borassodendron, Licuala*, as well as rotans), loose strips of bark and bite marks on tree trunks (notably of *Ficus*), and bite marks on dropped fruit-peels and fruits (Annonaceae, *Durio, Artocarpus, Strychnos, Neesia*).

However, the easiest way to establish the presence of orang-utans in an area is to look for the typical sleeping platforms or 'nests'. No other animal in Southeast Asia is known to make such characteristic arboreal platforms, although inexperienced observers may be confused when spotting the 'nests' of large raptors and the giant *Ratufa* squirrels, or the crude 'platforms' of the Malayan sun-bear in dense bushes of climbers. Because orang-utan nests remain visible for any time between a few weeks and several months (Rijksen, 1978), they are a long-term proof of an ape's presence in an area. In addition to surveys on foot through a forest, careful observations from a low-flying helicopter can also be used for crude orang-utan nest counts. In this way very large areas can be covered. The method was first proposed and tested by Bangan Empuluh of the Sarawak Wildlife Service, and was subsequently applied in the WWF survey for the establishment of the Lanjak Entimau reserve in 1982 (Kavanagh *et al.* 1982). Payne (1988) also applied this method for some areas in Sabah.

The bulk of the surveys in Kalimantan were conducted during the period 1994-1996. The investigations concerning the situation in northern Sumatra have considerably benefited from the inception of the Leuser Development Programme.

A total of 78 field checks involving 208 days in the field were spent in Borneo, covering almost 35,000 km by regional aeroplanes, cars, boats, motorcycles, bicycles and on foot. In Sumatra a total of 20 field checks were carried out, involving 118 days in the field, and covering some 21,000 km of travelling distance. In addition 69 days were spent in towns and villages for the official visits and interviews[69]. The presence of orang-utans was most commonly verified by the detection of indirect signs, in particular nests.

The presence/absence data collected for this report were subsequently plotted on the potential habitat maps; i.e. areas of known forest with reference to recorded topographic features (mountains, streams, river valleys, flood plains and swamps). The same was done with locations where illegal hunting and trade involving orang-utans had been reported. The resulting maps representing the current distribution range of the orang-utan were produced by means of on-screen digitising. This method, in additon to providing some crude insight into the size of the present distribution range, clearly reveals the fragmentation of the ape's range and its overlap with logging concessions and conversion areas. In this way it is possible to assess the survival prospects of the orang-utan subpopulations.

Forest cover of Borneo (WCMC, 1996). No data is shown for Brunei.

---

69 The Kalimantan teams were led by E. Meijaard. Two Sumatran teams were deployed, one consisted of Idrusman and Ibrahim under the guidance of M. Griffiths, the other of Yossa Istiadi, Suroso M. Leksono, and Guritno Djanubudiman. The first Sumatran team, under the supervision of C. van Schaik, has over a decade of experience in searching for and observing wild orang-utans.

Criteria for estimating orang-utan numbers

In order to assess the present status of wild orang-utans, the following section of this report will estimate the total orang-utan numbers over large areas, based on surveys of the known area of currently available forest and the outlines of the distribution range. Usually such numbers are calculated by multiplying an average density figure and the total area of the distribution range. However, the densities used in such an exercise cannot be taken directly from Table V, because the nest surveys are usually done in such small areas that they represent point densities, that is, a reflection of the orang-utan density in one limited part of a single habitat sector during a particular period of time. Hence, the present report has applied the following considerations:

(1) Nest survey densities are often estimated on the basis of a single visit to a particular site. In view of the extensive seasonal movements in many orang-utan demes, and the tendency of surveyors to more consistently record positive rather than negative findings, such estimates may be unreliable; in other words, they more commonly over- rather than underestimate the annual average density of apes in the area.
- *A conservative approach is required.*

(2) Orang-utans prefer the conditions of lowland forests. In Sumatra, mean densities are believed to fall below 0.5 individuals/km^2 in areas at altitudes above 1,000 m, whereas in Borneo such densities are already reached above altitudes of 500 m. Apart from exceptional occasions, orang-utans cannot cross rivers wider than ten metres and deeper than some 60 cm.
- *As regards suitable orang-utan habitat, the forest-cover 'polygons' are restricted by an altitudinal limit of 500 m for Borneo and 1,000 m for Sumatra, and by river courses.*

(3) Good-quality habitat is often represented in the field as small patches within a larger forest matrix. Such patches cannot be recognised clearly on any but the finest-scale satellite images and aerial photographs, and although they usually occur near watercourses, their extent and shape is so variable as to be almost entirely unpredictable. Hence, it is virtually impossible to estimate the total size of good orang-utan habitat from the imagery available.

Where a differentiation of forest (habitat) types is impossible, estimates for larger areas may be based on average densities for a usual composite of habitats. On the basis of more than 20 years of combined field experience in rainforest trekking after orang-utans it is possible to make a crude estimate of the relative frequency of the different extent of habitat patchiness in Borneo as compared to Sumatra, and to relate these frequencies to the corresponding differences in density data, as obtained by means of nest count surveys (Table V).
- *Table V provides the crude average estimates for orang-utan densities in the polygons on the basis of a weighed assessment of the percentages of distribution of different habitat types in the polygon.*

(4) When surveying large areas of forest inhabited by orang-utans, one often finds stretches of forest not occupied by the ape. Indeed, it seems that most of the forest provides low to very low-quality habitat. Due to the patchiness of segments of different quality, the living requirements for a deme seem to be concentrated in a zone up to 10-15 km away from the alluvial bottomlands, lakes or deep peat forests, where seasonality and therefore food availability is almost continuous. Usually such segments comprise the foothills or, in peat-swamp forests, the higher central domes.
- *In Borneo no more than 25% of a forest matrix comprises suitable orang-utan habitat in flat lowland, and some 30% in undulating hill country up to an altitude of 500 m. In peat-swamp forest the segments of quality habitat reach some 35%, and in heath forest are less than 10%. Due to the difference in geomorphological scale this percentage in Sumatra is estimated at 50% in swamp forests and 35% in all other kinds of the forest matrix up to an altitude of 1,000 m.*

(5) The survival chances of orang-utans are affected in habitats modified by selective logging (see section The impact of logging). As a matter of fact, the carrying capacity of selectively logged (secondary) forest is reduced by at least 50% (Johns, 1988; van Schaik, Azwar and Priatna, 1996; Rao and van Schaik, in press), and is sometimes reduced to virtual obliteration of the habitat conditions, in particular in the vicinity of watercourses. Regeneration of habitat to its original standard quality will take at least two ape generations, i.e. some 36-40 years.
- *In each forest polygon, an assessment must be made of the impact of logging and encroachment on orang-utan habitat, and the carrying capacity of the polygon must be corrected accordingly.*

The present distribution

Introduction

It has been known in Holland since the seventeenth century that the orang-utan occurs on the islands of Borneo (Bontius, 1658) and Sumatra (Tulp, 1641).

Borneo is the third largest island in the world and is divided into a complex of large provinces of Indonesia, collectively known as Kalimantan, two federal states of Malaysia, namely Sarawak and Sabah, and the independent sultanate of Brunei Darussalam. Its human habitation goes back for at least 12,000 and perhaps as much as some 35,000 years, according to prehistoric finds in the Niah caves in Sarawak (Bellwood, 1985). Thus, where sectors of the eastern lowlands of Sumatra and much of Java's alluvial valleys seem to have been inhabited since the early Pleistocene, initially by people of the *Homo erectus* type, the finds in Borneo concern exclusively people of a modern *sapiens* type and their implements.

Until the seventeenth century the island of Borneo was sparsely inhabited, with a few concentrations of human communities along its coasts and upstream of some of the larger rivers. Cultural affinities and ethnic similarities suggest that some of the ancient tribes of Borneo originated in Indo-China, notably Assam (e.g. the Kenyah/Kayan and related Punan/Penan), while others may have reached the island along the island chain of Taiwan and the Philippines (Hooyer, 1962; Bellwood *op. cit.*).

In the earliest civilisations of the Asian mainland, Borneo was known for its rich gold and precious stone deposits as well as for its forest produce, in particular rhino horn, hornbill casques, edible swift-nests, bezoar stones (from the gall bladder of monkeys), pepper, rattan, incense wood (*gaharu*) and resins. The earliest trading centre was the South-Asian Hindu colony of Kutai (between Balikpapan and Samarinda) on the east coast, which had been established sometime during the fourth century. Other early centres of a Persian- and Chinese-influenced Malay trading culture were Sambas, Sukadana and Martapura (Banjarmasin) on the south coast, and Brunei on the northwestern coast. The earliest Chinese trading records concerning Borneo date from the fifth century.

In order to present the results of the orang-utan field surveys, several spot locations have been highlighted on the provincial maps of Indonesian Kalimantan and the state maps of Sarawak and Sabah. Such spot locations provide an indication of presence but cannot in any way be interpreted so as to yield an impression of abundance. Such spot locations within a general forest area as depicted on the map are meant to indicate the presence of orang-utans within (the polygon) of that forest, taking wider rivers, roads and the altitudinal limits as boundaries of a polygon. It will be noted that several such spot locations fall outside the currently known extent of suitable forest. In some instances this concerns areas in which the forest has been so degraded and fragmented that it has been omitted from the forest cover map, in other instances it concerns forest at an altitude above 500 m.

TABLE XI

The total extent of forest and cleared/converted land in Borneo; data from a composite GIS layering derived from the latest WCMC coverage interpretation (1996), the TREES coverage interpretation (1996) and ground-check information collected during the present surveys. Note that these data include very large areas where orang-utans no longer occur, but could be re-introduced.

Forest extent in km^2	Kalimantan	Sarawak	Sabah
Lowland forest up to 500 m	162,093	35,542	19,011
Peat swamp forest	41,874	6,106	899
Limestone forest	1,577	40	24
Heath and degraded forest	38,742	665	4,735
Mangrove	6,084	1,005	728
Montane forest (>1000 m)	24,883	11,797	3,067
All highland forest > 500 m	79,213	26,637	16,068
Cleared and converted	229,713	52,292	35,933

The GIS interpretation based on recent satellite imagery reveals that in Kalimantan some 55% of the potentially arable land has already been converted, taking into account that peat-swamp, limestone, heath and mountain forests are virtually unsuitable for human land use. The government statistics indicate that 88.5% of the land area of Kalimantan is State Forest land (National Forest Inventory, 1996). In Sarawak this is 53% and in Sabah 58%.

Differentiation of the forest cover into forest above and below the 500 m altitude contour reveals that in Kalimantan some 23% of the forest cover lies above 500 m, and in the East Malaysian states anything between 33 and 36%. If one takes the 1000 m contour, only about 5% of the forest cover is involved in Kalimantan and up to some 7% in the Malaysian states. If the 1000 m contour is considered as the limit of orang-utan distribution, as in Sumatra, it implies that anything up to 18% may be added to the assessments of the distribution range.

In the early 1980s the eastern section of Borneo was affected by a long drought, towards the end of which a total of some 40,000 km^2 of logged-over forest and *ladang* fields was destroyed due to innumerable fires. One of the most badly affected areas was the Kutai wildlife reserve. Despite reports that orang-utans were able to survive forest fires (Suzuki, 1986; 1992), many hundreds of apes must have perished. However, no reports of the actual immediate effects of forest fires on orang-utans and other wildlife were published.

During 1997-98 a more extensive drought occurred in Borneo (and southern Sumatra). It is believed that there is a relationship between such a prolonged dry period and the occurrence of an accelerating global atmospheric cycle affecting the Pacific region, known as El Niño Pacific Oscillation.

Section II: orang-utan distribution

> Since the general behaviour pattern of rural people in Indonesia does not appear to change in the face of increased fire – or other – risks, and people are accustomed to setting fire to the debris in their fields and in shrub savannas towards what they believe will be the end of a dry season, an incredible number of widely ranging fires scorched the island for at least nine months. Furthermore, prospective plantation owners often seized the opportunity to demolish remnants of forest in order to issue a formal claim for conversion.

Hence, arson destroyed the already battered ecology of an estimated area of well over 50,000 km² in Borneo and Sumatra in 1997 and 1998. In 1997, the fires caused clouds of smoke and smog to cover much of Southeast Asia, so that international airports had to be closed for many days on end. In this period one aeroplane crashed, killing some 230 people, and a ferry collided with a freighter in the Malacca Straits. The immediate economic impact due to timber and crop losses, impaired traffic and a virtual drying up of a regular tourist stream, has been estimated to amount to some six billion US dollars (*Int. Herald Tribune,* April 20th, 1998), of which some US$ 625 million were in timber losses alone (e.g. *Jakarta Post,* April 18th, 1998).

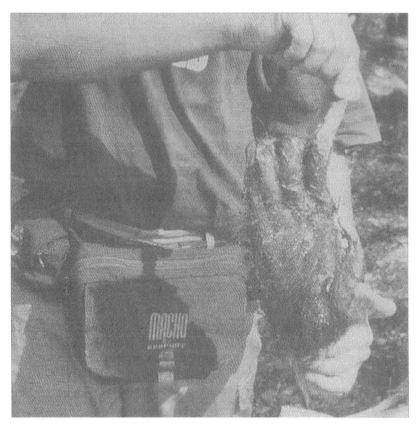

The remains of orang-utans, killed in the fires or butchered by local people, could readily be found all over Central Kalimantan in early 1998 (Meijaard).

The impact of these fires on the orang-utan population is unimaginable. Large sectors of the Tanjung Puting national park, the Gunung Palung national park and all of the Kutai national park have been badly affected. Particularly severely affected, however, are the peat swamps of the Kendawangan area in Southwest Kalimantan, of the coastal zone at Sampit, and the newly drained, reclaimed PLG 'mega-rice project' area between the river Barito and Kahayan in Central Kalimantan province. And where amazingly few people seem to have perished in the widespread and uncontrollable carnage, large numbers of orang-utans and other wildlife have reportedly been seen to flee the burning and smoking forest remnants, only to be slaughtered by local people lying in wait. During late 1997 and early 1998 every survey into fire-affected areas yielded at least three baby orang-utans for confiscation, freshly ripped from the bodies of their slaughtered mothers, and evidence of several butchered adults (e.g. meat, skin, heads, hands, etc.) was found in abundance.

A crude estimate of the loss of habitat after the 1997-98 fires will be made for every polygon of the range, in which fire hot spots have been detected and recorded.

West Kalimantan

General description

The province of West Kalimantan consists of approximately 146,000 km^2 of mountains, hills, plains and swamps, and is dominated by the great water catchment of the river Kapuas. Its human population was estimated to be 3.6 million in 1995, increasing at an annual rate of some 2.6%. The people live clustered in a few major dryland areas especially along the rivers (Cleary & Eaton, 1992). The region has a history of almost continuous human migration; it is believed that the large Iban tribal entity, which originated in eastern Sumatra, invaded Borneo along the Kapuas river sometime during the fourteenth and fifteenth century, and entered Sarawak in the early nineteenth century (Freeman, 1970). Originally known by Malay people as the 'sea Dayak', the Iban during the early nineteenth century were known as coastal pirates as well as shifting cultivators, and were particularly feared for their head-hunting tradition. This caused major movements of indigenous Dayak tribes during the nineteenth and early twentieth century. The Iban have, since the 1950s, modernised their lifestyle to that of nomadic slash-and-burn cultivators, and many of them have become skilled timber loggers.

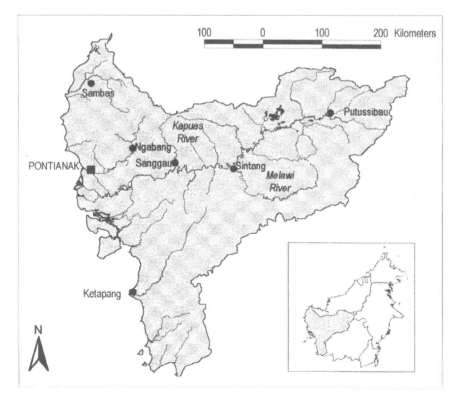

Basic geography of West Kalimantan province (Kalimantan Barat).

Other tribal entities in this region are, from west to east, the Selako (Klemantans), Land Dayak (Bidayuh) and the Maloh. For all these indigenous groups, gathering and hunting are important supplementary activities to their shifting agriculture; the bearded pig, monkeys and apes are their favourite prey (King, 1978). In 1981, it was estimated that 48% of the originally forested area of West Kalimantan had been demolished by slash-and-burn cultivation and commercial logging (MacKinnon and Artha, 1981).

West Kalimantan is of particular interest as regards conservation, because it is believed to have some of the richest and presumably most archaic forests on the island, particularly along the mountain chains. In addition, the extensive swamp forests, in which the valuable tree Ramin (*Gonystylus bancanus*) occurs, are of immense ecological importance. Unfortunately, the province has been very late in attracting any substantial conservation interest, although it has the highest number of established conservation areas in Kalimantan; the first PHPA office was established in 1979. An internal evaluation in 1981 reported that the PHPA staff was barely adequate for the task required of them (MacKinnon and Artha, 1981). This situation of understaffed and ineffective reserve management was repeatedly confirmed by local informants during the present survey.

Four major development zones have been identified for the province by the Ministry of Public works. These are Kabupaten Sanggau (allocated 3,343 km^2 of oil-palm plantations, reserved 10,328 km^2 of timber estates and allocated 590 km^2 for bauxite mining), Kabupaten Sambas, Teluk Air (sea port) and Kabupaten Ketapang (a belt of oil-palm estates). In addition, a wide zone along the planned and partly established arterial roads (i.e. a total extent of 1981 km) through the province will also be transformed by 'default' development.

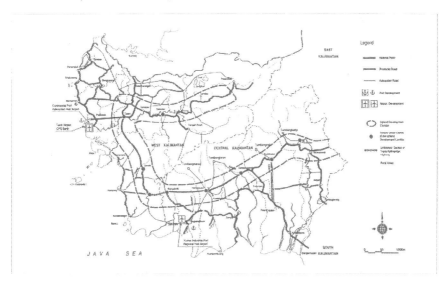

Official plan for the development of a road infrastructure in the southern half of Kalimantan. Note the extensive band of upland area earmarked for development.

Orang-utan distribution in West Kalimantan

1 Sambas area

Some 3,000 km^2 of coastal peat swamp and lowland area once stretched out in the westernmost corner of Kalimantan. Until the middle of this century the area was covered with dense, uninterrupted swamp and lowland forest, in which orang-utans were common. At present few orang-utans are believed to survive in the three fresh-water and peat swamp forest patches (27, 27 and 446 km^2) east of the township Sambas. Any surviving apes are virtually isolated from the populations further to the east by areas deforested through logging activities, massive timber stealing and reclamation development for transmigration. In 1981 MacKinnon and Artha proposed establishing the Sambas Nature Reserve, comprising 100 km^2 of peat-swamp forest and 430 km^2 of fresh-water swamp; the area was reported to be exceptionally rich in flora. A decade later, the area was reported to have an 80% overlap with a logging concession area (McCarthy, 1991) and logging and encroachment had seriously damaged the forest. The swamp- and hill-forest habitat of the Sambas region up to Gunung Niut is now fragmented into some 12 patches of less than 10 km^2. Virtually all the land in the Kabupaten Sambas has been allocated for development, including some 10,300 km^2 for timber estates. Poaching of orang-utans for meat and pets is reportedly common, with clear indications that several illegally captured young orang-utans are smuggled across the border to Sarawak every year. The proposal for official protection was apparently never processed, or was rejected, because the Sambas Nature Reserve is not indicated on any of the official maps of conservation areas at the DG PHPA. The river Sambas originates in the hill complex stretching northward into Sarawak (i.e. the Simitau-Kuching zone), of which Gunung Niut (1701 m) is the highest peak..

2 Mempawah water catchment

Due north of the main road between the coast (Pontianak) and the township Ngabang inland lies a small low hill complex in which orang-utans reportedly were seen during 1993: Of a group of three, one individual was shot (Suriansa, 1994 pers. comm.) when the orang-utans showed up in a fruit- tree garden. In 1936, the *Cagar Alam* (Nature Sanctuary – 20 km^2) Mandor was established in the foothills of this complex in order to preserve the area's extraordinary wealth of orchid species. The strict nature reserve contains some 5 km^2 of peat- swamp forest. The reserve has been under logging pressure since the 1960s; only the areas of *kerangas* and deep peat forest remained comparatively unaffected. The more spectacular orchid species have become extremely rare. The reserve is located in a relatively densely populated area, close to the large town of Mandor and closed in by two main roads. Nowadays it is far from other suitable orang-utan habitat, and in fact the habitat has been fragmented into three small forest sectors of 10, 12 and 23 km^2, and one larger sector of 129 km^2, adjacent to the Gunung Niut Reserve. Orang-utans are said to have migrated through the area, seeking refuge from the massive disturbance in the surroundings (Hadisuproto and Said, 1992). The remaining ape refugees are probably compressed into the few crowded habitat patches. The population is critically endangered, and given the present lack of protection cannot be expected to survive into the next century.

3 Gunung Niut

Orang-utans are reported from several sites in and around the approximately 1800 km² Gunung Niut mountain complex, which is the water catchment of the river Sambas and several rivers flowing into Sarawak. The apes are confined to the hill forest which is now broken up into three tiny, separate fragments (36, 61 and 98 km²); virtually all the lowland forest has been badly affected by logging, human encroachment and the impact of transmigration. Much of the once forested mountain complex (i.e. 1,635 km²) acquired a protected status in 1982, following the recommendation of MacKinnon and Artha (1981). The total area of cultivated land, including various stages of secondary forest, inside the reserve, was 200 km² in the late 1980s (Simons, 1987). It is assumed that orang-utan density is relatively low (Yanuar unpubl.; Simons, *ibid.*) in the reserve, due to poaching pressure: Poaching of apes for meat, skulls and for the pet-trade is taking place in and around the reserve.

The Gunung Niut Reserve has only one PHPA *resort* post, situated in Sanggau Ledo, 20 km from the western reserve boundary. Only one PHPA officer is deployed to guard the reserve (present survey, also Simons, 1987). A lack of transport means in combination with poor motivation hamper protection of the reserve. Under the present conditions the dwindling orang-utan population in the Gunung Niut wildife Reserve can not be expected to survive for much more than another decade. The latest official map of the conservation areas denotes the Gunung Niut Penrisen *Suaka Margasatwa* (wildlife reserve) as comprising 1,800 km²; it is noteworthy, however, that no more than 500 km² of forest cover remains in the region. Arson during the autumn of 1997 has affected the remaining forest patches.

4 Ketungau water catchment (Noyan) – Seluwah hills (Indonesian border with Sarawak)

Due east of the main road between the township Sanggau and Kuching (Sarawak) lies an extensive lowland flood-plain area with a few scattered hills. In this area the naturalist Selenka shot 217 orang-utans in the late 19th century. Due to the swampy nature of this region (approximately 2,000 km²) between the river Sekayan, and the upper stretch of the river Ketungau (both tributaries of the river Kapuas) it is inhabited by few people. However, some transmigration villages, e.g. Bonti, Balaisebut, are located in the middle of this region. Sightings of orang-utans were reported from several swamp forest locations in this area. According to the reports orang-utans were concentrated around the relatively unspoilt forested hills south of the border with Sarawak. Actually, the satellite imagery reveals that only three patches of forest remain in the Ketungau area, one of some

270 km², the others of 19 and 26 km². Whether the orang-utan populations in the Ketungau region can still exchange members between these pathetically small patches, or with the populations to the west, notably those around Gunung Niut, is unlikely. The busy road to Sarawak interconnecting adjacent agricultural development areas is a formidable barrier. Since 1990 a German technical assistance (GTZ) project for participatory forest management has been operating in the Noyan area. Its aim is to study the ecological integrity of the remaining forest patches under a regime of local resource extraction, and includes protection of the higher elevation forest. However, poorly controlled logging and forest conversion for transmigration is usurping and fragmenting the habitat in many parts of this region. Considerable numbers of orang-utans are killed for 'sport' and for consumption every year, according to local informants.

Between the lower stretch of the river Ketungau and the border of Indonesia and Sarawak lies an extensive hill country (approximately 2,000 km²) which must once have been covered with prime habitat for orang-utans. Satellite maps indicate that no more than some 430 km² of forest in some three major fragments remains. Orang-utans are reportedly common in the area, but they are under heavy poaching pressure from local Selako and Iban hunters, as well as from transmigrants and army personnel, and will in all probability soon become extinct. This orang-utan population was already distinct from the population in the peat-swamp forests around Danau Sentarum (and the population along the headwaters of the Batang Lupar in Sarawak), according to Selenka (1896).

5 Danau Sentarum – Bentuang dan Karimun mountain complex

The orang-utan in the easternmost corner of West Kalimantan province occurs in a large and almost uninterrupted area of mixed forest, which includes lakes, peat swamp and hill forest, and stretches into Sarawak. The Sentarum lakes (i.e. *danau*) and surrounding peat-swamp wetlands are located in the upper Kapuas basin, covering approximately 1,800 km² of interconnecting shallow lakes and peat forest. The Sentarum lakes area is unique in Kalimantan, as it has old inland peat swamps (i.e. at least 12,000 years) and is one of the two island-lake areas in Borneo. In 1986, therefore, MacKinnon and Artha proposed that the lakes and some surrounding fringe forest become wildlife reserve (*suaka margasatwa*), and in 1992 the Danau Sentarum Wildlife Reserve was said to have been established.

Orang-utans are found in the peat forests to the west, north and east of the Danau Sentarum Wildlife Reserve (Meijaard and Dennis, 1995). Russon and Erman (1996) found a considerable ape population surviving in the ecotone of two hills to the east of the reserve. A boundary extension has been proposed to include these hills in the reserve. Orang-utans are also common in the swam-forests to the north and east of the reserve. Local people reported that the ape is 'common' along the Sungei Embaloh. It is believed that the logging activities around the reserve cause considerable migration of the orang-utans to the less disturbed swamp forest in and around the reserve. Further field surveys and negotiations are required to redesign the Danau Sentarum Reserve so as to include all surrounding orang-utan populations and join it, by way of a wide forest corridor, to the population in the large, designated Bentuang dan Karimun National Park further north. During the later stages of the drought of 1997, arson seriously affected some of the peat-swamp forests northeast of the reserve.

The Danau Sentarum conservation area was, until 1997, the subject of a joint Asian Wetland Bureau (Wetland International) and Overseas Development Agency (ODA) project for improvement of its conservation through the participation of local people and sustainable use of resources. The area has been mapped in great detail and a management plan has been produced. It is remarkable that the Danau Sentarum conservation area has **not** been included in the latest (1996) official map of the conservation areas of Indonesia of the Directorate-General PHPA; again it has been stated that formal establishment is in progress.

The Danau Sentarum lakes (courtesy of R. Dennis).

The border between Indonesia and Sarawak in the northeastern sector of the province is formed by the watershed of the river Embaloh, a major tributary of the river Kapuas; the highest peak in this deeply dissected mountain range is Gunung Lawit (1767 m). East of this mountain, the range on the Indonesian side is divided into an eastern and a western section by the headwaters of the river Sibau, another tributary of the river Kapuas. The eastern section is called the Kapuas Hulu mountain range. MacKinnon and Artha (1981) proposed a nature sanctuary (*Cagar Alam*) status for some 6,000 km^2 of this watershed, and in 1992 the designated Bentuang dan Karimun National Park was gazetted. Little is known of the present conservation status of this largest reserve for orang-utans in Borneo, as there have been no recent surveys. Orang-utans have been reported to occur along the southern foothills and in the area's steep valleys all the way up to the headwaters of the Kapuas River in the east, as well as in the peat swamp and lowland Dipterocarp forest on the southern side. According to people living just south of the reserve, orang-utans are common, but live in lower densities in the mountains than in the swamp forest just south of the reserve. A recent (1998) 'biodiversity survey'by a WWF team with limited experience in ape-surveying has reported that orang-utans in low density are found only in the western half of the park (Yeager, pers. comm.) Orang-utans also still occur in the Malaysian Reserves of Batang Ai (Meredith, 1993) and Lanjak Entimau (Rijksen, pers. obs.; WCMC, 1995), which are contiguous across the border with the Bentuang Karimun Reserve. Inside the reserve there are no reported permanent settlements and the only people that regularly enter the area are Maloh and Iban hunters and *geharu (Aquilaria malaccensis)* collectors. These people do shoot and eat orang-utans during their forest forays. Recent satellite imagery of the Bentuang dan Karimun Reserve shows that the mountain forest is virtually untouched, and that most of the valleys are uninhabited.

The extent of this large mountainous reserve is unknown. Earlier reports mention 6,000 km^2 (MacKinnon and Artha, 1981), but the latest (1996) official map of the Directorate-General PHPA gives 8,000 km^2 for its extent. It is unknown whether this change or extension is officially gazetted; possibly some 2,000 km^2 of the contiguous Batu Tenobang Protection Forest in the Kapuas Hulu range has somehow been added to the designated National Park.

The designated Gn. Bentuang dan Karimun National Park and the adjacent Lanjak Entimau Wildlife Reserve are the subject of a US$ 996,000 project sponsored by the International Timber Trade Organisation (ITTO), to be carried out under the supervision of the Directorate-General PHPA by the WWF Indonesia Programme in two years. Technical assistance for the development of the conservation area as a National Park and a transfrontier management system, has focused on support for rural development in relation to the local people. It has been operant since 1994, although no information on the progress of the project is available as yet. The same applies to the Sarawak side.

6 Kapuas Hulu mountain range

The eastern section of the range is contiguous with the Müller mountain range to the south across the valley of the upper stretches of the river Kapuas and the Upper Mahakam water catchment to the east. The highest peak of this range is Gunung Cemaru (1681 m). Both the rivers Mahakam and upper Kapuas (Tajam) originate at this mountain. The valley of the headwaters of the river Kapuas (Tajam) ranges from some 500-1400 m altitude and has a Protection Forest status; the mountain complex south of the valley is known as the (8,830 km^2) Batu Tenobang Protection Forest. King (1985) notes that the foothills of the mountain complex on both sides of the provincial border are Punan territory.

Few data are available on the present conservation status of orang-utans in the approximately 2,000 km^2 area of the Batu Tenobang Protection Forest in the upper Kapuas area. In 1981, MacKinnon and Artha reported, 'It is a very extensive and well forested area of deeply incised hills, consisting mostly of hill forest with some montane forest on the higher peaks. Most of the area is very remote and uninhabited.' Recent reports from the region indicate that orang-utans are still present, but evidently 'rare.' The ape has been reported to occur as well in the adjacent forests just across the provincial border in East and in Central Kalimantan. The mountain range between Bukit Melatai (1554 m), Gunung Cemaru (1681 m) and Gunung Tibau (1565 m) forms the pivotal ridge of Borneo, where at least three mighty rivers, which divide the island, originate. These are the river Kapuas running into West Kalimantan, the river Oga-Mahakam running into East Kalimantan, and the river Baleh-Rejang of Sarawak. The southern valley flanks of the river Kapuas and the ridge along Mount Kerihun (1980 m) due northeast along Gunung Cemaru was the entrance into east and northern Borneo for orang-utans and several of the indigenous tribes. The area is still the major infrastructural 'bridge' for the hunter-gatherers ranging from the Müller range in West Kalimantan to the Piedmont zone of the Iran mountain range and central lowlands in Sarawak.

7 Madi-plateau – Melawi water catchment

The river Melawi is a tributary of the river Kapuas, which originates in the mountain complex known as the Madi plateau, linking the Schwaner and Müller mountain ranges; its three highest peaks are approximately 1390 m. Due east of the township Nanga Pinoh, the wide valley of the river Melawi is very sparsely inhabited. The foothills and bottomlands are excellent habitat, and orang-utans are reportedly common in the area. Much of the area is in the custody of timber concessions. This valley, and the adjacent Madi plateau, should in fact become part of the Bukit Baka conservation area, due to its unique flora and fauna and its vital water catchment function.

8 Bukit Baka

The northern water catchment of the Schwaner mountain range reaches its summit at Bukit Baka (1617 m). Some 10,000 km^2 of the forested area around Bukut Baka was proposed to become a nature sanctuary (*Cagar Alam*) by MacKinnon and Artha in 1981. In the same year 705 km^2 was gazetted as a wildlife reserve along the provincial boundary with Central Kalimantan, contiguous with the Bukit Raya conservation area. Although the area is mountainous and at an altitude of largely over 500 m, orang-utans reportedly occur in the Bukit Baka conservation area and nests were found in the logging concession north of this reserve. The once continuous forest cover north of the reserve has been fragmented into at least two major sections. The northern section is still contiguous with a large forest area east of the Bukit Raya reserve. The orang-utans in the area are very elusive; indeed, several botanical researchers reported no sightings of orang-utans, even after spending up to four weeks in the forests of Bukit Baka (J. Jarvey, pers. comm). Orang-utan skulls which were found in a village along the southwestern fringe of the reserve indicate that poaching does take place although the events could not be dated reliably. Considerable illegal logging is common inside the reserve; there is no protection of this conservation area whatsoever, and approximately 50% of the forest has disappeared.

On the official (1996) conservation area map of the Directorate-General PHPA the Bukit Baka reserve is mentioned as one unit with the adjacent Bukit Raya conservation area across the border in Central Kalimantan (see 15). If this complex of reserves could be redesigned to include some of the logged-over lowlands and be extended along with the adjacent Madi plateau across the river Malawi, it may become one of the most important strongholds for orang-utans in Borneo (but see Bukit Raya, no 26).

In 1992, the designated Bukit Baka-Bukit Raya National Park became the subject of the Natural Resource Management Project, sponsored by the government of Indonesia, the ITTO and the US Agency for International Development (USAID). The Kaburai Research station was established on the southern flanks of Bukit Baka, and therefore actually in Central Kalimantan province, but can only be reached through West Kalimantan province (see therefore 26, The central Schwaner mountain range).

9 Bukit Berangin – Sebayan complex

Due south of the township of Sintang a hill complex rises out of the flood plains of the river Kapuas and its tributaries Melawi-Sepinoh and Sepau. It is the northwestern extension of the Schwaner mountain complex. The forested foothill complex reaches a summit at Mount Saran (1758 m alt.) and is the major water catchment of the river Pawan/Kerabai, running south to release its waters at the coastal township of Ketapang. Some 2,600 km^2 of the steep forested terrain of Bukit Rongga (Rangga), and some 1,620 km^2 of the inland plateau of well-forested and relatively flat hill country of Bukit Perai, has a Protection Forest status. Orang-utan sightings are reported from logging concessions south of the protection forest and orang-utans were reportedly also seen along the river Pinoh. The present conservation status of the area is unknown, but much of the forest in the region is in the custody of timber concessions. MacKinnon and Artha (1981) proposed that the Bukit Perai Protection forest be upgraded to Nature Sanctuary (*Cagar Alam*). It is probable that control of logging operations and poaching is scant due to the remoteness of the area. It is even uncertain whether the Protection Forest status of the two areas has been maintained.

10 Kapuas swamps

Between Pontianak at the mouth of the river Kapuas and the coastal township of Sukadana on the east coast, a large area of swamp forest covers the delta-like river network of the Kapuas. Along the inland fringes of this swamp forest orang-utans reportedly do occur. Virtually nothing is known of this inaccessible region; it is possible that it contains one of the larger sectors of orang-utan habitat in Borneo. Remarkably, the swamp forest in this sector suffered hardly any arson impact during the 1997 drought.

11 Sukadana – Kendawangan coastal plains

Some time during the first decades of the seventeenth century the Dutch physician of the VOC, Jacob de Bondt, for the very first time described in his diary an orang-utan which originated from this area (Bontius, 1658), see Addendum I. Because de Bondt's notes were published some 27 years after his death, however, Nicolaas Tulp (1641) was the first to publish on the orang-utan, which originated from Sumatra.

The Sukadana area has one of the oldest nature monuments in Borneo, situated around the (then) sacred Gunung Palung. It was gazetted in 1936. MacKinnon and Warsito (1982) reported, 'It appears that the density of orang-utans in the G. Palung area is as high or higher than any other known population'. Indeed, an average density of 2.5 (Knott, pers. comm.) to 4 orang-utans/km^2 (Leighton and Darnaedi, 1996) was reported for the habitat around the Cabang Panti research station in the designated Gunung Palung national park (900 km^2). This indicates the presence of social arenas of average habitat quality for Borneo.

A forested region to the northeast of the conservation area has been proposed as an extension (MacKinnon and Artha, 1981; CBSG/SSC/IUCN, 1993). The area around the Gunung Palung shows a complete spectrum of vegetation types from sea level to peaks of 1160 metres and contains 300 km^2 of peat-swamp forest and 200 km^2 of freshwater swamp forest (MacKinnon and Artha, 1981). MacKinnon and Warsito (1982) believed that the presence of normal densities of wildlife was 'because the Gunung Palung lies in an Islamic rather than Dyak area and is reputed to be haunted, virtually no hunting has occurred there and excellent wildlife still exists'. A similar observation was made by Abott in the first decade of the twentieth century (in Lyon, 1911). Recent aerial pictures of the forest cover show degraded and fragmented forest northeast of the designated national park, but the results of the present survey indicate that orang-utans can still be found there. The Peabody Museum of Harvard University (USA) has established the Cabang Panti Research Station for long-term rainforest ecology at the banks of the river Air Putih within the Gunung Palung conservation area (Leighton, 1990). Long-term studies have been conducted to obtain insight into the dynamic ecological processes within the various forest formations, including floristics, phenology, growth and seedling regeneration of trees and lianas, the phytochemistry of fruits, and the population ecologies of larger birds and mammals. However, the most publicised results of the research concern the ways and means of so-called sustainable management of the area, and in particular seek to justify the exploitation of timber and forest produce through 'local participation' (Peart, 1996). Towards the end of 1997 the regional PHPA/BKSDA was alarmed to discover that the protected area was being demolished for quarrying of rock and earth for the building of nearby roads on the basis of a 1984 concession contract (Suara Karya, March 5th, 1998). The reserve is threatened by the increasing accessibility, since a number of roads through the region are being developed. The west coast road up to Teluk Batang and the new road between Sidang and Nanga Tayap have both facilitated encroachment and timber poaching in the west and along the southern boundary. PHPA (BKSDA) with the help of Untan university volunteers is active in seeking support for conservation from four nearby villages, pumping some US$ 160,000 per year into 'training, the provision of seeds and demonstration plots and a visitor lodge to be operated by the local communities'. Considerable enroachment into the park and massive timber poaching occurs along the logging road which branches off the southern route of the Trand Kalimantan highway near Sungei Dua, southwest towards the river Matan.

The coastal landscape in the Kendawangan region, due south of the coastal township of Ketapang, is dominated by swamp forest and very low forest-covered terraces which were deposited by the rivers Peusaguan, Kendawangan and Membuluh. Athough on satellite images the area shows up as being covered with forest, it is hemmed in by a major road network, and timber extraction, both scheduled and illegal, is in full swing to supply the port of Ketapang. The peat layer under the permanently wet swamp forest is extensive; it covers the coastal flood plains and extends into the narrow valleys which border the low terraces, developing even on the broad flat terrace surfaces. Nearly all of the western half of the notoriously infertile Kendawangan region is supposedly allocated as Nature Reserve (RePPProT, 1987a). The (ostensibly proposed) Muara Kendawangan Nature Reserve of 650 km^2 of peat swamp forest and 750 km^2 of fresh water swamp forest, reportedly still contains a relatively large population of orang-utans, according to Rusila and Widyanarti (1995). It was reported that the reserve has a low forestry value and is not much threatened because extremely few people live in the area. However, during the survey by Rusila and Widyanarti (*ibid.*), it was apparent in all areas 'that illegal timber extraction is a major threat to the reserve'. Poaching of orang-utans for food and pets was reported by both Rusila and Widyanarti (*ibid.*) and Sugardjito (pers. comm.). There is at present one resort guard for this supposed reserve. This is perhaps not surprising because the area is not included in the latest official PHPA map of protected areas. Indeed, during a recent (1997) meeting, PHPA officials proposed that the Kendawangan Nature Reserve 'should no longer be protected' as timber poaching by Java-based companies had destroyed large sections of the reserve, while the area was too remote to be protected anyway.

The northeastern fringes of the Gunung Palung National park, as well as the logged-over and encroached peat-swamp forests of the whole Kendawangan area, were very badly affected by arson during the 1997 drought. And although this area was among the most promising regions as regards offering the orang-utan a chance of survival, the feasibility of this prospect is now highly unlikely. Indeed, it must be feared that the combined effects of poor planning, unscheduled migration, encroachment and massive poaching of timber and apes, in combination with the widespread arson, have pushed a hitherto major orang-utan population to the very brink of extinction in less than half a decade.

The present distribution

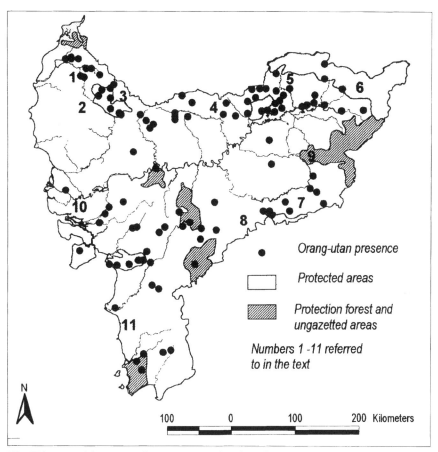

West Kalimantan. The presence of orang-utans is indicated in relation to the main rivers and protected areas. Numbers in the map refer to the text above.

Central Kalimantan

General description

Central Kalimantan is one of the largest provinces in Indonesia, fringed by the southern hills of the Schwaner mountain range in the west and the Müller mountain range in the north. It comprises the largest continuous peat swamp and flood plains of western Indonesia, encompassing the catchment areas of the rivers Seruyan, Sampit, Katingan, Kahayan, Kapuas Murung (or eastern Kapuas) and Barito, their alluvial swales and ridges, swamp-filled basins, coastal mangrove estuaries and sandy beaches. The province is generally of low elevation with no high mountains except in the northern watershed country, where the dry upland is deeply incised in the foothills and headwaters of the southward flowing rivers. The Schwaner mountain range and the central north-south watershed (the Müller range) belong to the pre-Tertiary geological core of the island. More than 65 million years ago the mountainous core of the island was formed through collision of two tectonic plates, while all its current lowland areas were still submerged (Smit Sibinga, 1953). This geological history meant that, on the one hand, much of the mountainous interior of Borneo is composed of substrate of low nutrient quality (e.g. sandstone), while on the other hand, some sectors of the mountain ranges have seams and mother lodes of valuable minerals and metals, for instance, gold. Volcanic activity in Borneo was greatest during the Tertiary.

In the lowlands, peat has developed to varying thickness (i.e. 3-18 m), and some of the deposits are among the oldest known in the tropics, i.e. circa 60,000 years. Closer to the coast a shallower peatlayer developed on loamy clays, mixed with laterite, so that reclamation of the land for agricultural development results in (toxic, unproductive) acid sulphate soils.

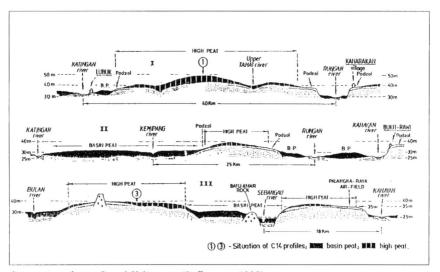

Cross sections of peat, Central Kalimantan (Sieffermann, 1988).

At maximum development, the thick layer of peat forms a convex or domed surface with a flattened top which is the substrate for a mature, high-rising rainforest. Shallower peat has developed around the major swamps or on younger, less developed inter-river basins. Between the major rivers, very large blocks of both peat types occur (Sieffermann, 1988).

The peat-layers and their living forest cover in this province are one of the best Carbon sinks in the world, with a sustainable accretion of anything between 0.5 and 1.3 cm of peat per year – or anything between 290 and 670 g $C/m^2/year$ or, on average the annual storage capacity is some 4.10^5 tonnes of $Carbon/km^2$. Being among the best in the world, this storage capacity will, in the near future, be of major economic value in the growing global concern for the Arrhenius effect (i.e. global warming).

Approximately 1.6 million people inhabited the 152,000 km^2 of this province in 1995, increasing at an annual rate of well over 3%. Until the 1980s, the people were clustered along the rivers, so that figures for populations density (as presented in official documents) are meaningless. Notwithstanding the relatively small number of people, land clearance as a result of wasteful slash-and-burn agriculture, timber poaching, arson and conversion for plantations has been extensive and has proceeded at an accelerating rate with the building of roads across the province.

A rapidly expanding road network effectively severs north-south connections between the major rivers for animal movement. Two major road networks traverse the province in a west-east direction, one along the coastal zone, the other along the foothills of the mountain range. In 1997 the public roads network amounted to some 7,000 km, in addition to some 12,000 km of logging roads. Thus, the patterns of human habitation are increasingly spreading over the province, as roads and a growing market facilitate massive unscheduled encroachment into the hitherto forested areas. Although roads have a tremendous environmental impact, especially in peat and sandy flood-plain areas like Central Kalimantan, no official Environmental Impact Assessments have as yet been conducted.

In addition to forest-based industry, the province is subject to a growing mining industry, mainly for gold and coal. In 1997, some 14 major coal reserves were identified, with a capacity of at least 500 million tons. Exploration for gas and oil are currently in progress, and production in the near future is anticipated.

The government has identified five major development areas in the province: The capital Palangkaraya and surroundings, the area northwest of Pangkalanbun, the area around Sampit, the Kahayan-Kapuas Muring-Barito floodplains (acronym Kabab, or *proyek lahan gambut* / mega-rice project PLG) and the area enclosed in the triangle Buntok-Ampah-Muara Teweh. The regions Pangkalanbun and Sampit are also often considered as one development zone (i.e. *kawasan andalan* Sambun). These five areas cover much of the entire province except for the upper water catchments of the rivers. Timber extraction and plantation development has concentrated mainly in the western half of the province; 60% of the applications and 77% of the awarded allocations for conversion (*hak guna usuaha* HGU) are situated in the Kotawaringin region (i.e. the area north of Pangkalanbun, and west of Sampit). Most of the estates are owned by either the Salim Group (PT Indotruba Timor) or the Astra group, with both of which the former President Soeharto had strong connections.

According to the *Jakarta Post* (November 18th. 1996) 163 major investors had submitted proposals to establish over 22,000 km^2 of plantations in Central Kalimantan; local licences had been issued for 15,000 km^2, and another 44 companies had just established almost 3,000 km^2 of plantations. During the 1990s, the area of reclaimed land – or removed forest – in the province increased from 119,000 to 136,000 km^2, while Protection Forests decreased from 34,000 to 17,000 km^2.

Illegal timber exploitation, followed by arson, is now so rampant in the province that regeneration of selectively logged forests has become unlikely if not impossible. The extent of this illegal activity reflects the clash of interests between the State and the private sector, and between centralised authority and regionalism, in which the regional government appears to play an inconsistent role. By removing the forest cover the State may be obliged to yield the land up to use, other than forest management.

Yet, in spite of the tremendous impact of timber extraction, demolition, fire and conversion, Central Kalimantan in the mid 1990s was one of the very few regions in western Indonesia still covered with some vast expanses of closed-canopy forest in which orang-utans have a home. The latest (1997) official data of the regional government (Bappeda) show that some 34,000 km^2 (22%) of the province is considered to be 'Protection Forest', and 119,000 km^2 as productive land, including some 30,000 km^2 of limited Production Forest land, 46,000 km^2 of permanent Production Forest land and almost 19,000 km^2 of plantations and agricultural land.

View of dead swamp forest after drainage and burning (Palangkaraya, 1997).

In December 1995 it was decided by Presidential Decree RI No 82, to initiate the reclamation of some 15,000 km^2 of swamp-forest flood plains in Central Kalimantan. This so-called mega-rice reclamation project (PLG – *proyek lahan gambut*) was planned to result in the complete conversion of some 11,000 km^2 of swamp forest in the Barito/Kapuas River water catchment and some 4,000 km^2 in the Sampit River area (McBeth, 1995). A regional EIA study by the Agricultural University Bogor (IPB) has revealed that no more than some 4,900 km^2 can possibly be prepared for wetland rice cultivation and some 1,000 km^2 for dryland arable crops, while some 8,500 km^2 should be kept inviolate, with some 1000 km^2 for 'alluvial reserves' and 7,500 km^2 for 'nature conservation'.

The public justification for the PLG project is to increase the potential for rice production in Indonesia. Expert evaluation of the plans has resulted in considerable opposition because it has been proven elsewhere in Kalimantan and on Sumatra (see for instance Whitten *et al.* 1987; AARD/LAWOO, 1992; Giesen, 1993) that reclamation of peat-land flood plains may cause the development of acid sulphate soils, disturb the hydrology, and result in die-off of the forest cover, thus altering the climate, while at the same time resulting in very low agricultural production. As a matter of fact one of the best studied reclamation projects in Southeast Asia, namely the Pulau Petak area, straddles the PLG.

The Pulau Petak project which was funded with Russian aid during the 1960s, may well be the best example of a premeditated failure and a deliberate ecological disaster. Notwithstanding three decades of continuous and expensive 'improvements' in drainage and soil fertility, the extremely acidic soils never yielded even a subsistence level of crops, while the surface and ground water became undrinkable and in the streams and canals all fish of relevance for human consumption disappeared (AARD/LAWOO, 1992).

Planning a mega-reclamation project to expand on this gross failure in economics and environmental stewardship suggests a profound ignorance of widely publicised scientific and empirical facts. It is, however, evident that reclamation would, in the first instance, facilitate the total exploitation of the hitherto poorly accessible swamp forests to supply the currently inactive timber and plywood plants in Banjarmasin, and, in the second instance, would provide space, no matter how miserable, for the planned transmigration of at least 300,000 Javan and Madurese families into the area. This was the situation until the drought hit and the area was set ablaze towards the end of 1997, to be followed by the breakdown of Indonesia's economy and the toppling of the government in early 1998.

Towards the end of 1997 it became increasingly evident that the project had opened up a road to disaster; the existing drainage and irrigation system failed to function according to plan, and arson accelerated the destruction process. The target for transmigration was reduced from over 300,000 families to some 100,000, and some 4,000 km^2 in the northern sector of the PLG was designated as 'conservation forest for ecological protection'. At present, development of the drainage network is not to proceed into this (still) forested area, although drainage of the flood plains to the south will inevitably have a devastating impact even here. It must be realised that this area of low forest and heath forest is in the geological zone of white sand with a shallow peat cover.

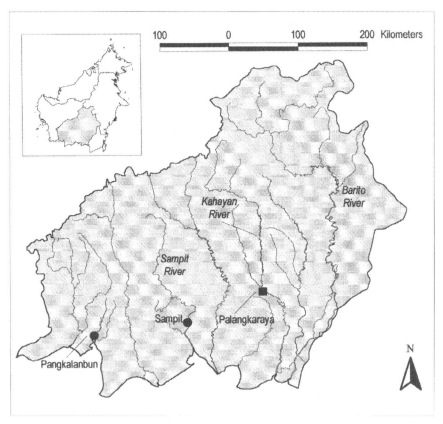

Basic geography of Central Kalimantan province (Kalimantan Tengah).

Major tribes in this province are the Ot Danum along the upper stretches of the large rivers and the Ngaju in the headwaters; to the east the Ma'anyan live on both sides of the provincial boundary with South Kalimantan.

It was generally assumed that the extensive swamp forests of Central Kalimantan are relatively poor in terms of biological diversity, probably because, until recently, biologists could not generate sufficient interest to overcome the physical challenges of field study in such a habitat. Recent surveys and studies indicate that this forest type actually contains a much wider range of floral and faunal diversity than previously expected (Rieley, pers. comm.). Further ecologically rich areas can be found in the lower foothills which surround the lowland flood plains along the western and northern boundaries of the province.

For the present study it is relevant to note that virtually all the alluvial habitats, including the swampy fringes of the rivers, have been affected by timber extraction, or are occupied by people. If these extensive alluvial zones were, until the mid 1990s, a prime section of the range of the largest contiguous sub-population of orang-utans in the world, in order to survive, the apes have now been forced to move into the higher central peat-dome regions between the larger rivers, and into the inaccessible coastal swamps. However, the plans (1994-1995) to drain, reclaim and convert virtually the whole of the flood-plain forests of Central Kalimantan province threaten to sacrifice anything up to some 50% of the present Bornean orang-utan population. Full implementation of these plans will result in the wanton destruction of the largest community of orang-utans and other wildlife in Borneo, unless much of the drainage can be timely aborted. During the extraordinary drought of 1997-98, the recently drained and reclaimed areas were most heavily affected by arson, causing the accelerated demise of at least a thousand orang-utans and destroying virtually all the benign environmental functions of this once densely forested region.

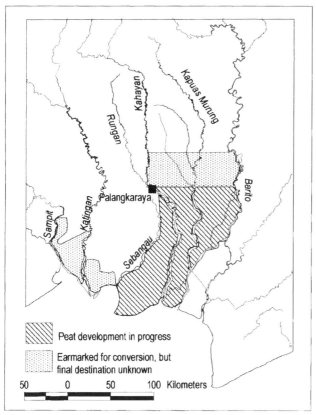

Map of Central Kalimantan province indicating the reclamation of its extensive peat-swamp forests.

Orang-utan distribution in Central Kalimantan

12 Jelai – Lamandau – Arui catchment

The southernmost extension of the Schwaner mountain range into Central Kalimantan province forms the major water catchments of the river Jelai-Bila, comprising the boundary of west and Central Kalimantan provinces, and of the upper tributaries of the river Kotawaringin; the major tributary is the river Lamandau which changes its name downstream after the village Nanga Bulik into river Kotawaringin. The banks of the river Kotawaringin-Lamandau are occupied as far as the upper reaches, and much of the area between the rivers Jelai-Bila and Lamandau has been occupied for agriculture. Although the area of Kabupaten Kotawaringin Timur (between the upper reaches of the river Pembuang, which, with its headwaters extends into the river Seruyan and the river Balantikan, or upper Lamandau), has a small human population and is covered by a large contiguous forest range in which orang-utans occur, the lower reaches of the rivers are densely populated. The distribution pattern of the orang-utan extends from the mountain valleys along the backdrop of alluvial fringes into the peat-domes and across the coastal flood-plain swamps of all the major rivers. It used to be contiguous with the ape population of the Kendawangan coastal swamps in the west.

The river Arui is also a tributary of the lower river Kotawaringin; like the river Lamandau its headwaters flow through a wide valley where the conditions for high densities of orang-utans must have been excellent, but which is nowadays badly affected by timber extraction, poaching and slash-and-burn agriculture. According to various sources, orang-utans still occur in abundance in the middle and upper reaches of the river Arui. Sightings of apes were reported especially from forests some distance from the populated river banks. Poaching of orang-utans for pets, and the killing of them for food and as an agricultural pest, is reportedly still common.

Rapidly expanding developments for transmigration, drainage for agricultural plantations and uncontrolled logging threaten the swamp forests in the coastal region. An area of 400 km^2 logged-over and degraded forest at Tanjung Penghujan has been proposed as a nature reserve, but it is of very limited value as a conservation area and is in fact mainly intended to function as a recreation area for the developing township of Pangkalanbun. The Borneo Orangutan Foundation planned to use the area for the rehabilitation of orang-utans (Galdikas, pers. comm.), in violation of the regulations. Large sectors of the logged-over forest land have were scorched as a result of arson during the 1997 drought, and no more than some 30% of the area contains orang-utans.

13 Tanjung Puting

Between the rivers Sekonyer and Pembuang the area of Tanjung (peninsula) Puting now comprises an isolated fragment of mixed flood-plain, swamp and heath forest. In 1939, the 3,050 km^2 Kotawaringin-Sampit wildlife reserve was established to protect some of the unique landforms, flora and fauna of this peninsula. About 980 km^2 was originally covered by peat forest and 780 km^2 comprises freshwater swamp. Another 700 km^2 of adjacent lowland forest was proposed as an extension for the reserve in the early 1980s, but subsequently suffered poorly controlled timber extraction, and open-pit gold-mining. In 1978, the name of the wildlife reserve was changed, and, in 1986, its status was changed so that it became the designated Tanjung Puting National Park. It is believed by the management that the park contains an orang-utan population of about 2,000 individuals (PHPA internal report, 1994), but no surveys have been conducted to substantiate this.

The area was brought to international attention by the activities of Birute Galdikas and her then husband Rod Brindamour, when they established an orang-utan rehabilitation project along a tributary of the river Sekonyer (1972-1991). Since the early 1980s, Galdikas has run a touristic 'research' programme on the feral apes; it is therefore hardly surprising that the scientific results are far below the quality one might expect from a research programme spanning two decades. In 1991 the Directorate-General PHPA refused her further admittance to her former research site, Camp Leakey, and to the rest of the national park. Since the mid-1990s she has tried to establish for tourist development a new rehabilitation site in a logged-over ex-concession area of some 900 km^2 near Pangkalanbun. She collected orang-utan juveniles illicitly and kept them on her private grounds in anticipation of a permit for the new rehabilitation site. In 1998 the Directorate-General PHPA was still trying to determine how this illegal stock of 89 apes could be confiscated for proper rehabilitation and reintroduction without attracting more criticism from Galdikas' powerful US constituency.

For the Ministry of Forestry the Tanjung Puting orang-utan population is considered one of the more important in Kalimantan, because it is acknowledged that 'it may be intermediate or even more closely related to the Sumatran population than to the Bornean populations of north and eastern Borneo' (PHPA, 1994). Despite more than 25 years of work by Galdikas, it is still unknown to what extent the influx of some 190 provisioned rehabilitant apes has displaced and influenced the wild population. Logging inside and around the eastern fringe of the park has considerably degraded the habitat, leaving behind large patches of poorly regenerating fern wilderness on sand (H. Djoko Susilo, pers. comm.). Due to the gold-mining operations, north of the park, in the water catchment of the river Sekonyer, the originally tea-coloured waters of the black-water river have turned permanently into a thin mud solution containing more than 200 times the toxic level of mercury. Poaching along the eastern fringe is not uncommon (Yanuar, unpubl.). The Tanjung Puting orang-utan population is now entirely isolated from any surrounding population. During the 1997 drought, at least three major, secondary-forest sectors of the national park were scorched as a result of arson, and it must be feared that the carrying capacity of the area has been reduced by at least 30%.

The designated Tanjung Puting National Park was brought to international attention by Birute Galdikas and her husband Rod Brindamour, who established an orang-utan rehabilitation station in the early 1970s.

14 Eastern Pembuang – Seruyan water catchment

The headwaters of the long river Pembuang are referred to as the river Seruyan, which originates on Bukit Tikung (1175 m alt.) in the Schwaner mountain range west of the Bukit Raya area. The valleys of the river Seruyan and other upland tributaries of the river Pembuang are wide and were supposedly covered with excellent orang-utan habitat, capable of supporting high densities. Three logging concessions in the area reported that orang-utans are common in the headwaters of the Seruyan River. Obviously the distribution of the orang-utan used to show a band of concentration in the ecotone (i.e. ecological transition zone) along the foothills of the continuous central Schwaner, Müller and Barito Ulu mountain complexes and the flood plain (swamps). The eastern side of the river Pembuang contained a very extensive area of forest habitat which ranged from the headwaters down to the Sembulu (Belajau) lakes over a distance of some 350 km. At present this area is traversed just north of the lakes, where the higher ground of the hills begins, by the main road between Palangkaraya, Sampit and Pangkalanbun, and all the highland area is being converted into mega-sized oil-palm plantations, along the expanding network of roads that cross the middle part of Kabupaten Kotawaringin Timur. It is estimated from aerial observations that no more than 40% of the forest remains. The opened up and pillaged forest in this sector suffered extensively from arson during the 1997 drought.

Large machinery can not be deployed in logging in swamp-forest areas. Use must be made of (railway) tracks on sleepers to drag the timber out by hand; such inroads are a convenient means of transport for biological surveyors, and for poachers.

15 Western Sampit flood plains

The Schwaner mountain range extends with long hill ridges southwards, separating major river systems, e.g. the Kotawaringin, the Arui, the Pembuang and the Sampit. From the Gunung Tikung – Bukit Baka mountain complex a major hill ridge extends all the way down to the coast of Sampit Bay. The once contiguous expanse of forest from the mountains down to the coast has been fragmented due to road construction, massive conversion and settlement. Due north of the mainroad Sampit – Pangkalanbun a large expanse of peat-swamp forest remains, in the water catchment of the river Seranau, which may be saved as permanent forest from conversion. The large expanse of swamp due south of this road has partly been occupied and will probably be reclaimed soon.

16 Katingan flood plains

East of the hill ridge separating the Pembuang and Sampit flood-plains begins a vast expanse (i.e. > 30,000 km^2) of swamps and flood plain stretching to the east as far as the banks of the mighty Barito River. This area is subdivided, by rivercourses, reclamation canals and roads, into a number of more or less isolated forest islands. The western Katingan (7,878 km^2) flood plain is the most western sector of this expanse. On satellite imagery it still looks like an impressive continuous expanse of forest but from the air (2 km altitude) the forest looks so uniform as to suggest that much of the southern section is *kerangas*, and therefore contains little, and extremely poor, orang-utan habitat. However, orang-utans were reported by various sources from this area, and it appears that the ape is still relatively abundant here. Chivers and Burton (1986), reported that orang-utans are often encountered in the area of the middle section of the river Katingan, and noted that there was considerable poaching pressure on the orang-utans and for timber. The southern section of the area is scheduled to be reclaimed, but it is probable that the common occurrence of potential acid sulphate soils in the region will preclude such reclamation. Very large tracts of recently pillaged forest in this sector suffered the effects of arson during the 1997 drought.

17 Sebangau catchment

The river Sebangau is the shortest of the major rivers of Central Kalimantan, originating in the ancient (i.e. > 60,000 years), more than 18 m deep, peat dome due west of the township of Palangkaraya. Interestingly, the river Sebangau is one of the very few river systems in Kalimantan which has until recently remained virtually free of human habitation. From the air (2 km altitude) a large area area due southwest of Palangkaraya looks like excellent swamp forest for orang-utans, showing only the typical straight extraction rails of the selective timber exploitation crisscrossing the vast expanse of diverse forest. Three logging concessions as well as the research project on the ecology of peat-swamp forest (Palangkaraya University in collaboration with the Universities of Nottingham and Leicester in the UK) reported the common occurrence of orang-utans in the swamp forests in the catchment of the river Sebangau (Page, et al. 1995). Many nests were seen when the area was visited for this survey, and orang-utans were encountered almost daily by the research team. Logging has been taking place at several locations, and the results of a study on disturbance in the area suggest that the restricted selective logging of swamp forest has had an impact on a local population of apes similar to that of dry-land logging (see Page *et al.* in press.). The research project applied for protection of a 50 km^2 peat-forest research area, and in 1996 expanded their request for protection of 500 km^2. Timber and ape poaching continues, however, as no protective structure is present. It is probable that the area will be reclaimed before the end of the century and that its orang-utan population is doomed. It is planned to strengthen the provincial government office for regional planning (Bappeda) with an ecological unit in order to conduct a prior review and to revise the province's planning process accordingly.

The southern (coastal) section of the western Sebangau flood plain was part of the planned LPG project by the Salim group (Mc Beth, 1995), for the region between the Sampit and the Sebanggau rivers. Close to the mouth of the

Sebangau river two large transmigration projects had already been established. The integrity of some 4,000 km^2 of swamp forest would have been destroyed if the reclamation project had been implemented. This development would have meant the end of a major population of orang-utans in Kalimantan, but a belated Environmental Impact Assessment convinced the group of devlopers that the planned reclamation would never be economical, and so the project was reportedly cancelled. Unless some serious conservation efforts are undertaken to help guide planning, land allocation and set-up of the extensive water catchments, while further expansion and timber poaching of the transmigration projects can still be contained, the unique Sebangau area will be doomed nevertheless. It is particularly significant that much of the west Sebangau peat-swamp forest largely escaped the general arson during the 1997 drought, except for some areas in the coastal zone near the transmigration project.

18 West Rungan flood plain

Closed in by the river Katingan in the west, the river Rungan in the east and two major roads (Kasongan-Tangkiling in the south and Tumbangsamba-Tumbangtalaka in the north), the gently undulating lowland area used to link the habitat on the slopes of the Gunung Liangmangan in the Schwaner range with the low coastal plains. Although few reports of orang-utan sightings have come from the area, the ape must have been common. The forest in this area is under heavy pressure of poorly planned and uncontrolled timber exploitation, timber poaching and conversion. The Bukit Tangkiling Nature Sanctuary (*Cagar Alam*) (21km^2) was established in 1977; it has mainly been used as a recreation area, is not effectively protected and the forest is almost demolished now that the area has been mined as a quarry for road-building material. In the swamp-forest area of Balunkun near this tiny nature sanctuary orang-utans were found to be common. The Balunkun area is designated Conversion Forest, and the apes in the area are reportedly poached 'for food' (Yanuar, unpubl.). Most of the pillaged forest in this area was scorched during the 1997 drought; hardly any trees remain in the Tangkiling 'Nature Reserve'.

19 Kahayan – Rungan catchment

The water catchment of the merging rivers Kahayan Ulu and Rungan once provided excellent conditions for orang-utan habitat, being the ecotone between the foothills of the northern extension of the Schwaner mountain range and the flood plains. The banks of the river Kahayan between Palangkaraya and its source upstream of the township of Tumbangmiri (i.e. the river Miri) are densely settled, which must have a tremendous impact upon the forest (timber poaching) for at least ten kilometres inland. However, reports of frequent sightings of orang-utans and their nests indicate that the ape is indeed still common in this region. Since the area reportedly has large expanses of virtually undisturbed swamp forest it is probable that a considerable number of orang-utans survived in this area.

20 Sebangau-Kahayan flood plain

The coastal flood plain between the lower reaches of the rivers Sebangau and Kahayan is fed by two rivers of different origin; the former originating in the peat dome southwest of the town of Palangaraya, the latter originating in the geologically ancient Schwaner mountain range. This geomorphological interface situation must have caused a particular, and possibly unique, ecological arrangement of vegetation elements and aquatic fauna. The southern part of the flood plain consists of open swamp with very low fringing vegetation. A few reports indicate the presence of orang-utans in this swamp-forest area. The region is scheduled to be drained and reclaimed as part of the PLG mega-reclamation project of the Central Kalimantan coastal zone; this reclamation will destroy all forest and will cause the development of acid sulphate soils. The swamp forest in much of this sector survived the widespread arson during the 1997 drought mainly because the drainage canals had not been established, and several timber concessions were still operating to remove much of the valuable forest cover. In the northern sector of the PLG, a zone of approximately two kilometres wide along the already excavated east west drainage canal in deep peat (> 9 m), was entirely lost as orang-utan habitat. Arson and massive forest die-off during the 1997 drought made survival of the resident orang-utan demes impossible. Poaching in this area has attained rampant proportions.

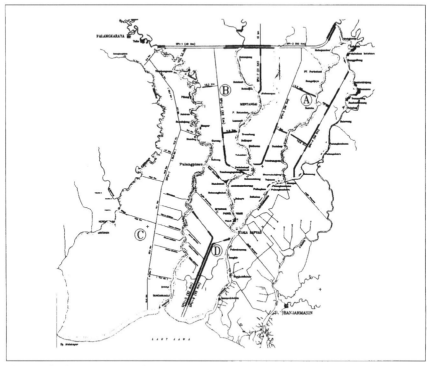

The scheduled drainage pattern in the mega-rice reclamation project; the essential first stage of an ecological disaster in the making.

21 The Mangkutup catchment between Kahayan and Kapuas

In this area of the lower and middle reaches of the rivers Kahayan and Kapuas orang-utans still occur in abundance, while in the northern section apes are reported to be seen regularly. However, poaching and encroachment seem to be pushing the ape back into ever smaller areas of undisturbed forest. Considerable numbers (65) of orang-utan nests were found along (a total of 3.5 kilometres) transects in the southern swamp forest, notably along the Sungei Arut, and 'very good habitat' can be found along the tributaries Sungei Hibur and Sungei Gadis (Yanuar, unpubl.). Along the S. Arut the ape is under heavy poaching pressure, while its habitat is being destroyed for transmigration, agricultural expansion and gold-mining. At the mouth of the river Kahayan, a 1,500 km^2 conservation area was proposed by MacKinnon and Artha in the 1980s, but has not been implemented. Although MacKinnon and Artha (1981; p 7) denote the Kelompok Hutan Kahayan area as 'approved *Cagar Alam*' it is not to be found on any of the official protected area maps at the Directorate-General PHPA and the Ministry of Forestry. Recent satellite imagery shows that this former large area of swamp forest, with about 900 km^2 of peat- swamp forest, 100 km^2 freshwater swamp forest and 500 km^2 of mangrove swamp (Silvius *et al.*, 1987) has now mostly been cleared by logging and is being converted for agriculture. However, just north of this once proposed reserve large stretches of peat-swamp forest can still be found, apparently still containing a considerable population of orangutans.

This area of approximately 10,000 km^2 of peat forest is at present being reclaimed, supposedly for wet-rice cultivation, being a part of the planned 37,000 km^2 of reclaimed land for industrial production in Indonesia (McBeth, 1995). In 1995, all of this planned development area still consisted of large stretches of peat-swamp forest, in which orang-utans are known to occur in 'normal' density (Page *et al.*, 1995). The large-scale drainage of the flood-plain swamps between the river Kahayan and Barito, which began in 1994, and the massive influx of transmigrants for the PLG land-reclamation project will obliterate the peat-swamp habitat across the entire Kahayan water catchment, affecting both the flood-plain swamps of the river Sebangau and the river Katingan. Hence the largest continuous population of orang-utans in the world, which inhabited this extensive swamp area until the early 1990s, will be driven to extinction within a few years. Perhaps careful planning, including ecological allocation of what areas to conserve and where to reclaim for agricultural production, and applying a ploder-technique with dikes to preserve remaining water catchments could have averted the disaster. The reclaimed sectors of this area have suffered badly from the widespread arson during the 1997 drought. Ridiculous ideas of 'translocating' all the doomed orang-utans have been proposed by the Directorate- General PHPA. The scope of such an operation is unrealistic in terms of funding and implementation, while surveys for reintroduction sites have not been carried out and the required investments have not even been contemplated. Some experts have suggested, however, relocating the doomed apes to the logged-over forests due north of the LPG project, and the Directorate-General PHPA has since invited foreign investment in this idea. It has apparently not even been considered that the carrying capacity of this area for the original population of surviving apes is already overloaded due to the logging, and the addition of more apes will result in a massive overshoot and starvation for all.

22 Kapuas-Barito plains and Negara swamps

An elongated area of some 10,000 km^2 of inaccessible swamp forest occurs between the rivers Kapuas and Barito, extending north as far as the dry foothills east of the township of Muara Teweh. The road between Sepinang and the Purukcahu airfield separates this vast area from the mountainous north and the valleys of the tributaries of the river Murung (29). Several reports of orang-utan sightings indicate that the ape still inhabits the vast area, although logging has extracted the most valuable timber from the inaccessible swamp forests. Together with the Sebangau Kahayan flood plain (20) the Kapuas Barito plain is the major target for the PLG reclamation project by the Sambu group (McBeth, 1995). Large-scale drainage was begun in 1996; the reclamation will cause the development of acid sulphate soils, which makes it unlikely that agricultural productivity will ever reach the desired targets and offset the investments. The large orang-utan population in the area is doomed and will become extinct in less than a decade. Being the primary focus of the PLG drainage and reclamation scheme, towards the end of 1997 this region was severely affected by deliberate burning of the drained peat domes. Much of the peat layer of the area has been burned away, and only dying forest patches remain in the sector. The orang-utans in the area have been and are being routed, slaughtered and eaten; or they die of dehydration, starvation and stress.

On the eastern side of the river Barito some patches of peat-swamp forest remain in the triangular area between the main course and the river Negara, its major tributary. Perhaps the orang-utan population in this area once split off from the main Central Kalimantan population west of the river Barito, helped by the dynamics of the river's main course. The population is doomed, because the Negara swamp-forest area is destined to be drained and reclaimed as a part of the one-million-hectare PLG 'mega-rice project'.

23 Kahayan – Kapuas foothills

The river Mangkutup separates the undulating dryland area between the rivers Kahayan and Kapuas from the swamp forests and flood plains between the two large rivers in the coastal zone (21). Local informants report regular sightings of orang-utans and it is assumed that the ape is common in this area because considerable patches of forest still look more or less intact on recent satellite imagery. It must be realised, however, that the foothills area of the mountain chain all along the central part of the province is traversed by a dense road network connecting hundreds of villages along the infrastructure axis between the townships of Kualakuayan (on the river Sampit-Mentaya) – Tumbangsamba (on the river Katingan-Mendawai) – Tumbangjutuh (on the river Rungan-Kahayan) – Kualakurun (on the river Kahayan) and Seihanyu (on the river Kapuas). This implies that much of the forest along the foothills has been fragmented and degraded due to human activity; this concerns the polygons 12, 23, 24, 25, 26, 27, 28 and 29. The pillaged forest areas in this region were badly affected as well by the widespread arson during the drought of 1997.

24 Schwaner foothills east

Few reports of orang-utan sightings are available from this remote region, but there is reason to believe that this area, being the ecotone between the upland hill country of the foothills of the Schwaner mountain range and the permanently wet flood-plain and swamp-forest country, represents orang-utan habitat of the highest quality. However, timber exploitation and subsequent timber poaching followed by deliberate burning have caused considerable damage to the region's forests.

25 Schwaner foothills west

Orang-utans are reportedly common in this ecotone between the upland hill country of the foothills of the Schwaner mountain range. As in the eastern section (24) this area must be considered to have been among the best quality habitat of orang-utans. Timber exploitation, poaching and arson, however, have caused enormous damage.

26 The central Schwaner mountain range

The Schwaner range is the oldest core of Borneo; it used to be the central ridge of the Sunda landmass stretching along the Karimata – Biliton range to Sumatra even before the lowland plains of Central Kalimantan had come into being through sediment deposit of its rivers and coastal accretion. The river Katingan originates on the watershed of the Bukit Raya at the border between Central and West Kalimantan provinces. The Bukit Raya mountain (2278 m) is the highest summit of the range and a major landmark in the region; it is of old, Tertiary, volcanic origin and was considered sacred in historical times; access to its slopes and summit was *tabu* for the indigenous Ot Danum and Ngaju (Dayak) people. Much of this mountain area on the Central Kailmantan side has recently been granted the status of designated national park. The Bukit Raya park covers some 1,100 km^2 of the southern water catchment of the Schwaner mountain complex, ranging in elevation from 100 – 2280 m, but largely above 700 m. Orang-utans occur in the area, although little further information is available regarding their abundance and status, and the altitude and forest type of much of the reserve make it unlikely that the area is important as an orang-utan sanctuary. However, it is mentioned in the Bukit Baka-Bukit Raya Preliminary Management Plan that in 1995 residents of the village of Batu Badak reported having seen orang-utans (in a group of some 20 apes) on the slopes of Bukit Raya.

In 1992, the designated Bukit Baka National Park in West Kalimantan, which is contiguous with the Bukit Raya conservation area, became the subject of the Natural Resource Management Project (Soedjito and Kartawinata, 1995). Its headquarters are at Kaburai, which can be reached most conveniently from the village Nanga Pinoh in West Kalimantan, at least until such time as the timber concessions – most probably before the end of the century – will have established a network of roads across the mountain range. The project is sponsored by the government of Indonesia, the ITTO and the US Agency for International Development (USAID), and was established to study 'technical forest management to improve natural resources management'. In reality it is a euphemism for subsidised timber extraction in a remote region, focusing on 'field testing of improved policies and practices for the management of production forests and protected areas.' Since research 'is not the primary focus of the project' (Soedjito and Kartawinata, 1995) it must be feared that the unique Schwaner range foothills forest in and around the established conservation areas will suffer considerable experimental damage, destroying the orang-utan's habitat, as has happened in a similar ODA-sponsored project at Sangai, adjacent to the Kaburai region. In any case, why such a destructive project was set up so near, or within, an extremely sensitive protected area is beyond comprehension.

27 Kahayan-Miri catchment

Like the foothill country of the Schwaner range, the upper reaches of the rivers Kahayan and Miri must have provided excellent habitat for orang-utans before the timber concessions moved into the region. The area is densely populated with Dayak settlements, and the forest has been considerably affected by shifting cultivation and scouring. Little is known of the present orang-utan population. Although not as extensive in the southern sectors of the province, the partially logged over forests in this sector have to some extent been affected by scorching due to arson during the drought of 1997.

28 Bundang east (Kapuas Murung Ulu)

The situation for the area in the upper reaches of the river Kapuas Murung must be similar to that in the Kahayan Miri catchment. Recent reports indicate that orang-utans are present in some sectors of the valleys and hill forest of this area. Satellite imagery indicates that also in this region arson has affected the regeneration of a forest cover, during the 1997 drought.

29 Upper Dusun

Orang-utans are reportedly rare in – if not altogether absent from – the area east of the river Barito. However, there are two remarkable sets of records indicating that the ape has dispersed across the first and second 'bottleneck' barriers between the headwaters of the river Barito and the river Mahakam. The records come from the Dusun area, between the tributaries called the river Murung and the river Maruwai, and from the northern bank of the river Ratah, a tributary of the Mahakam (see 31). Since the record has not been verified by a ground check, this entry (referring to a region comprising 1899 km² of potentially suitable forest below an altitude of 500 m) may be considered dubious. Satellite imagery indicates that during the 1997 drought this area contained a number of arson hot spots as well.

Ratah – west Mahakam

Due south of the Gunung Batuayan (1652 m), where the river Murung (i.e. a major headwater of the river Barito) originates, is a wide valley belonging to the river Ratah, a tributary of the river Mahakam. The Ratah originates from a steep ridge (at 1000-1500 m) covered with extensive heath forest, some 20 km wide. On the other side of the ridge the river Maruwai (a tributary of the river Barito) finds its origin. The watershed appears to be the last of the three physical barriers for orang-utan expansion into South Kalimantan province. Few orang-utan sightings have been reported from the northern side of the Ratah River valley, but absolutely none from areas across the watershed to the south.

30 Busang Hulu (Ulu Busang)

The northwestern corner of Central Kalimantan province is mountainous but of fairly low elevation (300-1735 m), comprising the watersheds of the Kahayan, Mirih and Upper Kapuas (of Central Kalimantan). Much of the western flanks area of the Busang valley has the formal status of Protection Forest, called Batikap, Block I, II and III (7,404 km²); the Protection Forest complex links the Schwaner mountain range (i.e. the Bukit Baka-Bukit Raya area) with the Müller mountain complex and the (West Kalimantan) Kapuas Ulu. The 720 km² Bukit Batikap I area was proposed as a Wildlife Reserve in the 1980s, but has so far not been established. The area is contiguous with the 7,000 km² Batikap II and III Protection Forests, as well as with the Bukit Baka-Bukit Raya conservation areas. All three areas of Protection Forest were reported to contain orang-utans (MacKinnon and MacKinnon, 1991). The present survey, however, was able to confirm the presence of orang-utans only in the upper Busang valley.

The provincial boundaries of Central, West and East Kalimantan meet on a watershed of relatively low (500-1300 m) elevation, between the origin of the river Busang (a major headwater of the river Barito which changes its name upstream of the village Muara Joloi), and the origin of a small tributary of the river Mahakam. Here, a narrow ridge (15 km) forms a 'bottleneck' through which the orang-utan has expanded into Sarawak, and westwards into the Upper Kapuas region, as well as into northeastern Borneo. Records from the area indicate that the ape has been seen in forests located at altitudes well over 500 m, and it may be surmised that the extent of forest through which the orang-utan has dispersed and possibly still traverses in this central Bornean corridor is considerably larger than the (119 + 373) 492 km² which shows up on the GIS map and is applied for this report. As a matter of fact, it is possible that the upper Barito forest still forms a contiguous link with the Liangpran (31) and upper Boh catchment area (32) in East Kalimantan.

In 1988, the University of Cambridge in collaboration with the Ministry of Forestry and the Centre for Biological Research and Development (LBN) established an ecological field research project with field locations at the headwaters of the river Barito, at the Joloi, Busang and Murung tributaries (Chivers, 1992). The project is to study forest dynamics due to plant-animal interactions and the regeneration of selectively logged forest.

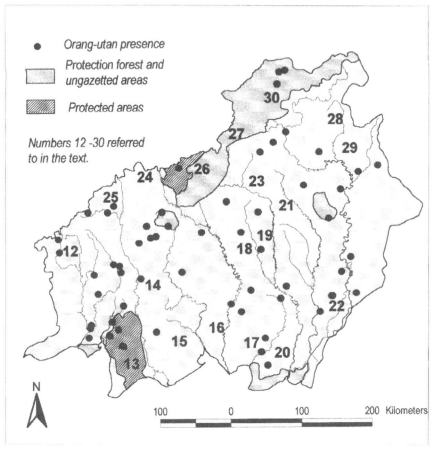

Map of Central Kalimantan. The presence of the orang-utan is shown in relation to the main rivers and conservation areas. Numbers on the map refer to the preceding text.

East Kalimantan

General description

In terms of socio-economic parameters East Kalimantan is at present (still) the most important province in Kalimantan. It contained 2.3 million inhabitants in 1995, increasing at some 4.7 % per year. The province covers 200,000 km² of land. As in the other provinces the human population has been clustered in a few sectors of the province, but is currently fanning out. The coastal zone contains rich coal seams as well as oil and gas deposits. Since the 1980s, this wealth in fossil resources, in addition to timber, has caused a surge in industry, attracting many people from Java, Sulawesi and other regions in the archipelago. In addition to local migrants (notably Iban from Sarawak) and to the large influx of people from outside Borneo, the indigenous tribes in this province encompass the Kenyah, Kayan, Punan and Lun Dayeh (Kelabit Murut, Potok, or Dayak). Many of the indigenous communities reside in what became State forest land in the 1960s, yet the provincial government explicitly denies the legality of community rights (*hak ulayat*) or traditional rights (*hak adat*). In practice this implies that only transmigrants have legal property rights (*hak milik*) over the land they (were) settled on.

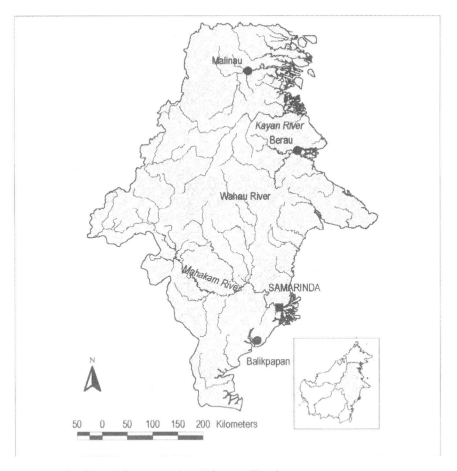

Basic geography of East Kalimantan province (Kalimantan Timur).

The province covers the catchment areas of several major rivers, notably the Mahakam and the Kayan, which discharge their waters on the east coast in the deep Makassar Straits. The area contains extensive hill and montane forests along the Sarawak border. In 1924, the colonial forestry service declared that 94% of the province was covered with forest; in 1981 the forest cover had been reduced to 61% (MacKinnon and Artha, 1981). The main forest types are hill, lowland Dipterocarp and heath forest, but a wide range of other forest types are also represented, including rare and valuable forests on limestone on the Sangkulirang peninsula. The peat- and freshwater-swamp forests in East Kalimantan are mainly of the coastal flood-plain type where the rivers Mahakam and Kayan enter the coastal flatland. This wide area of terrain includes the shallow Mahakam lakes, and is unsuitable for agriculture, consisting of some 1,250 km^2 of infertile sandy terraces, next to 5,672 km^2 of peat swamps. However, the areas of swamp forest are far less extensive than in Central and West Kalimantan.

The exploitation of timber, coal, oil and gas in this province plays a very important role in the economy of Indonesia. In the early 1980s, for instance, the annual timber exports were nearly twice as high as those of the rest of Kalimantan put together (MacKinnon and Artha, 1981). The coastal region, between Balikpapan, Samarinda (i.e. *kapet Sasamba*), and stretching north towards Bontang, Sangatta and Moara Wahau-Sangkuliran is the main development zone, with a perpendicular belt stretching inland from Samarinda along the river Mahakam being of lesser importance. The government considers Kabupaten Berau to be the second major prospective development zone. In 1997, the province had no more than some 3,000 km of public roads; all the roads have been upgraded from former logging roads. The eastern, coastal sector of the Trans-Kalimantan Highway cuts through the Kutai National Park, and a major planned extension of the present stretch of 'high-way', linking Sarawak and Sabah, will pass through the Kayan-Mentarang reserve.

Apart from the remote Kayan Mentarang Wildlife Reserve, the two other protected areas in East-Kalimantan are in poor condition and their current design and status should be revised. With a few notable exceptions, much of the PHPA staff is reportedly poorly motivated, and inadequate reserve management as well as an absence of protective measures were repeatedly confirmed during the course of the present study.

Just as during the 1982-83 drought, the 1997-98 drought also facilitated arson, so that many thousands of hectares were scorched. As a matter of fact, all the remaining lowland forest south of the Sembaliung mountain chain and the Sangkulirang peninsula has perished due to desiccation and fire.

Orang-utan distribution in East-Kalimantan

31 Liangpran (Long Boh)
The reported occurrence of orang-utans in the forest on the northern bank of the river Mahakam, and along its headwater tributary the river Boh (32), suggests

that the upper catchment of the river Mahakam was, and is, the main entrance for orang-utans to East Kalimantan province. Yet the upland forest is unlikely to support high densities of orang-utans.

32 Boh water catchment (southern Apo Kayan)

The Apo Kayan is the mountainous region between the headwaters of the rivers Kayan and Bahau on the Kalimantan side, and the river Rejang across the watershed of the Irang mountain range on the Sarawak side. Even before colonial times the Apo Kayan had been settled because of its fertile soils of volcanic origin. Forest hunter-gatherers (the Punan) occurred in the forests to the north and west (Piazini, 1957; Hoffman, 1986; Tillema, 1995), and engaged in a barter relationship with the agricultural tribes. In the Apo Kayan area, orang-utans are reportedly absent, probably due to the fairly intensive human traffic, the presence of subsistence hunters, and, more recently, massive encroachment.

However, around the rivers Kayaniut and Boh (the northern headwater tributary of the river Mahakam), in the southern sector of the Apo Kayan there are both historic and recent indications of the occurrence of orang-utans. This plateau-like mountain range of ancient volcanic origin and somewhat low altitude must have been the natural habitat corridor across the infertile central Bornean mountain range between the Eastern Kutai and West/Central Kalimantan, for both the orang-utan and the human forest dwellers. In the late nineteenth and early twentieth century it was the main entrance for the Iban raiding parties from Sarawak when they invaded the Kayan and Kenyah longhouses in Kalimantan. It is now evident that this connection has been severed for the orang-utan: Satellite imagery shows intensive settlements in the central valleys of the Apo Kayan, and extensive *ladang*[70] developments all the way up into the steep valleys of this former corridor. In the region there is a considerable amount of overland traffic and travel across the Indonesia-Sarawak border, and there are also numerous airfields used by small charter aircraft.

MacKinnon and Artha (1981) indicate that a 1,000 km^2 area in this region (Apo Kayan) was proposed to become a conservation area, and a UNESCO Man and Biosphere Reserve. However, the area is not on the official maps of the Directorate-General PHPA conservation areas, and there is no conservation office nearby. In early 1998 deliberately set fires destroyed large areas of the regional forest.

33 Pari – Sentekan catchment (Longbleh)

Although the mountain area between the upper reaches of the river Boh and the catchments of the smaller tributaries of the river Mahakam downstream is well over 500 m high (and hence is excluded from the GIS map concerning lowland forest with potential orang-utan habitat), there are indications that orang-utans did, and still can, range from the southern Apo Kayan to the lower reaches of the river

---

[70] I.e. a forest area which is cleared by means of slash-and-burn cultivation for (cash) crop production.

Mahakam, and to the headwaters of the river Belayan. South of the river Pari, a tributary of the river Mahakam, the terrain is undulating with peaks not exceeding some 750 m in height. The forest in this area has been, and still is, under heavy pressure of uncontrolled timber extraction, so that the habitat quality has been dramatically affected. The foothill country between the river Boh and the upper reaches of the river Kedangkepala once formed a contiguous range of suitable habitat for orang-utans, and undoubtedly was a main dispersion channel into East Kalimantan.

34 Belayan – Kedangkepala (the entrance into East Kalimantan) (Longnah)

Evidently the orang-utan population was distributed all across East Kalimantan through the lower altitude (100-650 m) mountain range on the northern side along the river Mahakam between the Apo Kayan and the coast of northern Kutai regency. Along the foothills of the upper reaches of the river Belayan and Kedangkepala the habitat conditions for orang-utans must have been excellent before the logging companies moved into this area. Some major relict populations still survive in the forest fringes of this corridor. Along the river Mahakam, downstream of the township of Long Bagun, orang-utans appear now to be absent, but they still occur across the watershed north of the river. Orang-utans were reported in the remote hill country due east of the township of Long Bagun between the rivers Belayan and Mahakam. One logging company, approximately 100 kilometres northeast of Long Bagun, reported that both nests and orang-utans were commonly sighted in the concession. MacKinnon and Artha (1981) mention that a proposal for a 3,500 km^2 Long Bangun Nature Sanctuary was agreed upon for gazetting; however, the area is not to be found on the most recent official Directorate-General PHPA maps of the conservation areas in Kalimantan. The nearest representative of the Directorate- General PHPA (resort) lives in Muara Kaman. The remnants of forest in the area suffered heavily from the widespread arson during the drought early in 1998.

35 The central flood plains of East Kalimantan (West Muara Kaman)

The northern tributaries of the river Mahakam form an extensive (13,000 km^2) large flood plain that is bordered by the central mountain range to the west and north and by a coastal range of hill ridges to the east. The coastal ridges are exceptionally rich in coal seams as well as oil and gas deposits. The flood plains close to the main river have formed into semi-permanent lakes, surrounded originally by high-rising peat-swamp forest. In the northern section of the lakes the 625 km^2 Muara Kaman Nature Reserve was established in 1976. The present survey, however, obtained no reports at all of orang-utan presence in this area. It is noteworthy that the Muara Kaman Reserve has little conservation value at the moment, because it was very seriously damaged by the 1982–1983 forest fires. It has also been badly degraded by extensive *ladang* cultivation, illegal logging and poaching. There has been no protective action in this region whatsoever.

Nevertheless, there are reports that a population of orang-utans occurs in the estimated 800 km² of relatively undisturbed swamp forest to the west of the reserve, close to the foothills in the region of the river Senyiur. This area of excellent orang-utan habitat is designated Conversion Forest. It is important to assess the exact status of orang-utans in this area of swamp forest as soon as possible. Furthermore, to the east of Muara Kaman in the Sebulu and Separi areas and to the northeast in the Menamang area, many sightings have been reported. This may reflect both high densities of orang-utans and a high degree of encroachment. There are many accounts of clashes between local farmers, both indigenous and transmigrant, and orang-utans that come to the gardens to raid fruit trees and other crops. This area is also one of the most important sources of young orang-utans which will be captured and sold illegally as pets; 75% of the approximate 150 orang-utans that were confiscated by PHPA (KSDA) in Samarinda, came from the Sebulu area (Agustiana, pers. comm.). In addition to the ongoing habitat degradation due to logging, poaching is the most serious threat to the remaining populations of orang-utans in this area. During the 1997-1998 drought several sectors of this area were demolished by arson, and virtually all the forest cover died due to the extensive drought conditions.

Coastal Mahakam plains

Between the rivers Mahakam and Santan a large area of approximately 30,000 km² has been turned into a mosaic of badly degraded secondary forest patches, grass plains and agricultural fields. Frequent fires have gradually turned the area into a patchwork of grassland and low shrubs, entirely unsuitable for orang-utans to subsist. According to a considerable number of reports orang-utans still occured in this region. It is unlikely, however, that these apes had any real chance of survival. Even if any did survive up to the drought of 1997, widespread arson set much of the remaining forest patches ablaze, or smoked out the ape inhabitants who were then readily slaughtered.

36 Coastal Kutai

The coastal region of East Kalimantan consists of low hill country. During the first half of the twentieth century all of this coastal area was still covered with dense rainforest, and inhabited by orang-utans. Halfway between the town of Samarinda and the Sankulirang peninsula lies the Kutai Conservation Area which was established in the 1930s at the instigation of the Sultan of Kutai. In the late 1960s, a large sector of the wildlife reserve was allocated as a timber concession and the coastal lowlands were destined to be crossed by a highway. Without any form of protection the coastal lowlands were soon occupied by illegal settlers, mostly of Bugis origin. In the 1970s, two large industrial complexes were established in the southern section of the area and over 1,000 km² was degazetted. Proposals for the extension of the conservation area in a northwestern direction, across the river Sangatta, where Sumatran rhinos and orang-utans reportedly occurred, were not granted. Instead, the area to the north was first deforested and subsequently given out as a mining concession to the international mining consortium PT Kalimantan

Prima Coal[71]. With an investment of over $60 million the area was turned into a huge open-pit mine, attracting thousands of people, and obliterating over 1,000 km² of prime orang-utan habitat along the river Sangatta.

The Kutai area has suffered badly from the massive forest fires that razed eastern Kalimantan during the early 1980s, while all its coastal lowland forest has been demolished by repeated logging and encroachment for slash-and-burn cultivation. The north-south road track – part of the trans- Kalimantan 'highway' – which traverses this coastal region attracts increasing numbers of illegal settlers to the area.

Uncontrolled logging has ruined many forests in East Kalimantan; orang-utans cannot survive in such poor 'secondary forest', even after a decade of 'regeneration' (Kutai National Park, 1995).

In 1990, the mining company KPC invested in the production of a Master Plan for improved conservation of the remainder of the park, which involved 'participation' of the industries in its surroundings. In 1995, a formal agreement was reached and the Committee to oversee management was finally established. There are as yet no indications that this new 'support structure' (convened in a body called 'The Friends of Kutai') has had any effect as regards protection; poaching of orang-utans within the designated National Park is still rampant, and the repeated emergence of orang-utan infants to be illicitly 'rehabilitated' at the park's own Teluk Kaba recreation site suggests that the park staff is implicated in the poaching. On the official map of the conservation areas in Indonesia the extent of the designated Kutai National Park is given as 1,986 km².

In 1993, the orang-utan population inside the designated Kutai National Park was

---

[71] KPC is owned by the Australian Mining Company and BP (UK).

The present distribution

estimated to amount to anything between 1,200 and 2,100 individuals (CBSG/SSC/IUCN, 1993). This is probably a gross overestimate. Since the 1970s, the orang-utan population inside the conservation area has been affected by extraordinary habitat degradation caused by drought, fire, logging, oil exploitation and human encroachment, as well as by poaching. It was reported recently that the Park management issues permits for the hunting of protected banteng (*Bos javanicus*) and sambar deer (*Cervus unicolor*); up until 1997, a butcher in Bontang reportedly prepared and sold orang-utan meat on a regular basis, and orang-utan infants were on sale in Bontang and Samarinda regularly until early 1997, when a major dealer was apprehended.

An extension of the southern boundary of the park in the direction of Menamang would link the conservation area with the Muara Kaman Reserve and thereby include some of the larger stretches of peat-swamp forest and its significant populations of orang-utans. This extension had already been proposed in the early 1980s (MacKinnon and Artha, 1981), but has so far not been taken into serious consideration. All the coastal areas of the National Park again suffered insurmountable damage from arson at the hands of the illegal Bugis settlers during the drought of 1997-98, thus putting an end to the regeneration of the pillaged forests many kilometres inland. In April 1998 it was estimated that some 40% of the park was scorched (*Jakarta Post* April 18th, 1998). An aerial survey of the area in 1998 at the onset of a wet season revealed that virtually all the forest in and around the designated national park had dried out, and was partially burned. Like a temperate forest in winter, it was entirely leafless, and is presumed dead. This implies that for many years to come the carrying capacity of the area for orang-utans is reduced to a fraction, and it must be seriously feared that the remnant population in this section of East Kalimantan will die out.

37 Telen-Wahau catchment

The river Belayan originates in a high mountain complex with at least three peaks over 2000 m altitude (e.g. Gunung Menyapa); it is likely that this river, and the more northern headwaters of the river Telen, posed a formidable barrier for orang-utans to cross from the Pari-Sentekan area north into Borneo. Reports of orang-utan sightings in the mountainous area along the southern banks of the headwaters of the river Kayan indicate that the apes occur here at elevations of over 500 m, and through this forest corridor have populated the forests in the Telen and Wahau river catchments, and the Sembaliung (Tindah-Hantung) range which forms the Sangkulirang peninsula.

38 Sambaliung (Tinda-Hantung foothills)

Where the lower reaches of the northern tributaries of the river Mahakam (Belayan, Kelinjau, Kedangkepala and Telen) pass the foothills of the central mountain complex to form the huge central flood plain and swamp areas of the central basin of East Kalimantan province, orang-utans once enjoyed excellent habitat conditions. In the north, the foothills between the upper headwaters of the river Wahau and the river Karangan (Sankulirang) are believed to be one of the main centres of orang-utan distribution in East Kalimantan. Many reports of recent sightings come from the areas on both sides the road Sepasu – Muara Wahau. The area is subject to poorly controlled logging, timber poaching, large scale conversion for plantations and transmigration and expanding habitat destruction by migrants. Poaching of orang-utans is common, although it seems that the illegal trade peaked in the late 1980s and has now subsided, due to the confiscation campaigns of the PHPA-Wanariset Samboja reintroduction programme. Logging pressure and forest conversion have forced many orang utans together in degraded forest blocks and have dramatically increased the incidence of hungry apes raiding gardens in search of food. As a consequence, many orang-utans have been killed. The orang-utan population is believed to extend up along the headwaters of the rivers Kelinjau and Telen although the incidence of reported sightings decreases rapidly with altitude westwards. MacKinnon and Artha (1981) proposed establishing a 2,000 km^2 nature sanctuary (the Mangkilat/Kelompok Kapur Sankulirang) in the northern sector of this area, but the proposal has never been processed.

The Sangkulirang area is one of the most important forested areas of Borneo in terms of endemic biological diversity; recently orang-utan sightings were reported from the limestone forests of the Gunung Sekarat (near Sangkulirang township) south of the river Karangan. The extensive limestone plateau is unique in Southeast Asia. It is an area of extraordinary rainforest that should have been protected on the basis of its unique flora and fauna – a new species of Rafflesia was recently discovered in its forests (W. Meyer, pers. comm.) – and vital water catchment function. Forestry regulations prescribe that in order to prevent soil erosion its steep forested slopes should not be logged. However, half of the 250 km^2 area of the Gunung Sekarat lime stone hills has been allocated for open-pit coal-mining, while the other half will be excavated for lime-stone mining and cement production. The orang utan population in the area is isolated, and will be obliterated along with the other unique faunal and floral elements, because dispersion and migration to adjacent areas is impossible.

39 Segah catchment

Across the rivers Longgi and Kelai, the Telen-Wahau area is contiguous with a large catchment area of the river Segah, consisting of the northeastern foothills of a mountain complex with a high peak of 2467 m (Gunung Guguang). The forest used to be an uninterrupted expanse as far as the banks of the river Kayan in the north, and continuing up to the coastal plains. Only few records of orang-utan sightings are known from this area, and the status of the population in this impressive expanse of primary forest is unknown.

Bulungan – Kayan-Mentarang

The Bulungan area is closed in by the large rivers Sesayap in the north and Kayan in the south; to the east it is bordered by the Bantama Abu – Apo Duat range across the headwaters of the river Baram in Sarawak. The water catchments of the river Mentarang and the adjacent river Bahau (a headwater tributary of the river Kayan) is traversed by a network of roads, leading across the Apo Duat range into Sarawak, at the townships of Longbawat, Pa Rupai and Longsia. Along all the major rivers and the road network at least one-hundred villages can be found; the human impact in all the valleys of this area was, and still is, tremendous.

In 1980, the mountain complex of the Bulungan area was gazetted as a nature sanctuary (SK Mentan 25/11/80 No 847/Kpts/Um/11/1980). All recent surveys into the 16,000 km^2 Kayan-Mentarang Nature Reserve indicate that orang-utans in the water catchment of the rivers Mentang and Kayan, if still extant, are extremely rare. The so-called traditional hunting of orang-utans has been reported several times in this area, although it was invariably said to have occurred one to several decades ago. No sightings have been recorded for the central and southern sectors. Surveys in the far northern sectors, which are contiguous with the Ulu Padas Forest reserve in Sabah, revealed the presence of the ape in some isolated upper catchments of smaller rivers. These orang-utans are presumably representatives of the northern population, of which larger fragments are known to survive in the proposed Ulu Sembakung reserve. There are plans to expand the Kayan Mentarang reserve along the boundary with Sabah, to cover all State forest land up to the coast at Moara Sebulu. In this way the entire Sesayap – Sembakung water catchment would become part of the conservation area, and hence give the northern orang-utan population a better chance of survival.

Be that as it may, there was a time when the lower reaches of the southeastern Bulungan area was populated with orang-utans, notably along the river Kayaniut (Heynsius Viruli and van Heurn, 1935), so that a continuous link existed between the population of the Apo-Kayan and that of northern Borneo (i.e. the Sembakung area and Sabah).

Thus, at present, this large mountainous reserve is unfortunately of little importance as regards the conservation of wild orang-utans. Nevertheless, it may become extremely significant as the only area suitable for reintroduction of (ex-captive) rehabilitants in Borneo, if poaching can be contained, if the reserve can be redesigned to exclude peripheral settlements and its inhabitants in the interior can be resettled.

The Kayan Mentarang conservation area is the subject of a long-term project by the World Wide Fund for Nature, in order to establish effective conservation. The major problem concerning the reserve is that it was established as a nature sanctuary (*Cagar Alam*) while being inhabited by whole groups of people of distinct tribal identity, notably Kayan, Kenyah and Punan. According to the legal framework, the nature sanctuary is the strictest of conservation areas, and should not be inhabited, or even utilised for the harvesting of forest produce. The WWF project seeks to establish a status quo, consolidating tribal habitation of the reserve, and promoting so-called sustainable utilisation of the protected area's resources. Whether this essentially European approach to the conservation of archaic (agri-)cultural landscapes will be applicable to a delicate tropical rainforest ecosystem in a developing country will be revealed only in a few years time by an ecological evaluation. It may be an indication, however, that the orang-utan and the proboscis monkey have already been hunted to extinction. In any case, it is certainly questionable whether such an approach is legally correct.

On the latest official map of the conservation areas, the Kayan Mentarang area is denoted as a wildlife reserve; it is unknown when – or whether – the status was formally changed, for it is also often referred to as a 'national park.'

40 Berau

A mountainous area of some 5,000 km², Berau is closed in by the rivers Kayan and Segah-Berau. The trans-Kalimantan highway cuts through the middle of this region, between the major townships of Tanjung Redeb and Tanjung Selor. Orang-utans were reported to be 'rare' or absent in the foothills and the wide valley of the upper stretches of the river Segah. Further to the west, around Long Bea and into the mountains no orang-utans were reported. The area has been inhabited by forest-dwelling gatherers of two distinct identities, namely Punan Malinau and Dayak Basap. The former have recently been settled, the latter still dwell in limestone caves; both subsist by means of collecting forest produce – for the market – and by hunting, and it is almost certain that they have hunted the orang-utan population in the area to extinction As this area is enclosed by the rivers Kayan, Segah and the headwaters of the river Kelai, any surviving subpopulation of apes is isolated from adjacent populations. MacKinnon and Artha (1981) proposed that 1,100 km² of this area should be established as a nature sanctuary, also because it contained unique outcrops of limestone. The proposal was apparently never processed, and virtually all the area has been allocated for commercial logging and conversion for transmigration. Much of the region is in the custody of the State-owned forestry company Inhutani I; the European Union provides technical assistance for sustaining the remaining forest cover.

41, 42, 44. Southern Sembakung swamp (Malinau), Northern Sembakung swamp, Sebuku – Sembakung water catchments

Between the rivers Sembakung and Sebuku the flood plains comprise extensive peat swamps. The ecotone between the foothills of the Apo Duat mountain range and the flood plains was once excellent orang-utan habitat. Several villages are to be found along the headwaters of all the major rivers. Recently a road network was constructed to traverse this ecotone, and increasing numbers of people are settling along it. Several recent orang-utan sightings are reported from the area west and north of Malinau, and reports by logging concessions to the west and northwest of Malinau indicate that the ape may still be common. Protective management in the area is absent. However, in September 1998, some 300,000 ha of the area was established as a wildlife reserve by the Ministry of Forestry, at the instigation of a team from the World Wide Fund for Nature Indonesia Programme (i.e. F. Momberg, P. Jepson and H. van Noord), after seven earlier recommendations for a reserved status for this area (by MacKinnon and Artha,

WWF/FAO, 1982; Silvius and Scott, IUCN, 1989; RePPProT, 1990; Min. KLH, 1992; Bappenas, 1993; Prianto, Wibowo, ODA, 1997; and K. MacKinnon, Worldbank, 1997).

43 Sembakung Ulu valley

The rivers Sebuku and Sembakung originate in Sabah, and only the lower foothills and flood plains are located in Kalimantan. The river banks are densely settled; in colonial times the area was known as *Tidoengsche Bovenlanden*. Orang-utans reportedly occur in two logging concession areas close to the border with Sabah, and various orang-utan sightings were reported in the lower peat-swamp areas around the rivers Sembakung and Sebuku. MacKinnon and Artha (1981) proposed establishing a 5,000 km^2 nature sanctuary (*Cagar Alam*) in the Ulu Sembakung water catchment, because of its rich wildlife, including Kalimantan's only herd of elephants, and a 1,100 km^2 nature sanctuary at the mouth of the river Sebuku in order to protect a unique expanse of undisturbed mangrove forest. Both areas are not denoted in the most recent official maps of the conservation areas in Kalimantan and presumably were never seriously processed for gazetting. Recent satellite imagery reveals that much of the forest in this entire area is badly degraded. If orang-utans survive here, their numbers are probably small.

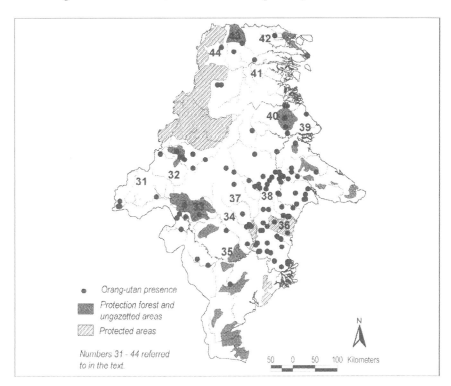

East Kalimantan. Reports of the presence of orang-utans are shown in relation to the main rivers and protected areas. Numbers on the map refer to the preceding text.

Sarawak

General description

In the early 1970s, when the impact of timber exploitation began to be evident in Sarawak, some 70% of the State's total area was still covered by relatively undisturbed rainforest (FAO, 1987). Since then the forest has rapidly been depleted and converted, while development for human purposes has advanced into the interior. Since the early 1990s, 37% of Sarawak's land area of 124,500 km^2 has been considered Permanent Forest Estate (PFE). Some 2,593 km^2 (2 % of the total land area) is gazetted as a protected area network, while another 7,958 km^2 has allegedly been proposed for a protected status. Well over 2,000 km^2 of forest is logged annually (Hurst, 1990). In the early 1990s, log production peaked at 18 million m^3, earning well over US$ 4,000 million; according to the World Bank, 'harvesting is carried out haphazardly with unnecessary and excessive damage to the residual forest, preventing regeneration.' All but 2.5% of the PFE remained unaffected (Collins et al., 1991). In addition to logging, slash-and-burn agriculture is also a major factor in the deforestation of Sarawak. On the one hand, encroachment into the forest estate along the logging tracks for slash-and-burn cultivation is virtually uncontrolled, but on the other hand, the aboriginal Penan and Kenyah of central Sarawak have repeatedly challenged the destructive intrusion of timber companies into their traditional domain (Hong, 1987; INSAN, 1989; Manser, 1992). A major issue in the challenge for some of the tribes, notably the Kenyah, is the lack of financial compensation for what they see as government-condoned piracy of the riches of 'their traditional forest'; for others (Penan) it is said to be mainly their desire to conserve a gathering culture in their traditional life-support system against the swell of a consumerist civilisation.

On its inland border with Kalimantan Sarawak shares the central mountain range, but much of the State consists of alluvial and often swampy coastal plains with gently undulating hilly country inland. The human population is relatively small; in 1990, it numbered some 1.7 million inhabitants, of which 30% were of Iban tribal identity (Cleary and Eaton, 1992). The annual population growth rate is over 3%.

Orang-utan distribution in Sarawak

Orang-utans have almost died out in Sarawak. Already in the 1960s 'the ape was confined to only a few areas of apparently suitable habitat, almost all in the southern third of the State' according to Bruen and Haile (1960). Reynolds (1967) compiled all evidence for the distribution and showed that during the 1960s the orang-utan occurred between the river Sadang and river Rejang, all along the foothills of the Iran range which forms the boundary with Indonesian Kalimantan, as far north as the headwaters of the river Baram. To these early statements Payne (1988) added, 'evidently the range has contracted over the past century as a result of human activity.' The survey carried out by Jane Bennett has confirmed this gloomy picture (Bennett, 1993).

The present distribution

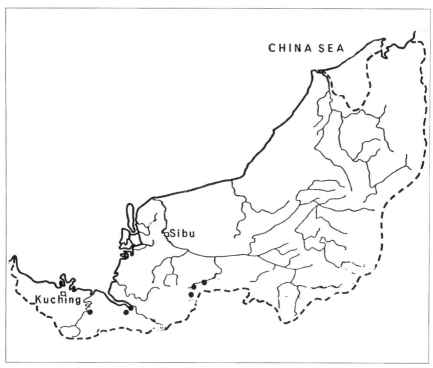

Orang-utan distribution in Sarawak in the early 1990s, after Bennett (1991).

Lanjak Entimau

Orang-utans have been found in the Lanjak Entimau Wildlife Sanctuary (approximately 1,700 km^2) and adjacent Batang Ai National Park (300 km^2) (Bennett, 1993). The surveys to prepare for the establishment of the reserve in 1981 (Kavanagh *et al.* 1982), which were carried out by means of a helicopter and on foot, did confirm the presence of the ape in the area. No attempt was made to assess the density. The incidence of nests along the helicopter's flight path indicated that the ape was patchily distributed. During the surveys on foot one fleeing ape was spotted, and the charred remains of a small orang-utan were found in an abandoned poacher camp, indicating that poaching in the interior was taking place at the time. Iban villagers at the periphery of the proposed Sanctuary readily admitted that the orang-utan was a favourite prey (Rijksen, unpubl.). After extensive surveys (i.e. 1,568 km of transects at 10 locations) through the area, Blouch (1997) believed the total population in Lanjak Entimau to be about 1,000 animals, with highest densities in the southern fringe, i.e. the water catchments of the rivers Engkari and Batang Ai, and very low densities in the northern third of the park (close to the heartland of the Iban). The method used by Blouch to assess the total number is a simple extrapolation of the limited nest-count survey method, and therefore probably yields inflated results.

Batang Lupar – Sebuyau swamp

The satellite images show that some pockets of swamp forest remain in the coastal region around the mouth of the rivers Lupar and Saribas. In her overview of the orang-utan situation Bennett (1993) does not give records for the area, but her maps suggest that the ape still occurs in the region. It is probable, however, that a surviving population in this region is doomed without management, for it will probably not exceed some 25 individuals. An Environmental Impact Assessment of the Sebuyau swamp in 1996 detected several orang-utans in the riparian forest of the swamp area; the swamp was destined to be reclaimed for conversion into oil-palm plantation, but on the basis of the EIA findings the decision was made to refrain from any further consideration involving reclamation of the swamp (A.C. Sebastian, pers. comm. 1996).

Sabah

General description

Sabah (73,710 km^2) occupies the northern 10% of land area of the island of Borneo and is Malaysia's second largest federal State (Payne and Andau, 1989). The State consists of alluvial and often swampy coastal plains with hilly inland country intersected by large rivers and mountain ranges in the interior. The central mountain ranges rise abruptly from the west coast to the granodiorite peak of Mount Kinabalu (4094 m), the highest summit in Southeast Asia. The river Kinabatangan is the largest of Sabah's rivers; it drains eastward, watering extensive plains, and is navigable for long distances (Collins et al., 1991).

The human population of Sabah was estimated at 1.5 million in 1990 (Cleary and Eaton, 1992), increasing at a rate of over 3% annually. In 1953, the natural forest cover of British North Borneo (i.e. Sabah) was 63,275 km^2 (Fox, 1978). According to the Sabah Forest Department the rainforest cover was 33,500 km^2 (45 %) in 1984 and the prediction was that this would fall to 29,110 km^2. (39 %) in 1990 (FAO, 1987).

Almost 8% of Sabah's land area is reportedly included in the system of national parks and other categories of existing reserves. In view of the fact that 'other categories' do not safeguard the forest from despoliation the percentage is quite meaningless. Moreover, on paper, the protected area system may be a fair 'representation of eastern Malaysia's ecological and biological diversity', but since the status for many of these areas is highly dubious, they are readily subjected to exploitation and conversion as demands for land and timber resources increase.

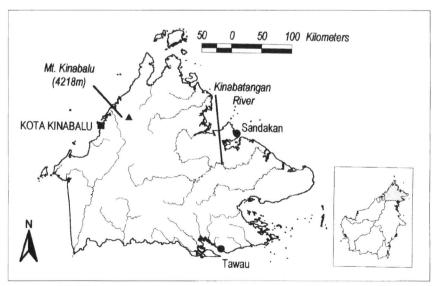

Basic geography of Sabah.

Orang-utan distribution in Sabah (after Payne, 1988)

The most comprehensive searches for the distribution of orang-utans during the second half of the twentieth century were conducted in Sabah. The first record was provided by Harrisson in 1963. In the 1970s MacKinnon (1971) reviewed the record. In the 1980s, only one-thirds of the State was still covered by primary forest unaffected by timber extraction. Nevertheless, Davies (1986) concluded, 'there are at least 4,000 orang-utans in eastern Sabah', although his calculations were of the simple 'extent of forest times average density' kind. Since the mid- 1980s the forest surveys in Sabah (and Sarawak) have been able to be conducted with the aid of helicopters. Commissioned by WWF, Payne (1988) applied this new technology in combination with ground surveys searching for orang-utans and counting nests. He also conducted inquiries among people living and working in rural areas. However, many of the historical records have become obsolete, due to massive forest obliteration and conversion.

At present, no more than a few forest areas are be suitable for long-term maintenance of wild breeding populations of orang utans in Sabah, although Payne still described at least ten such areas. These are described and assessed separately in the following sections. The information is largely derived from Payne's 1988 report. Where the original information included a somewhat confusing range of guesses concerning ape densities, the data have been simplified and adjusted.

45 Ulu Kalumpang Commercial Forest Reserve

The 511 km^2 of the Kalumpang Forest Reserve is badly affected by heavy logging; recent satellite imagery reveals no more than some 408 km^2 of more or less degraded forest. Orang-utans are reportedly present in this area, and Payne believed that the population which he assumed to be 50 apes is now isolated from forest reserves to the west and north, and is probably doomed.

46 Tabin Wildlife Reserve

The 1,225 km^2 Tabin Wildlife Reserve is badly affected by heavy logging, in particular in its lowland and hill dipterocarp forest. It also contains some swamp forest and a few small patches of unexploited dipterocarp forest. Davies (1986) noted, 'There are about 100 km^2 of primary forest in this reserve, where moderate densities of orang-utans have been recorded, i.e. less than one animal per square kilometre.' The satellite imagery reveals that four separate sectors of forest are to be found in the area, totalling some 374 km^2. Orang utans are present in the eastern and northern parts of the reserve, supposedly at population densities typical for the habitat, but are absent in the western sector. Payne assumed that orang-utans were present at a population density of 1.2 individual/km^2 in about one-third of the Reserve, and believed the total population to comprise fewer than 530 apes.

47 Melangking

The Melanking region (655 km^2) is mostly converted for palm-oil plantations, and all the remaining stands of forest along the river are privately owned, according to Payne. There are no prospects for a formal reserve status, but the swampy conditions and the private status of this area of some 298 km^2 of closed-canopy forest showing up on the satellite imagery may well provide better protection for the remaining orang-utan population than any other formally 'reserved' forest in Sabah.

48, 50, 51. Southern Forest Reserves

The 13,683 km^2 Southern forest reserve complex consists of parts of Sungai Pinangah Commercial Forest Reserve (about 1,955 km^2), and all of Sapulut Commercial Forest Reserve (2,440 km^2), Gunung Rara Commercial Forest Reserve (2,697 km^2), Kalabakan Commercial Forest Reserve (2,324 km^2), Kuamut Commercial Forest Reserve (1,165 km^2), Malua Commercial Forest Reserve (340 km^2), Ulu Segama Commercial Forest Reserve (2,476 km^2), Sungai Imbak Virgin Jungle Reserve (181 km^2), and a number of insignificant 'Virgin Jungle Reserves' (Batu Timbang, Sungai Sansiang, Imbok, Malubuk, Brantian, Kawang Gibong, Merisuli, Sepagaya), as well as the Sepagaya water catchment reserve (26 km^2). The area contains the most extensive remaining tracts of as yet unexploited dipterocarp forest in Sabah. On satellite imagery it shows up as three separate blocks of considerable size. Virtually all the Commercial Forest Reserves comprise the 9,728 km^2 logging concession of the Sabah Foundation. Large sectors are sedimentary rock, with scattered volcanic and ultrabasic mountains of rugged topography. Marsh and Senun (unpubl.) on behalf of the Foundation note that some 20% of the concession area is scheduled to remain untouched, including the Danum Valley and the proposed Maliau Basin (Gunung Lotung) Conservation Areas. Approximately half this area is 'unworkable' anyway due to 'non-commercial forest' (i.e. heath and montane forest) and steepness. Some 7 % of the area lies below 150 m altitude, and 10 % above 1,000 m altitude. Payne assumed orang-utans are present in 60% of the area, at a population density of 0.3 individual/km^2, and believed that up to 2,680 apes are still to be found in the region.

49 Middle Kinabatangan Forest Reserves

The 2,175 km^2 Middle Kinabatangan forest range consist of parts of the Sungai Pinangah Commercial Forest Reserve (about 530 km^2) and all of the Tangkulap Commercial Forest Reserve (276 km^2), the Deramakot Commercial Forest Reserve (551 km^2), the Segaliud-Lokan Commercial Forest Reserve (572 km^2), the Sungai Lokan Virgin Jungle Reserve (19 km^2) and the Tawai Protection Forest Reserve (227 km^2). Much of the area resembles Sepilok in topography, altitude, geology and forest types, although hardly any unexploited forest has remained. The exceptions are the Tawai Forest Reserve and about half of the Pinangah Forest Reserve, which are ultrabasic hills (300–1,300 m altitude) with some chert-spillite. At the eastern end of the area, the swampy Sungai Lokan Virgin Jungle Reserve seems to support a higher orang-utan population density than does the remainder. The Karamuak River and a strip of forested State land bisects the western end of the area, but this region is still mainly under forest cover and orang utans are able to cross the river at many points. Payne assumed that orang-utans are present at a population density of 0.5 individual/km^2 throughout the area except on the lower slopes of ultrabasic hills, and somehow believed that up to 2,320 orang-utans may have survived in the area. In 1996, Isabelle Lackman-Ancrenaz began preparatory studies to develop a management plan for the flood plain of the river Kinabatangan, supported by WWF; in the meantime, conversion of forest remnants for oil-palm plantations continues, so that a badly fragmented forest structures may remain. The WWF support has not been able to arrest or reverse any of this ongoing forest destruction, but is seeking to develop some snippets of remaining land for touristic development. It is unlikely that the remnant orang-utan populations can survive in this area.

52 Trus Madi Commercial Forest Reserve

Most of the forest surrounding this approximately 1,845 km^2 'reserve' has been destroyed by slash-and-burn cultivation; only 394 km^2 of forest below an altitude of 500 m can be found on recent satellite imagery. The southwestern section rises up to 1,000 m and the lower peripheral parts of the reserve elsewhere have been logged; the remainder is scheduled for logging. There are no reliable records of orang- utans in the more mountainous western half of the reserve, but the ape occurs at lower altitudes in the eastern half. Payne assumed a population density in the eastern half of the 'reserve' (i.e. 37% of the total area) of 0.1 individual/km^2, which suggests a maximum population of 190 apes.

53 Ulu Tungud Commercial Forest Reserve

Most of the forest surrounding the approximately 1,307 km^2 commercial forest reserve adjacent to the Bidu-bidu hills has been destroyed by slash-and-burn agriculture. About 60% of the 'reserve' is in an altitude range of 150–500 m, and is covered by forest on sedimentary rocks; much of the forest has been degraded through logging. No more than 163 km^2 of forest below an altitude of 500m can be discerned on recent satellite imagery. The remainder, on ultrabasic igneous rocks, has not yet been logged. The forest on the upper, steeper ultrabasic slopes is of low stature and is affected by landslips; it seems that this forest contained very few orang-utans. Payne assumed that the orang-utans may attain a density of 0.8 individual/km^2 in the better patches, and suggested that up to 420 apes may inhabit the area.

54 Bidu-bidu Hills

The Bidu Bidu Hills Protection Forest Reserve comprises a total area of about 161 km^2, yet no more than 74 km^2 of closed-canopy forest below an altitude of 500 m can be found on recent satellite imagery. The reserve covers an ultrabasic hill range rising to 680 m, which until the 1980s was surrounded by logged dipterocarp forest and scattered settlements. Payne assumed that orang-utans are present throughout the area at a population density of 0.8 individual/km^2, and believed that it harbours up to 130 apes.

55 Sugut South

The Sugut South area is about 1,000 km^2 in size, consisting of State land and comprising parts of the Sugut and Timimbang Forest Reserves and all of the Bonggaya Commercial Forest Reserve. About 5 % of the area is swamp forest, the remainder heavily logged dipterocarp forest, mostly below an altitude of 150 m; however, no more than 25 km^2 of closed-canopy forest is to be found in the area. In the 1980s Payne assumed that orang-utans are present in 50 % of the area at an average population density of 0.5 individual/km^2 and suggested that the orang-utan population comprises fewer than 580 individuals.

56 Sugut North

The Sugut North area is about 600 km^2 insize, consisting of State land and parts of the Sugut and Paitan Commercial Forest Reserves. About 40% of the area is swamp forest, the remainder heavily logged lowland/sandstone hill dipterocarp forest, much of which was burned in 1983. Recent satellite imagery reveals no more than 57 km^2 of closed-canopy forest in the area, possibly peat-swamp forest. Poor access to the region, extensive swamps and infertile sandstone soils preclude the likelihood of extensive agricultural development affecting the area in the near future. Payne believed that orang-utans are present in 30% of the area and that the whole population amounted to fewer than 190 individuals.

57 Lingkabau Commercial Forest Reserve

Payne noted that in the 1980s orang-utans appeared to be absent in much of the northwestern portion (perhaps one-third) of the 713 km^2 Lingkabau area. Most of the forest which surrounded the reserve has been despoiled by shifting agriculture. The reserve lies at an altitude between 150 and 1,000 m on steeply dissected sedimentary rocks, and the forest has been subjected to logging. Recent satellite imagery reveals no more than 84 km^2 of closed-canopy forest below 500 m altitude. Payne assumed that orang-utans are present in about two-thirds of the Reserve at a population density of 0.26 individual/km^2, suggesting a maximum number of 185 orang-utans.

58 Kinabalu region

The largest sector of the 754 km^2 Kinabalu Conservation Area lies well above an absolute altitudinal limit for orang utans. No orang-utans appear to be present at all on the southern and western sides of the park, but they have been found in the northern extension. On the eastern side of the park, orang utans were said, in the 1980s, to be concentrated almost entirely within the sub-montane zone between an altitude of 850 – 1,200 m, occurring in an area of some 79 km^2. In the northern extension, orang-utans seem to be concentrated at the more common altitude zone of 300–1,000 m, occurring in approximately 185 km^2 of mountain forest. Recent satellite imagery reveals that no more than 207 km^2 of closed-canopy forest below 500 m altitude occurs in the region, especially between the National Park and the Lingkabau area. Payne assumed a population density between 0.2-0.8 individual/km^2 in the eastern section and between 0.05-0.3 individual/km^2 in the northern, suggesting a population of fewer than 120 individuals.

59 Crocker Range

A large sector of the uplands of the Crocker mountain range has been gazetted as National Park. About 10% of the approximately 1,400 km^2 Crocker Range National Park is considered to have been affected by shifting agriculture; Payne described some tall forest outside and contiguous with the park boundary, especially in the upper Membakut River. Payne notes that, strangely enough, the orang-utans seem to be concentrated in the upper hill dipterocarp forest – lower montane forest zone, at an altitude of approximately 850–1,200 m (about 50% of the Park's area), except in the upper Membakut River and possibly the Kimanis River, where they may occur at usual density at lower altitudes. Recent satellite imagery reveals that only some 97 km^2 of closed-canopy forest below 500 m can be found in the southern parts of the Crocker range. Payne believed that the average population density is 0.2 individual/km^2, and suggested that the orang-utan population does not exceed 160 individuals.

Furthermore, Payne (1988) has enumerated a great number of forest areas which, during the 1980s, apparently still had relict populations of orang-utans. The satellite imagery of the mid-1990s indicates a considerable reduction of forest, while the standard exclusion of all areas above an altitude of 500 as being less suitable for orang-utan habitat has meant that for a large number of areas mentioned by Payne no quantitative assessments have been attempted for the present report. Some areas for which it may be held that orang-utans have become extinct during the last decade will not be described, such as the Sapi Region (600 km^2), the Lokan Region (about 1,220 km^2), the Ulu Milian Commercial Forest Reserve (about 777 km^2) and the Gomantong–Supu Forest Reserves (90 km^2). Others deserve at least a brief description because it demonstrates what the situation was just a decade ago.

Kulamba Wildlife Reserve

The 207 km^2 Kulamba wildlife reserve consists mainly of freshwater swamp forest. Payne believed that no more than 620 orang-utans survive in this area. In 1985, nest surveys yielded a population density estimate of 3.0 individual/km^2, while subsequent surveys in 1988 suggested that the density had dropped by 75%. On recent satellite imagery the Kulamba area can no longer be discerned as forest. Presumably its orang-utan population has faded into oblivion.

Pin region

Total area: some 1,145 km^2 including the Lamag Commercial Forest Reserve. Much of the land in this region is suitable for permanent agriculture, and was allocated for that purpose during the late 1980s. The Sabah Forestry Development Authority (SAFODA) has developed a rattan plantation within 4,300 ha of logged forest in the northern part of this region. Ironically, the plantation appears to have a considerable population of orang-utans, compressed to the extraordinary density of 5 individuals/km^2, the highest density recorded in Sabah, and no doubt the reflection of refugee crowding at the time. Since the apes are likely to cause considerable damage to the rattan stock, and must in any case have overshot the carrying capacity of the remaining forest, it is unlikely that any have survived in this so-called refuge.

Tenegang – Kretam Region

Total area: about 2,300 km^2. Virtually all the land in this region is considered suitable for permanent agriculture, and the forest was allocated for conversion during the 1980s. There are five isolated tiny Virgin Jungle Reserves in the region, none of which could possibly support even a single deme. Payne did not present estimates for the doomed orang-utan population in this region, and seemed to ignore their fate, though it may well have concerned one of the State's largest subpopulations. It is unlikely that the translocation programme of the Wildlife Department could or did handle the survival of such numbers of apes as were doomed in this area.

Sepilok Forest Reserve

This forest reserve has been the site of orang-utan rehabilitation since the late 1960s. Its total area is about 43 km^2, of which 65 % is predominantly lowland dipterocarp forest and 35 % mixed sandstone hill dipterocarp and heath forest. Sepilok is virtually isolated, and surrounded by gardens and plantations, but it is still connected to a narrow strip of dry land forest along Sandakan Bay where it links up with the Elopura Mangrove Forest Reserve, which is unsuitable habitat for orang-utans anyway. Payne assumed that orang utans occur in 65% of the reserve at a population density of 1.2 individual/km^2 and reported that the orang-utan population comprises fewer than 55 apes. It is noteworthy, however, that since the 1960s at least 200 and possibly up to 400 orang-utans must have been rehabilitated in this area, and one wonders where these apes could have gone, if they have not simply vanished forever.

Thus, in conclusion, orang-utans still occur in the lowlands of eastern Sabah, and the hill country in south-central Sabah. Yet, most of the apes occur outside the few conservation areas, and even if they manage to avoid being captured and killed by poachers, they will nevertheless slowly and silently starve to extinction as a direct result of the ongoing massive forest conversion.

Section II: orang-utan distribution

> Already during the 1970s orang-utans were reportedly absent in some lowland regions and in forest regions fringing areas of human habitation throughout western Sabah. Subsistence farming and hunting appear to have been the main reason for their absence. However, in two lowland areas of southeastern Sabah, Payne identified some 'apparently natural gaps' in the distribution of orang-utans. He noted that these gaps cannot easily be explained by human interference, because historically there is neither evidence of *ladang* disturbance nor of traditional hunting communities. For instance, the gap on the southeastern side of the lower Segama river, which has recently been gazetted as the western side of the Tabin Wildlife Reserve and the Silabukan Forest Reserve. However, archeological finds prove that the river Segama and many other rivers in Sabah have been systematically explored by Chinese gold-prospectors since the 18th century. As in other regions where exploration and digging for gold occurred, it cannot be ruled out that mixed bands of Chinese and indigenous gold-prospectors along the rivers had such a serious hunting impact that they eradicated the ape. Also in much of the Semporna peninsula and the Tawau region a gap is discernible in the orang-utan's distribution pattern. This area was under natural forest cover until the 1950s, and it is as yet impossible to assess possible human impact in this area.

It is noteworthy that orang-utans reportedly were found in 'remarkably high concentrations' in some patches of secondary forest, e.g. along the river Kinabatangan (Scott, pers. comm. 1992). This is not remarkable at all in view of the total obliteration of the surrounding habitat. Orang-utans will not immediately disappear (or die) after their habitat has been demolished by conversion. They will crowd together in any patch of vegetation, possibly in the vain hope that if anywhere in the newly created desert, a clump of trees and shrubs is the most likely place to find something to eat. They will simply waste away in such secondary forest patches,

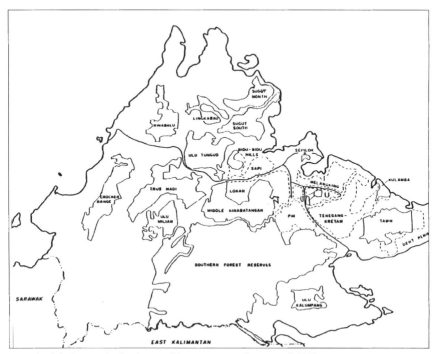

Map of Sabah showing the distribution of the orang-utan; after Payne (1988).

because there is no other option. It is truly remarkable, however, that they can hang on to life for several years under such miserable conditions.

In the late 1980s, Payne believed that between 9,800 and 21,000 orang-utans occurred in Sabah, of which the majority were to be found outside the permanent forest estates (Payne, 1988). Again Payne applied the simple 'extent of forest times density' calculation, and although he used conservative estimates of density, he seems to have made the common inflation-related errors. Since the publication of the report, very large areas of forest have been converted into plantations, and although the Wildlife Department has translocated more than one-hundred displaced orang-utans from cleared areas to the Tabin Wildlife Reserve since 1993 (Payne and Andau, 1994), it is obvious that many more apes have been and continue to be sacrificed to development, unbeknown to the public at large.

Brunei

General description

The sultanate of Brunei is a small (5,765 km^2) coastal settlement and its immediate forest surroundings, on the northwestern coast of Borneo, straddle the boundary of Sarawak and Sabah. It was the seat of the Malay sultans, who more or less ruled Sarawak's coastal settlements until the mid-nineteenth century. In 1841, the sultan was obliged to cede virtually all his coastal territory to an adventurous former officer of the English East India Company, James Brooke, who established a hundred-years rule of 'white rajahs' in Sarawak. Only the tiny coastal territory of Brunei was retained. Brunei has a predominantly urban population of some 300,000 people, deriving its income from a huge coastal oil reserve and international trade.

No orang-utan distribution

Much of the tiny state is still covered by rainforest (4,692 km^2). However, there are no reliable records to suggest that a population of orang-utans exist in Brunei Darussalam. The rare records of sightings, in the remoter parts near the border with Sabah, all concern solitary males (Payne, pers. comm.). The forests of Brunei have been, and still are, within the range of the nomadic Penan who freely enter the territory from Sarawak.

> In the Temburong district, the eastern lobe of Brunei's territory, several people have reported sightings of a large ape-like primate, other than the common Bornean gibbon (A.C. Sebastian, pers. comm.). The reports are reminiscent of the mysterious *orang pendek* of Sumatra. Among those who allegedly saw this ape were members of the Royal Geographic Expedition who studied the ecology of the Belalong area, in the Temburung district, under the leadership of the Earl of Cranbrook (formerly Lord Medway). Yet the excellent report of the explorations, entitled *Belalong, a tropical rainforest* (Cranbrook and Edwards, 1994) makes no mention of such mysterious sightings. Furthermore, the extraordinary high density of gibbons (5.3 groups per km^2), having an atypical 'harem' organisation rather than the common monogamous pair-structure, and of other primate competitors for food in the Belalong forest, suggests that the area has no resident large apes.

Sumatra

General description

In geological terms Sumatra can be differentiated into three major zones. The long Barisan mountain range stretches all along the western half of the island. Along the central axis eastwards it is straddled by a low, undulating, 40 km wide Piedmont – or foothill – transition belt which traverses the entire island. On the eastern side the Piedmont zone drops to an altitude below 50 m to extend in a wide Peneplain belt of swamps, flood plains and mangrove coast (Scholz, 1983). The consistency of this major geomorphological structure was interrupted some 75,000 years ago by the explosion of a large volcano, which left the Lake Toba and surrounding plateau behind.

A few regions of Sumatra have been inhabited by humans for a very long time. Paleolithic migrants settled along the east coast and along the larger rivers perhaps as many as 80,000 years ago; stone implements have been found buried beneath the ash of the Toba eruption (Whitten, *et al.* 1984). As a matter of fact, the finds of stone tools throughout Sumatra indicate that the early migrants predominantly settled where the large rivers (notably the rivers Baruman, Tembesi and Musi) entered the central foothill zone. Whereas the first migrants were typical hunter-gatherers of an archaic mainland type, they were soon displaced by a mixture of technologically more advanced peoples arriving by boats. Ethnic and cultural traces of many different peoples of Asia and the Pacific can still be recognised in the communities of Sumatra. The aboriginals retreated into the less accessible mountains to the west and the swamps to the east of the foothills.

The famous waves of Hindu and Buddhist colonial civilisation, around the millenium, again occupied the most accessible regions of the Piedmont zone. In the sixth century 'the empire was in full cultural development' (Loeb, 1972). The South Asian theocracies explored and developed trade routes across the island, opened up some suitable harbours along the west coast, and eventually facilitated the growth of cultural identities in some ethnic communities, e.g. the Minangkabau and Batak, through admixture and the enforcement of sedentary agriculture.

In about 1400, a new colonial wave swept over the civilised regions of Sumatra when seafaring Muslim traders from the Persian empire acquired a foothold in the major harbour settlements. Again some original inhabitants retreated into the inaccessible interior of the mountain range (e.g. the Gayo in Aceh). Finally, the Dutch colonial power gradually began to exert its influence through the same harbours towards the end of the nineteenth century.

The Dutch colonial influence in Sumatra was of short duration, but for some regions it brought about fairly dramatic developments. If, at the close of the nineteenth century, much of the island was inaccessible and virtually unexplored, the colonial influence soon initiated major conversion of the lowland forests for cash-crop plantations, developed an infrastructure and translocated and imported large numbers of people. After the 1960s, the Indonesian government expanded the development, causing a spectacular further demolition of the forest cover which is nearing completion at the end of the twentieth century. In 1975, less than 42% of the island was still covered with forest, but in 1991 the World Conservation Monitoring Centre noted, 'Sumatra probably continues to lose its natural vegetation faster than any other part of Indonesia' (Collins et al., 1991).

Now, in the late 1990s, almost all Sumatra's lowland forest east of the Barisan range has disappeared, while the few remaining forest fragments are being demolished, scorched and encroached. During the 1997 drought the pillaged State forest lands were affected heavily by arson, especially in the central and eastern lowlands south of the equator. The distribution range of the orang-utan has largely been spared, however, except in a few forest areas adjacent to transmigration projects in Tapanuli.

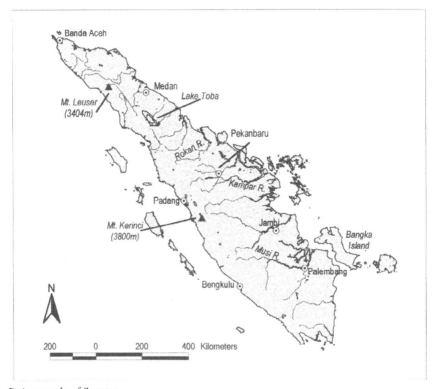

Basic geography of Sumatra.

Orang-utan distribution in Sumatra

Following the publications of the 1930s, it was generally held that the distribution range of the Sumatran orang-utan was restricted to Sumatra north of the equator, if not north of Lake Toba, and was concentrated particularly in the Gunung Leuser wildlife sanctuary. At the time this population was already fragmented into four major subpopulations: (1) the contiguous population of Aceh, west of the river Alas and the river Wampu, (2) the subpopulation of the protection forests Dolok Sembelin and Batu Ardan in Dairi district and the contiguous forest expanse east of the river Alas, stretching along the foothills of the westcoast down to Sibolga bay (3) the subpopulation of southeastern Tapanuli between the river Asahan and the river Baruman, and (4) the subpopulation of Anggolia, Angkola, and Pasaman, all along the western foothills of the Barisan range, from the lower catchment of the river Batang Toru stretching southward between Padang Sidempuan and the environs of Pariaman in West Sumatra province, some 50 km north of Padang.

1 The North Aceh population

Carpenter (1937) mistakenly reported that the most northern boundary of the ape's range was a rough line between the city of Lamno on the west coast and Sigli on the northeastern coast. The present survey corroborated earlier reports that orang-utans occur even farther north; sightings have been reported from Gunung Kulu, Gunung Geureunthe and Paro. The surveys have also confirmed that orang-utans still occur in the hinterland (Lengah forest) of Banda Aceh, and on the slopes of Gunung Seulawah; the Taman Hutan Raya at Jantho and on Gunung Seulawah had a high density of orang-utan nests during the survey of February 1997. The present survey revealed that orang-utans can also still be found in most of the forested lowland areas and lower slopes (especially below 1000 m) on the western side of the mountainous province; the total extent of the forest below an altitude of 1000 m is 1167 km^2.

2 The Woyla catchment

A large area of forest below an altitude of 1000 m (1793 km^2) is found in the large hill-forest complex east of the headwaters of the river Teunom; the forest has unfortunately been allocated entirely for timber exploitation, while control is virtually absent. The lowland area is fragmented by roads connecting at least eight villages; a major road traverses the coastal Barisan mountain chain between Meulaboh on the coast and Geumpang in the central highlands. Many timber concessions operate in the forest.

3 The Gayo mountains

The river Woyla separates the large forest area in the north from its southern extension into the Gayo mountains, due east of Lake Tawar. Also in this region many roads and tracks traverse the forest, and timber exploitation is having a devastating impact upon the habitat of the ape.

The central Gayo valley

Due south of Lake Tawar near Takengon in the Central Gayo mountains the ape was found in three locations where forest in wider valleys remained, as well as in the forested valleys of the Serbeujadi mountain chain, which is the northern area of the Leuser Ecosystem. The extent of habitat in the forest at this higher altitude is unknown and hence will not be included in this report.

4 The Bahbahrot swamp

Of the two 'centres of distribution' which were recognised by Carpenter (1937) in Aceh, the most important centre supposedly was the extensive coastal swamp-forest area between the city of Meulaboh and the township of Blangpidie on the west coast. The northern half of this swamp area was reclaimed in the 1980s, the southern half, from the river Teripa to Blangpidie, currently known as the Bahbahrot swamp, was still largely intact until 1991. However, since the early 1990s, the southern swamp forests are also being largely reclaimed for massive oil-palm plantations and transmigration projects, although some 40% of the area is included in the Leuser Ecosystem (since 1995) and should be destined for further protection. There is precious little chance, however, that the remnant populations of the Bahbaraot swamp forests can be saved from the ongoing drainage, reclamation and poaching.

Satellite image of the coastal region of the northwestern section of the Leuser Ecosystem (Bahbahrot swamp, Aceh); the coastal swamps have been pillaged and converted since the early 1990s; Mount Leuser in the lower right-hand corner.

5 The Leuser population

In the early 1970s, Rijksen (1978) estimated that the Gunung Leuser Wildlife Reserve complex held anything between 2,000 and 5,000 orang-utans, mostly around its fringes; in 1990, van Schaik and Azwar (1991) estimated that the population within the confines of the reserve amounted to anything between 5,000 and 7,400 individuals. Despite the consideration that only some 25% of the

9,000 km² large reserve was lowland between sea level and an altitude of 1,000 m (Rijksen, 1978; Laumonier, 1989), it was insufficiently realised that not all the 2,250 km² lowland forest represented habitat for which a standard density could be applied. In fact the orang-utans are concentrated in habitat patches which cover anything between 35% (dryland) and 50% (swamp) of the area, so that the total number of apes within the confines of the reserve probably did not exceed 2,000 individuals. Be that as it may, since the late 1970s much of the reserve's lowland areas have suffered severe forest degradation, encroachment and conversion. As a matter of fact, the original wildlife reserve was ill-designed for the conservation of the north Sumatran mega-fauna; orang-utans and elephants found their habitat mainly outside its boundaries.

Earlier surveyors overlooked one of the most important centres of distribution in southern Aceh, because it was situated deep in the, what was then, more remote southern region of the Gunung Leuser Wildlife Reserve. It is the extensive lowland plateau of the Bengkung River water catchment. Until 1992, it was a major centre of biological diversity containing at least five important nutrient licks (*uning*), where all the larger wildlife on the reserve used to convene at times. In spite of its long protected status, official requests to the Ministry of Forestry (Rijksen, unpubl.) for acknowledgment of this status, and intensive awareness campaigning among authorities, the area was nevertheless readily allocated for timber exploitation in 1992, at the recommendation of the then Directorate-General PHPA. No control of the timber operations has been carried out, and the area has been under serious threat of conversion and transmigration settlement by the small Aceh Tenggara regency, which is striving to enlarge its area of jurisdiction and population by usurping parts of the conservation area on State forest land. Hence, much of the forest in the Bengkung area has been badly demolished, and it will require a tremendous effort and large investments to revert the area back to its original function, as a crucial part of the Leuser Ecosystem. It will take at least forty years before the carrying capacity of the region will return to anywhere near the original capacity for orang-utans.

Until around 1990, the Bengkung plateau formed a continuous forest with the coastal swamps between the river Kluet (or Kandang) and the mouth of the river Alas (or Simpang Kiri) at the village of Singkil.

The highland plateaus

Two very important highland plateaus, at altitudes of over 1000 m, occur within the Gunung Leuser complex of wildlife reserves (i.e. the formerly designated national park), namely the upper Mamas valley and the Kappi plateau. Due to the *massenerhebung* effect, the ecological conditions of these plateaus resemble those at a lower (sub-montane) elevation and it is therefore perhaps not surprising that orang-utan sightings at altitudes over 1000 m have been recorded for some sectors of the plateaus. The total extent of these upland regions is approximately 1,300 km², and it may be surmised that together they contain less than some 300 km² of habitat. However, for consistency concerning the 1000 m altitude limit regarding orang-utan distribution in Sumatra, these extremely important highland plateaus are considered to be outside the scope of the present report. Although the areas are reasonably well protected due to their remoteness and poor accessibility, and timber poaching is virtually absent, illegal small-scale forest conversion for the production of *Cannabis sativa* (marihuana) is common, as well as the poaching of fish and rare birds.

Transmigration project at the river Bengkung in the Leuser Ecosystem; a deliberate attempt to demolish the unique expanse of one of the last lowland rainforest ecosystems in Sumatra (1997).

6   The Singkil swamp

The coastal swamp forests of the western bank of the river Alas – Simpang Kiri, known as the Singkil swamp, was until recently one of most pristine forest areas of North Sumatra. It is the region from which the orang-utan originated that was described by C. Abel in 1826. The approximately 1,000 km^2 area has been formed by the interactions of the river depositing its silt, and coastal accretion, while its forest cover has deposited a layer of peat of varying depth over this mixed basis of volcanic ash (from the Toba eruption), laterite, clay, sand and erosion sediment. The area is of tremendous ecological importance as a flood plain of the river Alas, and may in particular have an important function as a breeding ground for coastal fish and shrimps. Due to its complex soil composition and the almost permanent availability of water, the swamp forest is exceedingly rich in plant species, and appears to contain a relatively high incidence of social arenas, supporting the highest densities of orang-utans ever recorded (C. van Schaik and A. Sitompul, unpubl.).

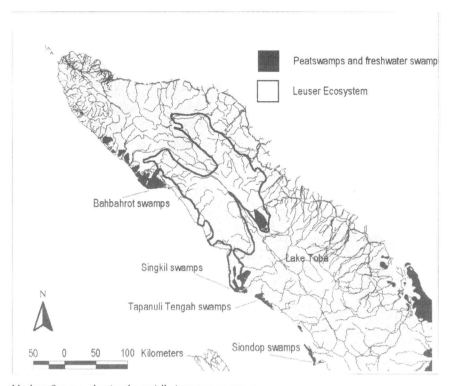

Northern Sumatra, showing the crucially important swamp areas.

The Singkil swamp is the most southern extension of the Leuser Ecosystem. However, recent developments, including the establishment of a major road which separates the swamp from the dry ground of the Kaparsesak and Bengkung plateau, ill-planned allocation of a large oil-palm plantation, the allocation of two timber concessions in the early 1990s, uncontrolled forest destruction, large-scale timber poaching and drainage and transmigration settlements, have effectively severed a major subpopulation of orang-utans from the dryland forests in the north. This demolition and conversion is condoned by the local government and will shortly also obliterate much of the swamp habitat if the Leuser Development Programme is unable to assert its management authority. Despite three years of awareness campaigns among authorities, which resulted in verbal agreements for support, officially sanctioned destructive developments continued to threaten the swamp ecosystem in 1998. It is hoped that the recently established status of wildlife reserve granted by the Minister of Forestry (1997) will help safeguard this unique swamp area and its considerable orang-utan population.

7 Peusangan catchment

Between the mouth of the river Peusangan and the river Jambu Aie is a wide bay, which is called the bay of the orang-utan, *Lhok Seumaweh* in Acehnese. Here the foothills of the Gunung Geureudong (2855 m) meet what used to be a shallow coastal flood-plain swamp, which must once have provided excellent habitat conditions for orang-utans. Now, the lowland forests along the northeastern coastal zone have mostly been converted into agricultural fields and plantations, and the ape has become extinct in this zone. Only two reports of the presence of orang-utans refer to the foothills on the northern side of the mountain chain between Banda Aceh and Bireuen; it was noted that orang-utans still occur on the southern slope of Gunung Bateekeubae (2756 m) south of Bireuen; the reference concerns the valley of the upper headwaters of the river Woyla and may suggest that this valley has been or still is a link between the west and east coast across the central Barisan mountain chain. On the eastern side of the river Peusangan (and the main road Bireuen–Takengon) a few reports indicate that the ape still occurs in the forested upland area between the two large rivers (Peusangan–Jambu Aie). Much of the forest has been given over to several timber concessions, and three small-scale transmigration projects have been planted in the middle of forest areas, apparently in order to further demolish the forest estate and facilitate 'development.' Across the watershed to the south the valleys widen out and wherever forest patches are left, orang-utans have reportedly been heard and seen. Much of the forested upland area between the road Bireuen–Takengon and the river Jambu Aie is included in the Leuser Ecosystem.

8 Jambo Aie catchment

Satellite imagery shows that much of the lowland forest of the river Jambu Aie catchment has been logged, but the canopy of the forest expanse is still largely intact; the timber concessions are currently moving into the higher slopes and highland valleys of the remaining 2,000 km^2 of forest. In the 1930s, Carpenter had already reported that orang-utan sightings were common, and recent reports indicate that the ape is still present. The Jambu Aie catchment area is included in the the Leuser Ecosystem, and bears some resemblance to the Bornean lowlands, a vast expanse of low undulating terrain, and hence it probably does not have the same density of top-habitat patches as many other places in western Sumatra. At least four major timber concessions are operating in the area, covering all the remaining forest, and there are plans to convert all the lowland areas into plantations. Since the area is included in the Leuser Ecosystem, the Leuser Development Programme will take into careful consideration whether any plan for conversion, and settlement will not reduce the opportunities for sustainable

9 Tamiang

In the 1930s, Van Heurn had already reported on the importance of the remote forest area south of the river Simpang Kanan (near Lokop), currently known as the Tamiang area. The forests of southeastern Aceh were contiguous with the large protected Sikundur lowlands across the river Besitang, along the Tamiang plateau. The forests also continued in a southwestern direction, across the volcanic Kappi plateau into the Alas valley. As a consequence the Tamiang area is one of the most important regions of the Leuser Ecosystem. Orang-utans still occur in the logged-over forest expanse of this vast area; logging continues in the area but the concessions are under close scrutiny of the Ministry of Forestry, through the vigilance of the Leuser Development Programme, and may be closed down if they transgress. A new road has opened up access to the area. This road, running west from Lokop, will soon link the township of Blangkedjeren in the Gayo highlands with the east coast.

10 The Langkat-Sikundur population

Until the 1970s, The Sikundur lowlands stood out as a protected lowland forest peninsula in the sea of plantations which occupied all the lowland area of Langkat and Deli. In addition to the Bengkung area, the Sikundur lowlands were one of the two major centres of distribution or strongholds for the former Gunung Leuser Wildlife Reserve orang-utans. The gently undulating area contained a great number of social arenas, and the unique lowland forest was renowned for its orang-utans. In the early 1970s, the Sikundur area was given out as a timber concession for 'habitat improvement' (according to the poster which was issued by PHPA) and in order to prepare the road alignment of the trans-Sumatran highway, the plans for which indicated that it happened to traverse the lowland forest area. After uncontrolled logging had ransacked the area, the plan for the highway was revised to re-align it along the existing main road. Orang-utans can still be found in small numbers, surviving in the pockets of regenerating forest in the foothills of the mountain chain to the west (van Schaik, unpubl.). The regenerating secondary forest of the Sikundur area is under continuous threat of encroachment, however; in 1994, the local press exposed a land speculation scandal with respect to some 6,500 ha of cleared secondary forest inside the designated Gunung Leuser National Park. A group of some 15 local companies had cleared the land on the basis of forged letters and sold shares for an oil-palm Nucleus Estate to at least 34 prospective owners (*Analisa*, Dec. 17, 1994; *Kompas*, Dec. 29, 1994).

The eastern slopes of the Serbolangit range in Langkat still contain orang-utan demes in the forested valleys; these sub-populations are contiguous with those of the western slopes on the eastern side of the Alas valley and the Kappi area in Aceh province. The Bohorok rehabilitation centre has reintroduced almost two-hundred rehabilitated apes into this population since 1973, and these rehabilitants have displaced the original population in an area of at least 200 km^2 around the Bukit Lawang rehabilitation site.

11 Alas valley

Until the 1920s, the orang-utan population in the Alas valley must once have been among the most important contiguous subpopulations, while the approximately 250 km^2 of alluvial forest must have contained a high density of social arenas. For all larger mammals, notably elephant, rhino, tiger and orang-utan, the valley was the major connection between the 'heartland' of their distribution in the southwestern corner of Aceh and the great eastern 'heartland' of the Sikundur-Tamiang-Jambu Aie catchment, across the Kappi plateau. At present, the once contiguous population in the centre of the Leuser Ecosystem, across the upper Alas valley, has been divided into a western and an eastern subpopulation, because since the 1980s the forested flanks of the valley have been almost entirely converted into *kemiri* (*Aleuritis moluccana*) and coffee (*Coffea arabica*) plantations by illegal settlers. No more than 5% of the ape's quality habitat in the valley remains, notably at the confluence of the river Alas and its tributary, the river Ketambe, which has been a well-protected international field-research area since 1970.

By the 1990s virtually all the alluvial forest habitat in the Alas valley had been converted.

> In northern Sumatra, within the supposedly protected confines of the designated Gunung Leuser National Park, almost all the (significant) lowland sectors containing high-quality habitat have been despoiled due to ill-planned and uncontrolled logging and timber poaching, as well as translocation and uncontrolled settlement during the past two decades. These include the Sikundur-Besitang and Bengkung-Kaparsesak regions, as well as the lower Kompas-Silukluk, and the lower Kluet-Leumbang areas. Even the narrowing alluvial valley of the river Alas above an altitude of 300 m has been occupied by squatters. In 1987, the Directorate-General PHPA notified the local government that the valley would be considered a buffer zone in the modified designated national park status of the area (SK Directorate-General PHPA 48/Kpts/DJ-VI/1987). This resulted in a massive influx of settlers in the narrow valley, who cleared the forested slopes for their *ladangs*, and established several new villages. In the entire Alas valley, which crosses some 80 km of the conservation area, orang-utans have only one single remaining social arena (Ketambe), of the approximately eight which existed in 1970 and the more than twenty which must have existed around the turn of the century.

12 East Alas

The large lowland area between the eastern bank of the river Alas (Simpang Kiri) and the Barisan mountain chain has been a poorly known stronghold of orang-utan distribution in Sumatra. Until the 1960s, it was quite extensive, covering well over 3,000 km^2 of continuous, uninhabited rainforest, which in turn was linked with the greater Aceh population of the Gunung Leuser Wildlife Reserve at the lower end of the Alas valley in the north and proceeded southward into the currently severed Tapanuli Tengah forest (15). In 1971, J.R. MacKinnon studied apes belonging to this sub population at the river Renun in the Dolok Sembelin Protection Forest which used to be part of the East Alas system. Since the early 1980s, the once continuous lowland forest has been fragmented and largely converted for large-scale industrial plantations and transmigration settlement, so that only about 175 km^2 of forest remained in 1995. Its orang-utan population is doomed. The Sembabala I and II Protection Forests have already been ransacked by uncontrolled logging and have been overrun by unscheduled encroachment. A small number of orang-utans have managed to survived in a narrow strip along the river Alas.

During the drought of 1997, this area was one of the few sites within the distribution range of the Sumatran orang-utan that was significantly damaged by arson. As in Kalimantan, in particular the forest areas adjacent to major transmigration sites were badly affected.

13 Dolok Sembelin

Some orang-utan demes of the once continuous expanse of the East Alas forest survive in a few isolated, still forested fragments in the foothills of the Barisan range, in particular some sections of the southern half of the Dolok Sembelin Protection Forest. In colonial times the Dolok Sembelin forest block was proposed repeatedly for a strictly protected status because of its extraordinary value in terms of biological diversity; this eventually resulted in the area being accorded a weak Protection Forest status. In modern Indonesia, Protection Forests are often not exempted from timber extraction, and encroachment is usually unhampered.

However, it is probable that due to the total lack of protection the remaining 148 km² of unique Protection Forest will be deforested by encroachment or further unscheduled invasions for timber extraction. It is a hopeful sign, nevertheless, that in 1996 the Diari regency authorities applied for inclusion of the Protection Forest into the Leuser Ecosystem, in order to have better opportunities for conservation. Also along the fringes of the Dolok Sembelin Protection Forest, arson resulted in some damage during the 1997 drought.

14 Batu Ardan

Like the Dolok Sembelin forest, the 200 km² Batu Ardan area has been accorded a Protection Forest status, but it is equally vulnerable to further timber poaching and encroachment. Orang-utans are still said to be present in this area. Some sections of the area were badly affected by arson during the 1997 drought.

15 Tapanuli Tengah

Between the coastal townships of Singkil and Barus lies an inaccessible coastal swamp area. The forest still covers the foothills inland, across the watershed to the river Simpang Kanan, a tributary of the river Alas, which runs parallel to the coast, and across the river up to the remote, deeply incised hill country of the northwestern Toba plateau.

Orang-utans were found in the area in the early 1970s (Rijksen, 1978), and are still reportedly common. With its 1,564 km² of forest cover, the area is the largest section of the fragmented range of what was, until the 1970s, the Western Tapanuli population. Small separate fragments of this section in the north (e.g. the 56 km² large, degraded Siranggas Nature Reserve, along the river Simpang Kanan), come close to the Batua Ardan Protection Forest; fragmentation is due mainly to timber roads and uncontrolled encroachment. Parts of the Tapanuli Tengah region have been proposed for protected status (*Cagar Alam*) by the regional PHPA office, namely some 21 km² of the coastal flood plain of the Aek Tapus river and 91 km² of the Barus Barat hill range. Notwithstanding the good intentions, however, it must be noted that such areas are far too small to afford any orang-utan deme the remotest chance of survival. Yet ample opportunity still remains for the authorities to establish a protected nature reserve of significant size and quality for orang-utans to survive in this region.

16 Kalang – Anggolia

Just north of the Bay of Sibolga some 188 km² of hill and mountain forest below an altitude of 1000 m remains in which orang-utans still occur. It comprises an almost continuous forest patch in rather inaccessible country which, until some two decades ago, formed a link between the East Alas -Tapanuli Tengah range and the upper Batang Toru valley extending into the Angkola range. At present the link is severed by several roads and an expanding network of logging tracks. Logging in the area is continuing, although some hillsides have been proposed as Protection Forest or wildlife reserve by the regional PHPA office.

17 Simonangmonang – Sibualbuali

The mountain complex due east of Sibolga Bay, as far as the valley of the river Batang Toru, has peaks which just reach the 1000 m altitude; much of the area has suffered poorly controlled logging as well as encroachment, yet orang-utans were found to occur in the forested valleys and hillsides. This is perhaps the *Silindoeng* area for which van Heurn reported the presence of orang-utans during the 1930s. The area is known as the *Pegunungan* Simonangmonang. Some 350 km^2 of the region within the triangle of roads between Tarutung, Sibolga and Padangsidempuan has been proposed by the regional PHPA as strictly nature reserve (*Cagar Alam*) Batang Toru. The area is under concession contract and logging operations are proceeding, with the extraction of over 20 truckloads of logs per day from the mountain slopes. The proposed area is separated by the river and a road from the tiny and degraded 50 km^2 *Cagar Alam* Sibual-buali. For conservation of the orang-utan population in the region, it would be of great importance to control the logging operations very rigidly, and to join the two conservation areas across the river, while adding a considerable stretch of alluvial valley forest along the river Batang Toru.

18 Sipirok

Due north of the township of Sipirok lies the catchment of a small tributary of the river Batang Toru. Much of the area is over 500 m high and rises up to an altitude of over 1800 m. Yet, some 50 km^2 of the area is covered with forest below 1000 m, and orang-utans reportedly do occur in the area. Approximately 70 km^2 of the area is gazetted as *Cagar Alam* Dolok Sipirok. During the 1996 survey, bird traders near the township of Sipirok offered to catch an orang-utan infant for sale within three days.

19 Angkola wilderness – Siondop

Formerly the Angkola population was contiguous with the East Alas population along a narrow strip of coastal plain- and foothill forest between the hinterland of Barus and the lower stretches of the river Batang Toru due south of Sibolga. The present surveys revealed that a fair population of orang-utans survives in the extensive forests of the coastal flood plain and foothills of the Tapanuli Selatan regency, notably in the 2,327 km^2 of partially logged-over Protection Forest of the Siondop area along the Batang Gadis River. The Siondop area is a unique ecotone between the mountain forest of the Barisan range and the coastal plains; its geomorphology has probably been influenced considerably by the Toba eruption, and it contains a unique tectonic *graben* structure which has developed into an upland swamp in the valley of the Batang Gadis River. Some 60% of the forested area has Protection Forest status, while the remainder has been formally allocated as a permanent forest estate for sustainable timber extraction. Yet virtually all the forest area has been subjected to selective logging by some six timber concessions, and one small area has been converted for timber production (HTI Siondop Jati Lestari) on the illegal basis of a temporary permit. Along the coast some people from the island Nias temporarily settled before moving inland in search of work. Nevertheless, the Siondop-Batang Gadis catchment and its coastal plain comprise some 3,500 km² of unique forest (Meijaard, unpubl. report 1996). This area should soon receive a fully protected status, because it harbours, along with a considerable selection of other spectacular wildlife, the southern Sumatran type of orang-utan. The Siondop area offers realistic opportunities for development into a national park of world standard.

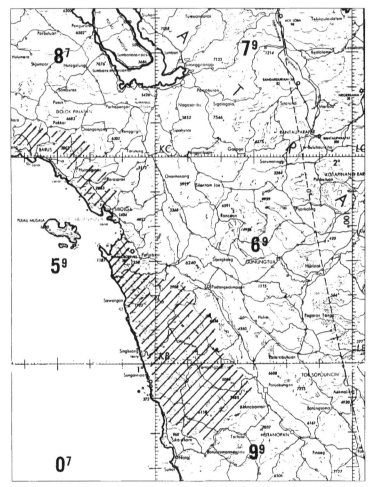

The unique Angkola-Siondop area on the western side of the Barisan mountain range could become a nature reserve of world standard; the crosshatching indicates the distribution of the orang-utan.

20 West Pasaman

At the border of the provinces of North and West Sumatra lie two major fragments of forest in which orang-utans occur. First the extensive coastal flood-plain swamp of the river Bantahan, in which some 190 km² of forest has survived, and second, another important fragment (281 km²) further east in the upper valley of the river Bantahan in the (southern) foothills of the Dolok Soposalak (753 m) west of the upper reaches of the river, and extending across the river eastwards into the (western) foothills of the Gunung Malinting (1983 m). These areas should be made a wildlife reserve, despite the logging damage already inflicted. Further into the Pasaman region another three small patches of forest in which orang-utans reportedly occur (total 242 km²) remain in the coastal zone; of these fragments only the swamps (some 170 km²) between the coastal villages of Sasak and Tiku may be considered relevant for the present survey. Since this coastal swamp is less than 100 km away from the city of Padang and a good coastal road exists, the area may still bear great potential for development into a national park.

21 Baruman range

Due east of the Angkola valley lies the Baruman mountain range with peaks of just over 2,000 m. Some 403 km² of this range has been established as wildlife reserve (*Suaka Margasatwa*). The forest area extends southwards into West Sumatra province, so that a more or less contiguous forest block of some 3,000 km² (below 1000 m altitude) stretches across the equator. These forest areas have the status of Protection forest. Only few reports on the presence of orang-utans were collected, but because the Baruman range and its southern extension of intact protection forest is situated in an Islamic region, it is probable that a significant population of orang-utans has survived in this area. The inaccessible eastern side of the Baruman range is the water catchment of tributaries of the river Rokan. Dammerman (1932) describes the mountainous Rokan district as one of the places from which reports of the *orang pendek* are most frequent.

22 Rimbo Panti – Gunung Talamau

In the late 1980s the French botanist Y. Laumonier saw an orang-utan and numerous nests during his botanic survey of the slopes of Gunung Talamau (Ophir) due west of the township Lubuksikaping. The place is less than 80 km away from the caves where Dubois found the first semi-fossilised teeth of the ape in his search for the missing link between man and apes. Already in the 1970s, the head of the regional forest division, K.S. Depari, insisted that orang-utans were still to be found in the Rimbo Panti reserve, where he had seen them during the 1960s. The originally strictly protected nature reserve (*Cagar Alam*) Rimbo Panti has since been degraded by illegal timber extraction, the cutting through of a wide irrigation channel, the Trans-Sumatra 'highway' and ill-planned touristic development for access to a hot-water spring. Some 147 km² of more or less degraded forest remains for orang-utan habitat patches. Large numbers of nests provided evidence that the apes were still present in the battered forest in 1997. The reserve is contiguous with protection forests to the west as well as to the northeast. However, the Protection Forest in the west covers a steep and high ridge (1400 m) which is apparently an effective barrier to keep orang-utans from crossing and wandering in a westerly direction, along the slopes of Gunung Talamau to the lowlands on the west coast, and in a northerly direction towards the Angkola region. Across the irrigation channel and main road to the east the reserve is contiguous with the large mountainous Baruman range in the north, along a very narrow strip of hill forest below 1000 m.

23 Habinsaran (southeast Tapanuli)

Of the once very extensive forests on the fertile foothills between the river Asahan and the river Rokan in Riau, southeast of Lake Toba, only a few patches are left. According to van Heurn, orang-utans still occurred in this region during the 1930s, and in the early 1970s an orang-utan was seen in the Asahan gorge (Rijksen, 1978). Large-scale developments in this area, in particular forest conversion for industrial oil-palm and rubber plantations, as well as (trans) migration and extensive cattle ranching, have resulted in the obliteration of much of the forest in this area. Only one wildlife reserve was established in the area, namely the Dolok Surungan Wildlife Sanctuary (238 km^2). It is under heavy pressure of encroachment, and its lowlands have been converted into an oil-palm plantation. The reserve is supposedly protected by one guard. Yet towards the south, satellite imagery reveals that the reserve is still contiguous with a much larger area of more or less degraded forest stretching all the way into the swampy flood plain of the Baruman River. However, it is doubtful whether any significant number of orang-utans have a chance of survival under pressure of the ongoing developments in the region. The local Batak people are known to have hunted and eaten the ape.

It is possible that a relict population survives near the provincial border with Riau province, in particular in a few inaccessible remnants of peat-swamp forest. During the 1996 survey, an informant in Langgapayung on the Baruman River insisted that orang-utans could be hunted in the swamp forests along some stretches of the water.

Thus, the once contiguous population of orang-utans north of Lake Toba has been fragmented into at least eleven rapidly shrinking subpopulations. The subpopulations of the immensely important coastal swamps are doomed if further reclamation and forest destruction for conversion cannot be halted and the forest links with the larger Leuser Ecosystem population cannot be restored. If the destructive developments on the Bengkung plateau cannot be stopped and reverted, the most important original Leuser population will also be decimated.

Fragmentation of the meta-population south of Lake Toba is still insufficiently known because the precise whereabouts of all surviving relict populations along the flanks of the Bukit Barisan mountain chain and in the extensive, yet badly demolished eastern plain swamp-forests have not yet been discovered. However, already more than thirteen fragments are known, of which only three may be large enough to offer the ape any chance of survival, if properly protected. Since some of

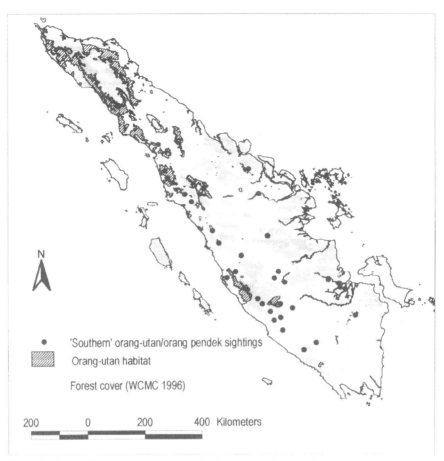

Present distribution of orang-utans in Sumatra, also shown are the forest patches in central and southern Sumatra, which may still contain relict populations of orang-utans.

the swamp-forest fragments of Sumatra's eastern flood plains are still fairly extensive, it is of particular importance to survey these areas for the possible occurrence of relict populations of orang-utans. Notably the coastal swamp forests between the rivers Bila and Rokan, the swamp region due north of Pekanbaru and the swamps northeast of Palembang deserve special attention, as they were once known as regions where the *orang pendek* or alternatively named orang-utans occurred.

During the 1996 survey, it became clear that the historical reports of orang-utan presence even farther to the south should not be dismissed as incidental sightings of stray individuals, or, even of released ex-captive apes. Indeed, orang-utans may still occur along the whole west coast wherever forests on the swampy coastal plain and the adjacent foothills are still extensive.

It is highly remarkable, however, that local people in the coastal areas and valleys where the ape occurs seem scarcely aware of its presence. Local inquiries with direct reference to orang-utan or *mawas* usually yielded negative answers, although nests in the nearby forests betrayed the ape's occurrence. Perhaps this is not surprising, considering that the ape used to be known by an alternative name and only a few modern Indonesians venture deep into forest, as it has been demolished from the fringes inwards. The few positive reports insist that the ape is extremely shy, and is usually encountered on the ground, apparently despite tigers being reportedly common in the area. Most interesting, however, is the suggestion that the apes are mostly active at dusk and at dawn, and may have shifted to a predominantly crepuscular lifestyle (van Schaik, unpubl. report of the University of Indonesia survey team, May 1997).

Considering that the orang-utan occurred all over the island of Sumatra, and in particular along the central foothills zone, it is probable that the present orang-utan populations along the west coast and in the few uninhabited valleys of the Barisan range are the survivors who escaped persecution at the hands of the tribes of hunter-gatherers (Kubu, Mamaq and others; Loeb, 1972) that roamed the extensive eastern coastal plains and foothills east of the mountain chain, and who still have an impact on the fauna. Consequently, the behaviour of a surviving relict population may be expected to be elusive. As a matter of fact, the reported behaviour and physical appearance of the southern population of orang-utans strongly suggest that what has been known as the legendary *orang-pendek* in the region of central and southern Sumatra is in fact the orang-utan, and in particular a darker, less arboreal, kind than the predominant type of northern populations.

Until further surveys can be conducted, however, the extent and numbers of these remnant populations will remain unknown.

Southern Sumatran populations?

At least four isolated population fragments of orang-utans may survive in southern Sumatra. A Ministry of Transmigration survey team in 1988 found evidence of orang-utans in the Mount Kaba region (S. Sculley, pers. comm.) due east of the township of Bengkulu. Both Westenenk (1918) and Maier (1923) describe the environs of Mount Kaba, i.e. the water catchment of the tributaries of the river Musi, as one of the major concentration areas for reports of the *orang pendek* in southern Sumatra.

The surveys to discover the legendary *orang-pendek* have revealed the presence of large numbers of nests, apparently made by orang-utans, in the Air Hitam forest complex along the coastal fringe of the Barisan range (between Benkulu and Mokomoko), more than 3°south of the equator (Yanuar, pers. comm.). Orang-utan nests are characteristic, and no other large animal in Indonesia is known to make a similar nest (sun bears and binturongs may make use of orang-utan nests but are not known to be able to construct such typical platforms). These findings also make it plausible that special surveys into the upper water catchment of the river Komering, between Gunung Nanti and Gunung Patah, in the province of South Sumatra can confirm the few serious reports of orang-utan sightings (Tantra, pers. comm. 1979) in this southern part of the Barisan mountain range (i.e. 5° south of the equator). The obligatory Environmental Impact Assessment report of P.T. Bengkulu Raya Timber, which concerns a forest concession area in the South Bengkulu regency (S.E. of the coastal township of Mana), explicitly mentions the presence of orang-utans in the concession, with an additional note on the occurrence of nests. The possible relict populations in southern Sumatra occur in forest blocks, altogether comprising some 3,697 km^2. All of this forest is under the severe impact of logging, encroachment and conversion.

In conclusion, in recent years the Sumatran orang-utan population has suffered the fragmentation as well as the obliteration of some of its subpopulations. At present the major refuge is the approximately 25,000 km^2 large Leuser Ecosystem, which is a recent extension of the former Gunung Leuser complex of wildlife reserves, which was also known as the designated Gunung Leuser National Park, *calon* TNGL (Rijksen and Griffiths, 1995). It must be realised that with proper land-use planning, protection and some forest restoration the Sumatran orang-utan may have a chance to survive in a main ecological structure comprising three large sections. The first could stretch all the way from northern Aceh, along the west coast down to the township of Singkil. The second, across the river Alas, could stretch from the Dolok Sembelin Protection Forest all the way down to the proposed Batangtoru protection forest. And the third could stretch from the mouth of the river Batangtoru, along the Siondop area south to Rimbo Panti and Gunung Talamau.

Evaluation of survey data

Introduction

A summary of the descriptive data relating to the major forest regions already indicates that a current meta-population of orang-utans is badly fragmented and no more than a relic of the past. The future looks ominous. Continued, minimally controlled timber exploitation, large-scale conversion, encroachment, arson and the poaching of timber and apes are expected to increase their combined impact, unless major counter-actions are undertaken.

The following tables give an overview of the survey results, and an evaluation of the situation for the six major geographical regions. The evaluation is meant to provide information on the situation of the orang-utan in major sectors comprising the more than 100 forest fragments found in Borneo as well as major sectors in Sumatra. It must be emphasised that many of the sectors no longer represent continuous areas of forest; many have been subdivided by roads, drainage canals, logging tracks, plantations and settlements.

For a more detailed evaluation it is necessary to assess realistically the ape's living conditions; it is illusory to presume that a tree cover as observed on a satellite image, or outlined on an official forestry map, represents orang-utan habitat. Therefore it is necessary to take account of (1) the estimated percentage of habitat within a forest area, and (2) the estimated reduction of quality of that habitat.

The present report has applied the following procedures to estimate the carrying capacity of a forest block, which supposedly reflects the number of orang-utan inhabitants under equilibrium conditions:

(1) Within the general forest sectors for which the occurrence of orang-utans was confirmed, the actual forest cover was determined on the basis of all available information, as polygons of a Geographic Information System (GIS). This actual forest cover disregards all grasslands, cultivated areas, villages and roads, as well as all forest at an altitude of over 1000 m in Sumatra and 500 m in Borneo, and is denoted as 'orang-utan forest' or simply 'forest.'

(2) Since the general type of forest in the GIS database is known, it is possible to apply the assessment for habitat percentage with reference to orang-utans (i.e. 20-35% for dryland forest, and up to 50% for Sumatran swamp forest) for each polygon, and arrive at the extent of habitat.

(3) Adding the interpretation of logging impact for every area of calculated habitat yields a measure of habitat quality, which is called a 'carrying capacity unit'. The 'cc unit' is considered to reflect habitat of a quality such that it can support the average orang-utan population density of each island.

(4) From these carrying capacity units one can estimate the minimum number of apes by multiplying the value by the relevant density estimate (see Survey Methods).

> One can arrive at value categories analogous to the IUCN criteria for endangered species (Appendix 4). Although the IUCN categories refer to species or meta-populations rather than populations, it is felt that the criteria can be adopted for this report at the population level.
>
> The criteria of the World Conservation Union (IUCN) are qualitative, and to a large extent speculative (see Appendix 4) and, when applied for a specific case, require a quantitative basis. In order to estimate the population reduction, the present study has focused on criteria A-c, namely 'a decline in area of occupancy, extent of occurrence, and/or the quality of the habitat' in combination with criterion A-d, namely 'actual or potential levels of exploitation.'

For this report, the evaluation of the major fragments of the found orang-utan range is differentiated into three categories, namely 'vulnerable', 'in danger' and 'critical'. In this evaluation a population is considered critical if:
(1) it inhabits an isolated forest fragment with less than 100 km^2 of habitat, or less than 185 cc units for Borneo and 120 cc units for Sumatra;
(2) its estimated size is below 350 mature individuals and declining;
(3) its range is part of a planned conversion process or subject to planned further fragmentation.

A population is considered 'in danger' if:
(1) it inhabits a forest fragment with less than 500 km^2 of habitat or less than 263 cc units for Borneo and 166 cc units for Sumatra;
(2) its estimated size is below 500 mature individuals and declining;
(3) its range is part of a planned conversion process or subject to planned further fragmentation.

A population is 'vulnerable' if:
(1) it inhabits a forest fragment with less than 1,000 km^2 of habitat;
(2) its range overlaps with timber concessions and/or that of traditional forest hunters.

The facts

The satellite imagery for forest cover, when entered into a GIS basis containing the locations from which orang-utans have been reported, provides a crude overview of the forest blocks in which orang-utans occur, the distribution polygons. The surface area of such polygons are readily computed. For Borneo all forest areas at an altitude above 500 m, and for Sumatra all forest areas at an altitude above 1000 m, were discounted from the estimated forest cover area, although it is realised that in particular the 500 m contour in Borneo is not an absolute limit. The rivers, some 10-50 km from their source downstream, are considered to be barriers preventing orang-utan dispersal. The results are shown in Table XIIa–f.

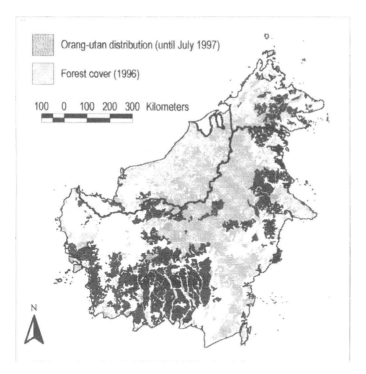

Current (1997) distribution of the orang-utan in Borneo.

TABLE XII (A)

Overview and evaluation of the survey results with reference to major forest sectors: Forest cover (Forest) in km² is not identical with habitat coverage; the forest cover within any sector is not uniform but consists of a mosaic of forest types, each with its 'own' habitat content. An assessed percentage (Hab %) is applied to calculate habitat coverage. Further application of an estimated quality reduction factor (Red %) for destructive disturbance in the habitat yields 'carrying capacity units' (Cc unit) as a measure of habitat quality.

Re	W. Kalimantan polygons	Forest – km²	Hab %	Red %	Cc unit	Value
1	Sambas (27+27+446)	500	30	75	37	Critical
2	Mempawah (10+12+23+129)	174	30	80	10	Critical
3	Gunung Niut (36+61+98)	195	25	70	15	Critical
4	Ketungau (19+26+268)	313	30	80	19	Critical
5	D.Sentarum-Bent. Karimun	5,869	25	50	734	Vulnerable
6	Kapuas Hulu 1962+595+41)	2,598	25	70	195	In Danger
7	Madi plateau-Melawi	2,856	30	70	257	In Danger
8	Bukit Baka	474	30	75	35	Critical
9	Berangin-Sebayan	4,047	30	70	364	Vulnerable
10	Kapuas swamps	6,798	20	70	407	Vulnerable
11	Sukadana-Kendawangan	7,768	30	75	583	In Danger

Table XII (B)

Re	C. Kalimantan polygons	Forest – km²	Hab %	Red %	Cc unit	Value
12	Jelai-Lamandau-Arui	12,516	25	75	782	In Danger
13	Tanjung Puting	3,518	25	65	308	Vulnerable
14	East Pembuang-Seruyan	7,500	30	70	675	In In Danger
15	W.Sampit floodplains	4,327	25	85	162	Critical
16	Katingan floodplains	7,402	20	50	740	Vulnerable
17	Sebangau catchment	5,878	30	60	705	Vulnerable
18	W.Rungan floodplains	2,217	30	70	199	In Danger
19	Kahayan-Rungan catchment	2,099	25	75	131	Critical
20	Sebangau-Kahayan floodplains	1,932	20	65	135	Critical
21	Mangkutup (3575+307)	3,882	25	80	194	Critical
22	Kapuas Murung-Barito plains	7,745	25	85	290	Critical
-	Negara floodplain	930	25	85	35	Critical
23	Kahayan-Kapuas Murung hills	3,115	30	85	140	Critical
24	Schwaner foothills east	1,426	30	75	107	Critical
25	Schwaner foothills west	1,460	30	75	109	Critical
26	Central Schwaner range	6,241	25	70	468	In Danger
27	Kahayan-Miri catchment	1,478	25	85	55	Critical
28	Bundang east	1,285	25	75	80	Critical
29	Upper Dusun	1,899	25	75	119	Critical
30	Busang Hulu (119+373)	492	25	50	61	Critical

Table XII (C)

Re	E. Kalimantan polygons	Forest – km²	Hab %	Red %	Cc unit	Value
31	Liangpran (633+765)	1,398	30	70	126	Critical
32	Boh catchment	1,755	30	65	184	Critical
33	Pari-Sentekan	1,870	30	75	140	Critical
34	Belayan-Kedangkepala	1,565	25	75	98	Critical
35	Central plains	617	25	80	31	Critical
36	Coastal Kutai	3,237	25	85	121	Critical
37	Telen-Wahau	888	30	70	80	Critical
38	Tinda-Hantung hills	5,729	30	80	344	Critical
39	Segah catchment	6,619	25	80	331	In Danger
40	Berau	2,105	25	-	-	Extinct
41	South Sembakung swamp	577	25	60	58	Critical

Table XII (c) (vervolg)

Re	E. Kalimantan polygons	Forest – km²	Hab %	Red %	Cc unit	Value
42	North Sembakung swamp	2,202	25	70	165	In Danger
43	Sebuku	2,415	25	75	151	Critical
44	Sembakung Ulu	499	30	75	37	Critical

Table XII (d)

Re	Sabah polygons	Forest – km²	Hab %	Red %	Cc unit	Value
45	Ulu Kalumpang	408	25	65	36	Critical
46	Tabin (120+132)	252	25	65	22	Critical
47	Melangking	298	30	70	27	Critical
48	Southern Forest east	3,710	30	75	278	In Danger
49	Kinabatagan (610+977)	1,587	30	70	143	Critical
50	Southern Forest north	982	30	85	44	Critical
51	Southern forest west	1,481	30	65	155	Critical
52	Trus Madi	394	30	70	35	Critical
53	Ulu Tunggud	163	30	70	15	Critical
54	Bidu-bidu hills	74	30	70	7	Critical
55	Sugut South	25	30	40	5	Critical
56	Sugut North	57	25	40	9	Critical
57	Lingkabau	459	30	60	55	Critical
58	Kinabalu	690	25	50	86	Critical
59	Crocker range	97	30	50	15	Critical

Table XII (e)

Re	Sarawak-Kal. polygons	Forest – km²	Hab %	Red %	Cc unit	Value
60	Batang Lupar coastal swamp	1,065	25	80	53	Critical
61	Lanj. Entimau/Batang Aie	812	30	35	158	Critical

Table XII (f)

Re	Sumatra polygons	Forest – km²	Hab %	Red %	Cc unit	Value
1	North Aceh	1,176	35	65	144	In Danger
2	Woyla	1,793	35	75	157	In Danger
3	Gayo Mountains	1,301	35	60	182	In Danger
4	Bahbahrot swamp	192	50	85	14	Critical
5	Leuser	4,445	35	40	933	Normal
6	Singkil swamp	981	50	20	392	Vulnerable
7	Peusangan catchment	1,086	30	70	98	Critical
8	Jambu Aie catchment	2,014	25	75	126	In Danger
9	Tamiang	1,826	30	75	137	In Danger
10	Sikundur-Langkat	1,548	30	70	139	In Danger
11	Alas valley	444	35	95	8	Critical
12	East Alas	174	30	70	16	Critical
13	Dolok Sembelin	148	30	50	22	Critical
14	Batu Ardan	206	30	50	31	Critical
15	Tapanuli Tengah	1,564	50	50	391	Vulnerable
16	Anggolia	188	30	65	20	Critical
17	Simonangmonang	350	30	70	31	Critical
18	Sipirok	190	30	60	23	Critical
19	Angkola-Siondop	2,327	40	65	325	Vulnerable
20	West Pasaman	405	35	80	28	Critical
21	Baruman	2,167	30	60	260	Vulnerable
22	Rimbo Panti	147	25	75	9	Critical
23	Habinsaran	1,283	30	70	115	In Danger

In Borneo the orang-utan is distributed over a total area of forest land of at most 150,000 km², which is fragmented into more than 100 fractions in 61 major sectors. In 1996 the sectors together comprised approximately 40,000 km² of habitat. However, the quality of this habitat has been so degraded that no more than 11,000 carrying capacity units (i.e. 28%) were still available before the massive forest fires of 1997. Virtually all this habitat is covered by concessions for timber exploitation or forest conversion, except for some minor sectors representing (parts of) the 14 conservation areas (i.e. established wildlife reserves, and nature sanctuaries)[72], although these have also been degraded. In Indonesia, the status of protection forest offers no guarantee whatsoever for the conservation of habitat quality; indeed, large sectors of habitat even within the conservation areas have suffered the degrading impact of uncontrolled logging, timber poaching, encroachment and arson.

---

[72] Although former reports often included areas which were 'proposed' for conservation, these are considered here to be meaningless because they may give a false impression of security.

An overview of the evaluation of the Bornean surveys prior to the fires of 1997, as represented in Tables XII (a-f), shows that:
- 34 sectors (i.e. 77%) of the range in Kalimantan, and three sectors (20%) in Sabah, represent fragments well over 1,000 km^2 and would apparently have been large enough to sustain surviving populations, if only habitat destruction had not been so rampant;
- in 45 sectors (74 %) of the fragmented range the Bornean orang-utan is critically endangered; for each of these 45 fragments, it is feared that a decline of 80 -100 % of the current population will occur within the next 75 years, implying virtual extinction;
- In ten fragments (16 %) the orang utan is endangered; this would imply that a decline of at least 50% is feared to occur within 75 years (see Appendix 3).
- In six fragments (10%) the orang utan is vulnerable, indicating that a reduction of at least 20 % is likely to occur within the next 75 years;
- the situation can nowhere be considered 'normal' anymore; that is, a condition in which a large area is still suffering only minimal habitat degradation, so that a normal population structure of apes may be preserved.

In Sumatra, the orang-utan is currently known to be distributed over a total area of almost 26,000 km^2 of forest land, which is fragmented into more than 40 fractions in 23 major sectors. Altogether this forest cover offers no more than some 8,000 km^2 of habitat. Due to timber extraction, *ladang* encroachment and ongoing conversion, this implies that no more than approximately 3,500 'carrying capacity units' (38%) are available. Some 80% of this remaining habitat area is covered by timber concessions, which may, under the current lack of control, turn at least 50% of the forest area into land fit for conversion to oil-palm, rubber, candlenut (*Aleurites moluccana*) and coffee plantations or cattle ranches[73]. Eleven sectors reflect critical conditions (48%); seven sectors contain populations in danger (30%); four populations are vulnerable, and only one of the populations may be said to be still living under 'normal' conditions (4%).

The disaster which struck Indonesia in the form of drought and arson in 1997-98 has devastated large areas of logged-over and pilfered forests as well as vast expanses of shrub and grass savannas in Borneo and Sumatra. A crude estimate of the damage in every affected range-polygon (information from the Ministry of Forestry and the Ministry of Environment – Bappedal, regarding hot spots, can be found on the Internet at *mofrinet.cbn.net.id/noaa* and at *www.bappedal.go.id*) has been made to show the reduced extent of possible forest habitat, although no satellite imagery was as yet available to measure the reductions. These estimates are represented in a separate set of tables (XIIIa – f). The fires which were still raging at the time that this manuscript

---

[73] Since 1994, American fast-food corporations have been establishing themselves in Southeast Asia, and are looking for a local 'hamburger connection', thus facilitating the conversion of rainforest into cattle ranches, as happened in Central America and the Amazon during the 1980s.

Section II: orang-utan distribution

was finalised may already have virtually eradicated the last remaining orang-utan populations in the Kutai (36) and the Telen-Wahau areas (37).

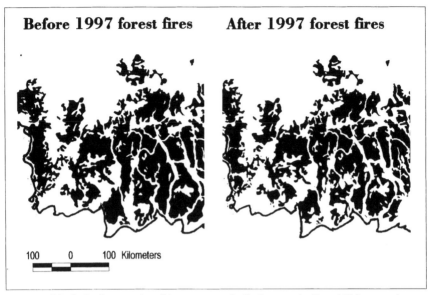

Overview of the further fragmentation of the orang-utan distribution range in Central Kalimantan due to arson towards the end of 1997.

TABLE XIII (A)

The estimated reduction in habitat due to arson during the 1997 drought; the polygons affected for more than 30% are indicated in italics.

Re	W. Kalimantan polygons	Forest 96 – km²	fire % red.	1997 Cc unit
1	Sambas	500	65	13
2	Mempawah	174	80	2
3	Gunung Niut	195	65	5
4	Ketungau	313	70	6
5	*D.Sentarum*-Bent. Karimun	5,869	30	514
6	Kapuas Hulu	2,598	-	195
7	*Madi plateau*-Melawi	2,856	40	154
8	Bukit Baka	474	-	35
9	Berangin-Sebayan	4,047	30	255
10	Kapuas swamps	6,798	-	407
11	Sukadana-Kendawangan	7,768	65	204

The areas indicated by an asterisk on page 275 (*) are being affected by the mega-rice reclamation projects, notably the drainage which causes forest die-off and invites arson during the dry seasons; it is probable that all orang-utans and other wildlife in these areas are beyond rescue, unless the drainage is reversed.

Table XIII (b)

Re	C. Kalimantan polygons	Forest 96 – km²	fire % red.	1997 Cc unit
12	Jelai-Lamandau-Arui	5,678	50	177
13	Tanjung Puting	1,185	35	70
14	East Pembuang-Seruyan	5,307	35	439
15	W.Sampit floodplains*	1,233	60	65
16	Katingan floodplains*	5,286	40	444
17	Sebangau catchment	5,834	15	599
18	W.Rungan floodplains	2,217	35	129
19	Kahayan-Rungan catchment*	1,286	50	65
20	Sebangau-Kahayan floodplains*	860	65	47
21	Mangkutup*	875	60	78
22	Kapuas Murung-Barito plains*	1,532	80	58
-	Negara floodplains*	930	80	7
23	Kahayan-Kapuas Murung hills*	1,513	50	70
24	Schwaner foothills east	1,426	-	107
25	Schwaner foothills west	1,460	-	109
26	Central Schwaner range	6,241	-	468
27	Kahayan-Miri catchment	1,478	10	49
28	Bundang east	1,285	15	68
29	Upper Dusun	1,899	20	95
30	Busang Hulu	492	-	61

Table XIII (c)

Re	E. Kalimantan polygons	Forest 96 – km²	fire % red.	1997 Cc unit
31	Liangpran	1,398	-	126
32	Boh catchment	1,755	10	166
33	Pari-Sentekan	1,870	15	119
34	Belayan-Kedangkepala	1,565	30	69
35	Central plains	617	80	6
36	Coastal Kutai	3,237	35	78
37	Telen-Wahau	888	10	72
38	Tinda-Hantung hills	5,729	-	344
39	Segah catchment	6,619	10	298
40	Berau	2,105	20	-
41	South Sembakung swamp	577	5	55
42	North Sembakung swamp	2,202	10	148
43	Sebuku	2,415	-	151
44	Sembakung Ulu	499	-	37

A comparison of the records on the fire hot spots (e.g. from the EUFREG GIS / Remote Sensing Expert Group / MoF) up to December 1997 reveals that arson affected 36 of the 45 major forest fragments in Kalimantan to a significant extent. This does not mean that all the forests in these sectors have burned down, but that the immediate and after-effects of the fires have seriously reduced the ecological quality of the affected areas and greatly increased the risks of further degradation. The immense heat radiation has often desiccated and 'cooked' many more trees than were directly scorched by the flames. In other words, the estimated reduction refers to the carrying capacity of the affected forest fragments. Even if more orang-utans have survived the conflagration than the simple calculation suggests, it is highly unlikely that any higher numbers of surviving apes can be supported during the coming years above the numbers related to the carrying capacity units.

This overview of the data indicate that, in order to give the orang-utan a fighting chance of survival, the governments of Indonesia and East Malaysia should:
(1) increase the efficaciousness of protection;
(2) establish new reserves of significant size and redesign existing reserves;
(3) keep the wildland forest within the orang-utan's range for wild timber 'production', and keep all protection and production forests free of people;
(4) stop further conversion of State forest land;
(5) immediately stop all drainage of swamp forests, and revert the hydrological situation.

The widespread arson during the droughts of 1982-83 and 1997-98 has demonstrated conclusively that human habitation and uncontrolled human access to the wildland forests is incompatible with the production of timber and the maintenance of ecological conditions of such forest, including its structure of biological diversity. One person with a box of matches can cause a catastrophe. It is therefore utterly unrealistic to try to implement 'awareness' programmes among squatters in order to prevent arson, or to invest heavily in modern fire-fighting equipment and other sorts of treatment of the symptoms, especially in drained peat forests. The trendy hope for 'participatory sustainable forest management' is also ill-conceived. Only an effective exclusion of unauthorized people from forest areas, through strict law enforcement, may help in the regeneration of timber and in salvaging the biodiversity of the wildland forests of Borneo and Sumatra.

Taking into account the currently isolated populations of the major types of orang-utan, it is sensible to pay special attention to the largest, and most important contiguous sectors where an improvement in the conservation situation can still be achieved. These are:

1. For the West Kalimantan/Sarawak population the crucially important region remaining is the northern corner of the province, extending into Sarawak. There the trans-frontier reserve Pegunungan Bentuang dan Karimun (some 6,000 km^2),

on the Indonesian side links up with the Lanjak-Entimau Wildlife Sanctuary (1,688 km^2) and the adjacent Batang Ai National park. Notwithstanding a national boundary and a network of strategic roads, this bi-national conservation area can become particularly important if the Danau Sentarum Wildlife Reserve can be linked up with the Pegunungan Bentuang and Karimun reserve.

The population of the western coastal zone comprising the Kapuas coastal swamps, and the Sukadana-Kendawangan region, as well as the foothills of the Beranging-Sebayan catchments, in fact harbour apes of the southern Bornean population, with its stronghold in Central Kalimantan province. Unfortunately, this area, in which the Gunung Palung Conservation Area is situated, has been affected very badly by encroachment, poaching and arson; nevertheless, it remains of prime importance for conservation of the ape.

2. The stronghold of the southern orang-utan population of Borneo is in Central Kalimantan; the (still) inaccessible, waterlogged peat forests in the catchment and flood plains of the rivers Sebangau and Katingan are extremely important for the ape's survival. However, most of the peat-swamp forests in this area are badly suffering the impact of drainage, reclamation, timber extraction and arson, e.g. the adjacent one-million-hectare PLG mega-reclamation project. Important sectors for conservation in the province are also the east Pembuang-Seruyan catchments, and the central Schwaner range, extending across the provincial border with Central Kalimantan including Bukit Baka, Bukit Raya, the Madi plateau and the Batikap I, II and III protection forests, as well as some of the lowlands (e.g. the Melawi catchment in West Kalimantan, and the catchments of the rivers Mendawai, Sampit and Kahayan). In 1997, the southern sectors of the range were affected heavily by arson, causing the death of thousands of orang-utans.

3. For the rapidly shrinking eastern Bornean population in East Kalimantan, the western and northern foothills fringing the central flood plains west of Muara Kaman, and including the water catchments of the rivers Belayan, Senyiur, Kelinjau and Telen, are important potential areas for forest conservation. Due to arson during the 1997-1998 drought, all the lowland and hill forest south of the Sangkulirang peninsula and the Tindah-Hantung range was destroyed by fire.

4. For the northern Bornean population the Sebuku-Sembakung water catchments along the boundary with Sabah (and extending as a transfrontier reserve into the Southern Forest) are of particular importance. Fortunately, much of the remaining forest in this area acquired a protected status in 1998. In Sabah the Southern Forest reserves of the Sabah Foundation, including the Danum Valley Conservation area, as well as some 13,000 km^2 of timber production forests, a the major stronghold of the ape.

5. In Sumatra, the Leuser Ecosystem and possible extensions north-westward, to include all the uninhabited mountain range, is not only essential for water

catchment and erosion control, but also crucial for conservation of the northern Sumatran type of orang-utan.

6. The Angkola–Batang Gadis water catchment, with its unique swamps and coastal flood plains, is probably the area best suited to afford the southern Sumatran type of orang-utan *(orang-pendek)* a chance of survival. A proposal to change the status of this area to that of wildlife sanctuary' has been submitted (1997).

Few of these crucially important areas have a protected status, apart from the transfrontier complex, the designated Bukit Baka/Bukit Raya national park, the recently established Sembakung-Sebuku reserve and the Leuser Ecosystem.

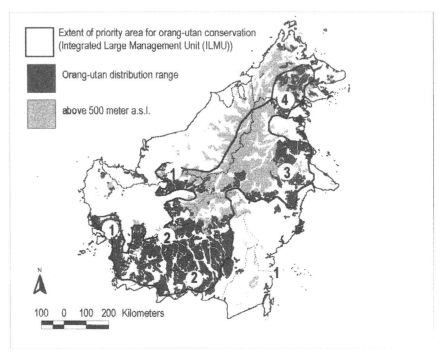

Map of priority areas – or Integrated Land Management Units (ILMU) – for the conservation of significant populations of orang-utans in Borneo.

Evaluation of survey data

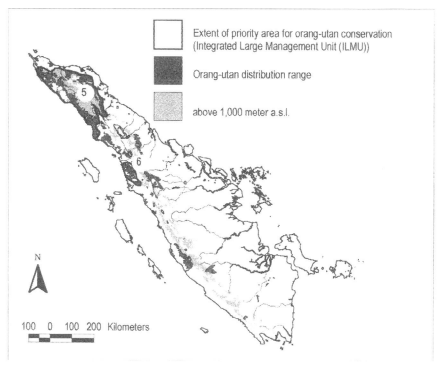

Map of priority areas (ILMUs) for the conservation of significant populations of orang-utans in Sumatra.

Section III

The Decline

Thousands of square kilometres of orang-utan habitat are being destroyed every year; forest removal for 'development' of the vast open-pit mine belonging to Kalimantan Prima Coal in East Kalimantan (1994).

Displaced wild orang-utan male on a roof, about to raid a papaya tree, in the new expatriate settlement of Tanjung Bara (Coal Cape) just north of the designated Kutai National Park; his habitat was obliterated to make way for KPC's vast open-pit mine. Like many of his deme, the male was searching for food in peoples' gardens (1996); (photograph courtesy of D. Eygendaal).

Summary of survey results

The decline

The incidence of media reports on orang-utans has increased dramatically during the last decades. The public is now able to see and hear of the ape more frequently than ever, and the numbers of orang-utans confiscated from illegal captivity is staggering. Even primatologists studying the ecology of the ape are finding higher densities than had been thought possible for a 'solitary ape.' For people of limited vision, this might indicate that there are apparently more orang-utans than had been believed, and that the ape is far from rare. In reality, however, the high incidence of reported sightings may simply reflect that the modern world, by way of the rampant exploitation and conversion of the forest is expanding in such a way that it cannot avoid confrontations with the last, desperately compressed populations of wild animals, in particular in the previously ignored vestiges of wilderness. Irrespective of their hiding skills, few orang-utans can avoid encounters with the many thousands of people constantly intruding into their domain; many orang-utans are forced out of devastated forests that used to be their home, to roam in search of food, either crowding into the last relict patches of habitat, or entering the private gardens of hostile people. Large numbers of orang-utans were massacred while fleeing the flames and smoke during and after the extensive forest fires of 1997-1998 in Borneo.

The survey results show that up until mid-1997, due to massive habitat degradation as a consequence of timber exploitation and encroachment the Bornean orang-utan occurred in a range of anything between 140,000 and 150,000 km^2 of forest land, which was equivalent to a maximum of only 11,000 carrying capacity units. Because Indonesian Kalimantan was not the only area to suffer from arson during the 1997 drought, it is feared that the conflagration reduced the total of this already battered carrying capacity in Borneo by an additional 30% over a period of some six months. This is due mainly to the disproportional destruction of peat-swamp and lowland forests, the hitherto most important habitat for the ape in Borneo. Applying the estimated densities for the different habitat mosaics in Borneo (Appendix 4), and the effects of the drought and fires of 1997, this implies that the present meta-population of Bornean orang-utans is believed to have suffered a decline from an estimated 23,000 in 1996 to some 15,400 individuals in 1998, i.e. a reduction of some 33% in one year.

If the approximately 150,000 km^2 of forest in Borneo had still been in its natural state, unaffected by timber extraction and subsequent demolition by encroachment, conversion and arson, this would have implied that the extent of the habitat (i.e. approximately 42,000 km^2) could be multiplied by the average density figures for the particular forest types in order to arrive at a figure for the total population. In this hypothetical case the meta-population of Bornean orang-utans would have amounted to more than 75,000 apes. Unfortunately, this is no longer so.

Applying the same basic criteria for the situation at the beginning of the twentieth century, one can conclude that some 230,000 orang-utans must have lived in the approximately 450,000 km² of forest in Borneo (excluding the area between the rivers Mahakam and Barito, Central Sarawak and some coastal zones). **Considering that the present population amounts to no more than anything up to some 15,000 apes, a mere 7% of the 1900 population has survived.**

It must be realised that the extent of the forest estate on official maps, as used for planning procedures and policy, has little bearing on reality. Table XIV illustrates the differences. In the Table, the column 'real cover' comprised, until 1997, all more or less closed-canopy forest remaining on the extent of State forest land designated Protection Forest, Conservation Area, Production Forest and Conversion Forest, excluding montane forest above an altitude of 1,000 m. For the three provinces in which orang-utans occur in Kalimantan, up until mid-1997 this still amounted to 266,457 km² or 74% of the estate, and for the two provinces in Sumatra to 54,506 km², or 81% of the estate. Thus, the State forest estate in Kalimantan was in reality covered by 'forest' for anything between 60 and 86% prior to the 1997 fires, and in Sumatra for some 80%. Of this 'forest' cover, anything between 45 and 78% comprised the current distribution pattern of the orang-utan, a discrepancy which mainly refers to forest areas at altitudes above 500 m.

TABLE XIV

A comparison of the figures relating to the formal extent of the forest estate, the observed forest cover in 1996 (prior to the fires), the extent of the range of the orang-utan within the observed forest cover, the respective percentages of real forest coverage (%), and the percentage of the orang-utan range within the real forest coverage (cov/ou.f).

Province	For. land	Real cover	%	Ou.forestCov/ou.f
W.Kalimantan	83,120	49,600	60	32
C.Kalimantan	114,089	97,722	86	76
E.Kalimantan	161,885	119,135	74	31
Aceh	38,176	31,220	82	15
N.Sumatra	28,941	23,286	80	11

It is relevant to consider in this respect that the PLG mega-reclamation programme in Central Kalimantan (McBeth, 1995; Presidential Decree RI / 82–26 Dec. 1995) does affect some 15,000 km² of peat-swamp forest in the flood plains between the rivers Sebangau and Barito, as well as another 4,000 km² of coastal floodplains near Sampit. The canals drain the deep peat swamp (3-18 m) over a very extensive area, causing a massive die-off of the forest cover during the dry season, and inviting arson. This implies that anything between 5,000 and 7,000 km² of orang-utan habitat will soon be utterly demolished for timber extraction, oil-palm plantation development, transmigration and rice and sugar production if it is not first largely destroyed by wildfire. One wonders whether the government has taken any note at all of the warnings given by professional national and international advisors with reference to this case, and has considered a belated obligatory Environmental Impact Assessment (AMDAL). If so, then how can it account for the fact that the forementioned development, if it proceeds unhindered, will not only be at the cost of several thousands of – fully protected – orang-utans (see Appendix 4, Table XXVI no 20-23), but will also result in an ecological disaster which will affect the entire southeastern corner of Borneo? If not, how then can a modern government at the close of the twentieth century be so short-sighted and irresponsible?

For Sumatra it is currently held that the meta-population of orang-utan occurs in an accumulated total area of approximately 26,000 km^2 of forest, i.e. roughly equivalent to some 3,500 carrying capacity units. Applying the estimated densities of apes in the different mosaics of habitat types in Sumatra, this implies that the current meta-population of surviving apes amounts to approximately 12,500 individuals.

In view of the recent discovery of hitherto unknown populations south of Lake Toba, it is possible that in the near future more small relict populations may be found in Sumatra south of the equator. However, these populations will probably add no more than a few hundred extra apes to the present assessment at most.

If the 25,955 km^2 of forest had not been degraded, the meta-population would have amounted to over 23,000 orang-utans. It may be surmised that at the beginning of the twentieth century the ape was to be found in at least 82,000 km^2 of forest in Sumatra north of the equator only, and amounted to no fewer than 85,000 individuals. **This implies that Sumatra's orang-utan population has been reduced to some 14% of its original size during the twentieth century.**

Together these facts demonstrate that the orang-utan must be considered an endangered species, while some of its regional types are critically endangered.

According to the IUCN criteria for categories of threatened status, a species is considered to be critically endangered if (a) its meta-population is estimated or suspected to have suffered a decline of at least 80% during the last three generations, or (b) its distribution range is found to be less than 100 km^2, is known to be severely fragmented and continues to decline, and/or (c) its population fragments of mature individuals number fewer than 250 individuals, showing a decline rate of 25% in one generation or a continuously declining trend (see Appendix 3).

In relation to an assumed original early Holocene situation, it is apparent that the meta-population of Sumatra has suffered a much longer, steady decline than the Bornean population. Whereas the Sumatran decline seems to reflect the history of cultural colonization of the island following the arrival of the Hindu and Buddhist South-Asian civilisations, the dramatic decline of the Bornean population set in with the twentieth century's technological revolution in resource extraction, and has been accelerating during the 1990s.

TABLE XV

Overview of the estimated reduction in orang-utan populations in Borneo and Sumatra during the twentieth century.

	No. of apes in Borneo	No. of apes in Sumatra	Total est.
(Early Holocene)	(420,000)	(380,000)	(800,000)
around 1900	230,000	85,000	315,000
in 1996	23,000	12,000	35,000
1997 after the fires	15,000	12,000	27,000
Remainder in % of 1900 population	7%	14%	9%

Discussion

Direct (proximate) causes of the decline

Habitat loss

Some human activities result in the entire loss of orang-utan habitat, because regeneration of the forest is unlikely or impossible, for instance, in the case of conversion for agriculture or plantations of cash-crops (oil-palm, rice, maize, sugar, pulp-wood, rubber, coffee, tea, etc.), and open-pit mining.

A comparison of early and recent satellite imagery and land-allocation maps shows a considerable reduction in forest area, both in Borneo and Sumatra. This is not surprising when one considers that even the official target of the Indonesian Forest utilisation development policy for *Pelita* VI (i.e. the Five Year Development Programme) indicated an annual clearing activity of some 6,860 km^2, and the conversion of some 2,100 km^2 of natural forest in 1994-95. The clearing of forest, however, is not in the least controlled or constrained by official planning: In 1988, the World Bank estimated that deforestation in Borneo alone amounted to some 7,000 km^2 per year (Davis and Ackermann, 1988); in 1990 a re-assessment made by the World Bank showed 9,000 km^2. In the same year, the FAO assessed 13,150 km^2 of deforestation in Indonesia, which was formally accepted by the Ministry of Forestry in 1992. In 1994, the World Bank undertook a new re-assessment and arrived at 13,000 km^2, while Indonesian NGOs stated that Indonesia had suffered an annual loss of 24,000 km^2 of its forest estate since 1986 (*Jakarta Post*, December 23rd. 1994). This enumeration of some assessments clearly indicates that the rate of deforestation has been accelerating during the last decade.

Over a period of only two-months, a ten-thousand-hectare area of primary forest is cleared and traversed by roads for conversion into an oil-palm plantation; a growing market for soap, margarine and cooking oil is destroying the habitat of (e.g.) orang-utans.

The area under impact of slash-and-burn cultivation in Kalimantan is estimated by the Indonesian Land Resources Development Centre to be 112,000 km^2; approximately 87% of this area has been prime orang-utan habitat.

In Sarawak and Sabah some twenty years ago this was already 30,000 km^2 and 11,000 km^2, respectively (Collins et al. 1991).

> Ethnological studies of the impact of indigenous tribal communities on their forest environment in South America could not find any conservation ethic in such tribals (Edgerton, 1992; Ridley, 1996). In East Kalimantan ethnological research in an ecological context (Vayda et al., 1980; Pierce-Colfer, 1981) has also shed a realistic light on the image of the 'noble savage' supposedly living in archaic harmony with nature. It has demonstrated that rural people of a traditional tribal background are as responsive to economic opportunity and consumerist lures as anyone else in the so-called developed world, and they make the best possible economic decisions under the given circumstances. From an ecological viewpoint, most important, however, is that under current conditions such rural people reproduce at the optimum rate of increase, and exert a disproportionate impact upon the natural environment due to a combination of their specialist knowledge of the forest and the ready availability of powerful technology. Colfer found that the resettlement village of Long Segar, on the river Telen, had an average household size of 9.3. Comprising 129 households (some 1200 people) the village occupants, in their 17 years of existence there, had cleared 11,600 ha of primary forest, creating larger rice fields and investing almost 30% more input in rice production than had their kin in the original village. In addition, the roles of the sexes shifted from equality towards male dominance, and to an exploitative attitude towards women. Virtually all the men worked in the timber industry, either as hired labour or as timber poachers.

Official information is often incomplete, biased, difficult to interpret and frequently outdated[74], but with reference to the types of forest which may contain orang-utan habitat in Kalimantan, a compilation of recorded changes gives the following indications:

TABLE XVI

Official records revealing the loss of forest estate in the distribution range of the orang-utan (source Statistik Kehutanan, 1990); reduction (Red 1) reflects the loss in the period 1960-1980), reduction (Red 2) reflects the loss after three decades.

Province	- Total area	1960 km^2	1980s	Red 1	1990s	Red 2
E. Kalimantan	- 202,440 km^2	134,390	97,660	27 %	58,769	56 %
C. Kalimantan	- 154,198 km^2	124,920	97,870	22 %	90,153	28 %
W. Kalimantan	- 146,760 km^2	108,150	65,590	39 %	55,000	49 %

It will shortly become evident that these records are little more than a crude indication of what has happened in reality. Nevertheless, the comparison implies that,

---

[74] The WCMC Conservation Atlas (1991) notes that the Permanent Forest Estate in Sabah, in spite of the Forest Department policy to seek a 50% increase, dropped from 35,700 km^2 to 33,500 km^2 between 1982 and 1984. Furthermore, during these years some 10,000 km^2 of logged forest was burned, while 'substantial areas ... are in the process of conversion to settled agriculture' (Collins, et al., 1991: 206).

even according to official statistics, more than half the forest cover of East Kalimantan has disappeared during the last three decades. West Kalimantan province has also lost half of its forest cover, while Central Kalimantan has lost only somewhat more than a quarter of its forests, but is currently striving to keep up with the trend.

> In view of this observation it is of more than just academic interest to note that different official sources give different figures for the extent and the planned conversion of State forest land. Most striking perhaps is the discrepancy between supposedly independent FAO data from 1980 and 1996, which indicate an *increase* in wildland forest cover of some 30,000 km² under a regime of 16 years of exploitation (see NFI, 1996, and FAO/UNEP 1981).
>
> A difference in official figures in Indonesia appears to be dependent on whether one refers to the central planning or to the regional planning process. Comparing provincial and central government documents and the records emerging in the newspapers yields the following picture: In West Kalimantan province the central planning for the period 1981-85 had allocated 15,000 km² (10% of the province) for conversion; in 1987 it had allowed an increase to 16,250 km². However, the regional planning process in early 1987 had already claimed 76,640 km² of converted forest, while Landsat imagery of 1991 shows that some 90,000 km² had actually been converted for plantation and non-forestry use. This would imply that as much as 78% of the State forest land has been 'lost' to the *laissez-faire* attitude towards the private and commercial semi-government sector. It certainly suggests a deficiency in central authority as regards the land-allocation procedure. Such findings make one wonder to what extent the government can guarantee any status of its State forest estate (Hurst, 1990). However, most disturbing perhaps is that in this age of technological might, and given the ease with which one can obtain precise and up-to-date thematic geographical information, some forces appear to be at work to effectively frustrate any such accuracy as would be required for proper government.

A further comparison of recent official forestry statistics reveals interesting trends, apart from discrepancies and an obvious overall decline in the forest estate. It demonstrates that Kalimantan may have lost at least 130,000 km² of its forest (25%) over a six-year period, including 38,500 km² of unscheduled clear-felling in its permanent forest estate, and 41,300 km² due to conversion. Sumatra may have lost approximately 214,000 km² (45%).

But it also depicts a marked decline in **Protection Forest** of anything between 9–13% for Kalimantan, and between 34–40% for Sumatra. More significant, it reveals as well a planned further reduction, over a six-year period, of anything between 74,000 and 106,000 km² of wildland forest in the three provinces where orang-utans occur in Kalimantan and some 10,600 km² in the three provinces in Sumatra. These plans concern the development of plantations, and are on top of, and partially overlap, earlier plans for conversion.

Because all these forests in Kalimantan contained orang-utan populations, the Ministry's nature conservation service as well as international conservation organisation could have been alarmed, because these data were commonly available. After all, such a tremendous loss of forest must also somehow reflect the crude loss of apes over the last three decades.

TABLE XVII

Summary of the changes of forest estate area according to official TGHK records (source RePPProT, 1990); these changes do not directly reflect the loss of forest but refer to the government's changing allocation policy regarding the use and the fate of State forest land. Note that the figures concern the whole of Kalimantan and Sumatra, and thus include forest areas outside the range of the orang-utan.

Island		Protection forest	Lim. Prod. forest	Prod. forest	Convers. forest	Total estate
Kalimantan	- 1984	69,237	118,292	142,345	115,325	445,119
	- 1990	60,082	101,695	110,311	74,027	346,115
Sumatra	- 1984	70,936	75,785	69,319	86,760	302,800
	- 1990	42,753	46,919	48,995	36,290	174,957

Concerning the official planning procedure at the central level, the official figures of the National Forest Inventory show that in Kalimantan the area of Protection Forest (59,000 km^2) had been reduced almost insignificantly (1%), the area allocated for Limited Production Forest (104,000 km^2) had increased somewhat (2%), Production Forest (104,000 km^2) had decreased 6%, and of the convertible forest land (112,000 km^2) only 66,000 km^2 was supposedly still under forest in 1996.

Of the official Protection Forest estate, some 8% lacks forest cover due to unscheduled or illegal timber destruction. Of the total official Production Forest estate 18% or some 47,000 km^2 is indicated as having lost its forest cover entirely. These opened up areas are set ablaze every year, precluding forest regeneration, and providing an ongoing potential fire hazard for surrounding forest and agricultural areas.

During the 1980s, the total area of the wildland forest estate in the three provinces of Kalimantan, where orang-utans occur, was still around 350,000 km^2. In the mid-1990s, between 19% and 38% of all lowland forest areas of prime habitat in Kalimantan became destined to be converted for plantations or transmigration, especially the alluvial areas. In Sumatra the percentage of forest, i.e. some 67,000 km^2 in the provinces of Aceh and North Sumatra (World Bank, 1988), which is destined to be converted is 11% for Aceh and some 16% for North Sumatra province (see Table XVIII).

Irrespective of unscheduled collateral developments, the trend clearly discerned in the plans referring to the wildland forests in Borneo and Sumatra already spells doom for the orang-utan outside conservation areas, unless stringent conservation measures are quickly integrated into sustainable wildland forestry practice.

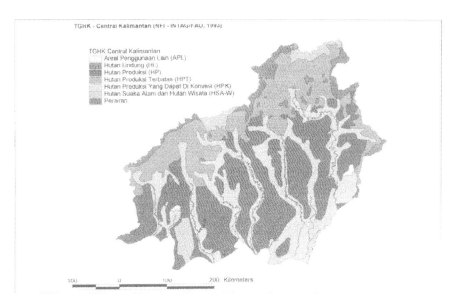

The State forest land allocation map (TGHK) of Central Kalimantan indicating that the most important sectors containing high quality habitat, i.e. riverine forest, is earmarked for conversion; however, the planning reflects a far from realistic outlook (NFI, 1993).

TABLE XVIII

Overview of the official differentiation of the forest estate into Production and Conversion Forest in the Indonesian provinces in which orang-utans occur, as established in the 1996 National Forest Inventory, revising the 1986 land-allocation (TGHK) procedure. The 'forest land' class does not reflect actual forest cover, but represents land which is in the custody of the Ministry of Forestry, and was once covered with natural forest; Conv% is the percentage of state forest land of a lowland and swamp forest status which is officially destined to be converted for plantations and transmigration.

Province	For. land	Prot for.	Prod. f.	Conv. f.	Conv%
W. Kalimantan	83,120	22,967	48,150	15,510	19%
C. Kalimantan	114,089	8,401	93,973	43,143	38%
E. Kalimantan	161,855	28,669	98,884	47,163	32%
Aceh	38,176	9,723	18,123	2,826	11%
N. Sumatra	28,941	15,434	22,798	3,603	16%

A comparison of the satellite imagery data relating to observed forest cover on with the planned allocation (see Table XVIII) reveals two facts, namely, (1) that the official figures reflect only between 60 and 86% of the photographed situation with respect to forest cover, and (2) that even if all conversion took place in areas currently already deforested – which is **not** the case – then there would not be enough real tree cover for either the planned protection function, or that of a sustainable timber production. In other words, if local rural people and private entrepreneurs continue to exert their tremendous impact on the forest cover, it is evident from the official data with particular reference to Conversion 'Forest', that the government continuously facilitates, rather than thwarts, reduction of the permanent wildland forest estate. This is not due to policy, but because of *laissez-faire* management without effective control of a bureaucracy which is plagued by petty corruption on the part of some individuals, and which generates only consistently biased information.

It must be realised that the Protection and Production Forest categories have been, and are continuously subjected to habitat degradation due both to scheduled timber extraction and to poaching, while the Conversion Forest category implies utter demolition of habitat. Thus, the official land-allocation procedure, when one analysed over the years – from one five-year plan to the next – provides little hope for the sustainability of wildland forestry in general, and for the conservation of orang-utans and other sensitive species within the structure of wildland biodiversity in particular.

That the TGHK procedure reveals higher figures for the coverage of conservation areas than other sources is explained by the fact that 'Recreation Forests' are consistently added in the official record; it must be realised that the figures for supposedly protected forest areas, i.e. protection forest, designated national parks, wildlife reserves, nature sanctuaries and recreation forests lumped together, are hardly relevant for the survival of the orang-utan.

TABLE XIX

The extent of conservation area, including recreation forest, and protection forest coverage from the latest (1996) official National Forest Inventory of the Ministry of Forestry, as compared to the conservation area coverage which is relevant for orang-utan conservation (which has been consistently used in the present report); % refers to the percentage of the extent of conservation areas of relevance to orang-utans regarding the total extent of forest cover supposedly protected, i.e. Conservation, Recreation and Protection Forests.

Province	Conservation area	Protection forest	Ou cons. area	%
W. Kalimantan	12,794	22,967	10,655	30
C. Kalimantan	6,327	8,402	4,156	28
E. Kalimantan	17,837	28,669	1,986	4
Aceh	8,324	9,723	7,472	41
N. Sumatra	2,535	15,434	816	5

The figures in Table XIX indicate that the official picture with respect to the conservation of wildlife forests in general, and the protection of orang-utan habitat in particular, is badly inflated. In reality no more than anything between 4 and 41% of the officially allocated Conservation and Protection Forest estate could have any significance for the formal protection of the ape's habitat, if such forests were effectively protected.

Despite the general trend in decentralisation, the government has reverted to centralised planning without further consensus from the regions for the latest forestland allocation procedure, as happened with the TGHK procedure in the 1980s. Perhaps this is no longer required, since a comparison of the latest maps of the National Forest Inventory presenting the land-allocation picture (1996) and other thematic maps reveals that virtually all accessible, lower altitude and irrigable land of the State forest estate has now been converted or is already scheduled for conversion. The remaining forest-estate will be on land which is of little interest to local governments or industrial entrepreneurs. Important in the context of conservation, however, is that it is also of scant survival interest for the region's unique wildlife.

Habitat degradation: the crucial role of timber concessions

Specific human activities can result in degradation of the habitat quality of natural or wildland forests, for short or longer periods, for instance, when trees are extracted or minor forest produce is harvested, or when small fields (i.e. *ladang*) are cleared for agriculture. Among these causes of temporary degradation, however, the one of most consequence is the exploitation of timber, and in particular its collateral (or side-) and after-effects. A review of orang-utan ecology, highlighted against measurements of the impact of timber exploitation, has demonstrated that selective logging ravages the ape's habitat by anything between 50% and total obliteration.

Map of a sector of Central Kalimantan showing the coverage of timber concessions in 1996.

Section III: the decline

Altogether some two hundred and seventy official timber concessions almost completely overlap the fragmented distribution range of the orang-utan in Kalimantan, and in northern Sumatra twenty-eight concessions overlap with the range[75]. On the formal land-use maps, not even the established conservation areas are entirely free of exploitation impact[76]. As noted earlier, major sectors of such areas have already been logged over during the last two decades, and in several areas timber extraction and encroachment continues.

Accurate logging data are difficult to obtain, and the information available is scattered, and inevitably of a crude or parochial nature, because it concerns exclusively the formally registered log yield in a particular area. Additional assessments of non-registered extraction in combination with interpretations of satellite imagery should complement this weak database.

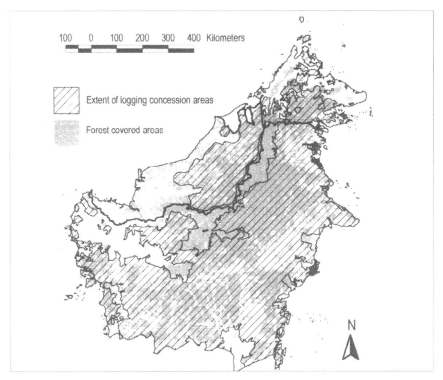

The accumulated coverage of all timber concessions in Borneo depicts more area than the actual forest cover of the island, and considerably overlaps with the few conservation areas; no data collected for Brunei.

---

[75] The number of concessions is subject to change because the Forestry Department may cancel, re-arrange and re-allocate a contract, dependent on possible violations of the regulations; until recently, concession rights could also be subdivided.

[76] It is noteworthy, however, that the official *Peta Saran RTRWP*, showing the extent of the timber concession areas in each province, clearly indicates, by means of a special legenda item, where the concession area (HPH) overlaps with conservation areas (*Kawasan Lindung*), demonstrating that the Ministry's planning department concurs fully with the exploitation of conservation areas (see also Barber *et al.*, 1995).

Be that as it may, it has been noted that, up till the mid-1980s, logging impact in Kalimantan affected some 14,000 km^2/year (Davis and Ackermann, 1988: 93). A total annual yield of anything between 9 (registered) and 15 million cubic metres of timber was extracted from the Kalimantan forests during the 1980s (*Indonesian Tropical Forestry Action Plan*, MoF/FAO, 1991). In the early 1990s, it was assessed that all remaining areas of Production Forest in Kalimantan amounted to anything between 170,000 and 280,000 km^2, of which some 108,000 km^2 had not (yet) been logged.

In Eastern Malaysia, logging intensity reportedly has also increased considerably during the last decade (Repetto, 1988; Manser, 1992): In Sabah it almost tripled from some 1,570 km^2/year in 1980 to 4,263 km^2/year in 1990; in Sarawak it increased from some 1,400 to 4,500 km^2/year. In the early 1980s, virtually all of Sabah's forests (i.e. 33,000 km^2; FAO data in Collins *et al.* 1991) were under timber concession. Since the early 1990s, much of the logged-over forests have been converted into plantations. By 1986, 86% of the forested land area of Sarawak (i.e. 84,000 km^2 in 1980; FAO data in Collins *et al.* 1991) had been conceded for timber extraction; at present, virtually all of the formerly forested state is under concession, and large areas are being converted into plantations. Only the most inaccessible secondary forest areas, in the mountain ranges and on swampy ground, remain.

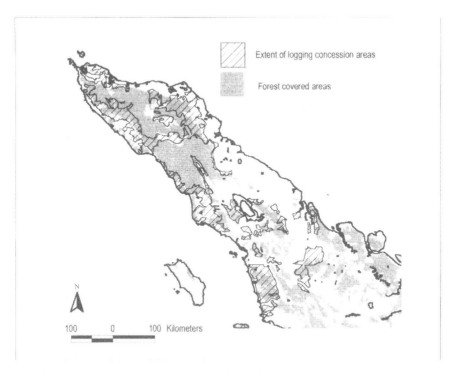

The timber concession coverage in northern Sumatra; all of the forest, except at high altitudes and on the steepest slopes, has been allocated for timber extraction.

In Sumatra the peak of logging impact (4-8 million cubic metres/year) lagged slightly behind Kalimantan, notably in the inaccessible regions of Aceh and the western flanks of the Barisan range, where the orang-utan still survived. Forest demolition gained momentum in the late 1980s, especially for large-scale plantation development and transmigration schemes. Although in 1990 the Ministry of Forestry announced that it would stop issuing new forest concessions in Aceh (*Indonesian Observer* October 10th., 1990), new concessions contracts were nevertheless signed, even for steep areas within the Gunung Leuser wildlife reserve, and, what is worse, several important areas of State forest land were demolished and spirited away from State custody to be converted into oil-palm plantations.

Considering the curve in the so-called 'production' of timber extracted from natural forests in Borneo and Sumatra, which reached its zenith during the late 1980s and early 1990s, it is ironic and troubling to realise that the increase in logging instantaneously followed the growing international and national concerns about the destruction of the rainforest. One wonders what role the government plays when increasing, rational concern for 'sustainability' appears to have spurred, rather than reduced, the destruction. One also wonders to what extent the deployment of fuzzy terms and concepts like 'sustainability', participation, integration and biodiversity has contributed to this acceleration. It is important to realise, however, that at the root of the problem have been misconceptions, deficient control, the absence of law enforcement, graft and collusion through the sharing of dividends, as well as the concomitant emergence of what may well be called a culture of timber poaching and sophisticated land grab for smaller or larger plantation estates. Very few logged-over State forests appear to have any chance of regeneration, since arson is usually one of the major tools in this illicit use of the land.

In any case, the demolition of the forests in Southeast Asia is perhaps the best example of a 'tragedy of the commons[77]', brought about by hit-and-run *laissez-faire* economics in places around the world (*The Economist* July 25th. 1994: 69 – see also Repetto, 1988, 1992; and Hurst, 1990).

Habitat fragmentation

Habitat fragmentation is the process of leaving behind more or less degraded forest 'islands' in what was originally a continuous expanse of primary wildland forest. The remaining forest fragments usually soon stand out amidst what has essentially become an environmentally hostile 'ecological desert', notably plantations, agricultural fields and fallow land. Fragmentation breaks up populations of plants and animals into

---

[77] The 'commons' are all natural resources which have been created without specific investment by any human or by a society as a whole, including public goods (e.g. roads, a lighthouse, etc.); commons are owned by no one, yet commonly used by everyone; the problem of sustainability of a commons was first described in mathematical terms by the economist Scott Gordon in 1954, and was subsequently publicised as a theory called the Tragedy of the Commons by the ecologist Garrett Hardin in an article in *Science* 162: 1243-1248 (1968).

smaller units which are more likely to become extinct due to a complex of processes in which genetic erosion through inbreeding (Soule, 1986) and increased exposure to fatal factors may play a major role. Turner (1996: 200) emphasised that especially 'animals that are large, sparsely or patchily distributed, or very specialised and intolerant of the [changed] vegetation surrounding fragments, are particularly prone to local extinction.' Thus, one may expect forest fragmentation to have an extraordinary impact on the survival chances of the orang-utan.

> The ecological effects of forest fragmentation and insularisation in tropical rainforest have been studied by Lovejoy (1980) and Lewin (1984) in a series of experiments in Brazil. Major effects are (Soulé, 1986; Turner, 1996):
> - edge effects – desiccation, radiation impact, greater vulnerability to wind and fire
> - relatively greater impact of hunting and gathering of forest produce (internal exposure effect)
> - genetic erosion due to decreased population size and small-scale calamities
> - displacement of climax species by pioneers from the edge inwards and invasion of alien species
> - changed regulatory function with reference to regional climate (dorsal effect)
> - more distinct seasonality in trees and lianas
> - crowding of animals and hence increased competition and overshoot of carrying capacity
> - extinction of larger mammals, notably predators and frugivores, due to starvation and genetic erosion
> - reduced chances for immigration and genetic exchange (distance effect)
> - shift in composition of animal species towards smaller omnivores and herbivores

The orang-utan needs an extensive area of suitable habitat in order to contend with seasonal variations in fruit production. It was elaborated earlier that the ape is extraordinarily sensitive to common logging practice, because the displacement of a few apes and the reduction of habitat quality for commuters can cause a shock wave of refugee crowding in adjacent forests. The migrants will overshoot and hence reduce the carrying capacity, even if such forest remains unaffected by logging. When one seriously considers the effects of logging, it is easy to imagine the devastating impact of inevitable forest fragmentation on such a sensitive animal. After all, dividing the typically elongated habitat structure of an orang-utan deme into two or more effectively separated fragments spells starvation for any ape which suddenly finds itself stranded in an isolated sector. In view of the extent of the ape's natural ranging pattern, and the estimated incidence of habitat patchiness, it is estimated that an area of less than 1,000 km^2 in Borneo (i.e. lower than 260 cc units in undisturbed forest), and an area smaller than 600 km^2 in Sumatra (i.e. lower than 165 cc units in undisturbed forest), must be considered insufficient for a surviving population (of 500 individuals).

On the basis of these criteria, as is shown in Table IX, some six orang-utan populations in Sumatra and thirteen in Kalimantan can still be seen to inhabit forest blocks large enough to be relatively free from a direct fragmentation impact. However, only three populations in Sabah and only one in Sarawak inhabit areas of the size required. In Sabah eleven blocks are so small that their orang-utan inhabitants are suffering from the fragmentation effects.

Section III: the decline

It must be realised that the present report can only reflect the macro-fragmentation of the distribution range and populations. Poorly controlled logging also causes fragmentation at the level of home-ranges, with equally serious consequences. In particular, the desiccation along logging roads and skid-tracks during a prolonged dry season (such as in 1997-98) can cause all forest up to some 300 m away from the tracks to dry out, die off and become extremely vulnerable to burning. In East Kalimantan the high density of (old) logging tracks and skid-roads has thus caused the die-off of virtually all the remaining forest cover. For the ape, the fragmentation of its home-range (i.e. anything between 1 and 60 km^2, comprising any number of habitat patches) into clumps of trees scattered amidst an expanse of regenerating shrubs and grass poses a serious physical constraint on its energetic economy, and exposes it to all sorts of dangers on the ground, in particular armed humans. In short, any manner of forest fragmentation will seriously affect the ape's survival possibilities.

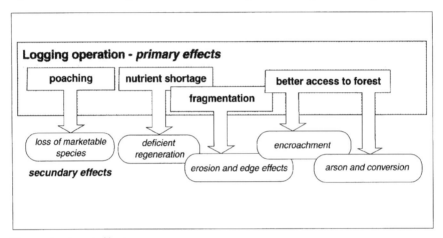

The secondary impact of logging.

Amplifier effect

The present reduction in habitat area and the fragmentation of the orang-utan (meta-) population, through timber exploitation, forest conversion, transmigration, drainage works and road building, has five major consequences:
(1) it seriously affects the typical habitat requirements in quality and extent;
(2) it splits up demes into smaller, unviable units obliged to survive in fragmented and hence insufficiently extensive habitat patches;
(3) it isolates demes, disrupting a regular genetic exchange;
(4) it causes shock waves of refuge crowding, first overshooting and subsequently depleting the carrying capacity while degrading remaining habitat; and
(5) it increases the risk of burning and of intolerable heat radiation during the dry season, further degrading and destroying remaining forest patches.
Therefore the reduction and degradation of habitat in combination with

fragmentation into ever smaller subpopulations has a complex impact with a fatally synergistic effect. This will be aggravated further when more displaced apes move into the forest refuges.

On the basis of acquired insight into the nature of the orang-utan, it is easy to predict that even temporary crowding under conditions of food shortage has a detrimental impact on the social conditions for breeding and infant care. The closer relative proximity of unfamiliar, immigrant apes of both sexes, and the breakdown of the social arena structure, is likely to disturb the social order continuously, while immigrant sub-adult and low-ranking adult males will have more opportunities to harass females. Such extraordinary conditions may, even in orang-utans, lead to infanticide, as is known to be typical of, for instance, Hamadryas baboons (Zuckermann, 1932; Rijksen, 1981).

Then, if a much reduced subpopulation eventually survives, the isolation will cause increased inbreeding and subsequent genetic erosion. Thus, a higher sensitivity of such a reduced local population to the combined effects of demographic, environmental and genetic chance events will soon result in the population becoming extinct (Shaffer, 1981).

If one adds the risk of common unscheduled impacts, like poaching of timber, encroachment and forest fires due to arson, the end result is what may well be called desertification, with no hope for the survival of the typical rainforest wildlife in general, and the orang-utan in particular. Numerous examples of local desertification in Borneo and Sumatra already exist.

The value of the conservation area network?

It is considered that the lower threshold of a minimum viable meta-population for orang-utans is above 5,000 individuals. Since the habitat of the ape has become fragmented and subpopulations can no longer exchange members, the fragmented orang-utan populations, in forest areas smaller than 5,000 km^2, must be monitored regularly in order to detect major demographic changes and to prevent local extinction. Since at present only one conservation area in the whole distribution range of the ape may be able to harbour anything up to the required minimum viable population, i.e. the Leuser Ecosystem in Sumatra, it is necessary for practical reasons to apply a lower absolute threshold of significance for surviving subpopulations of 500 individual apes/area, or some 1,000 km^2 for Borneo.

Many of the conservation areas in Sumatra and Borneo were established in colonial times. The majority of conservation areas (relevant for orang-utan conservation) that have been proposed since the 1980s (e.g. by MacKinnon and Artha, 1981, and several nature conservation PHPA/BKSDA offices in the provinces) have not (yet) been established. All the established conservation areas of any significant size reflect conditions which were (and are) of minor development interest

Section III: the decline

to humans, for example, poor soil, inaccessibility and, usually, location at a high altitude.

In Borneo, of the 14 reserves some eight may seem to be of significance for orang-utans because they cover an area larger than 1,000 km^2, namely, Gunung Niut, Danau Sentarum, Bentuang Karimun, Tanjung Puting (Kotawaringing-Sampit), Bukit Raya and Kutai in Indonesia, Lanjak Entimau in Sarawak and the Crocker range in Sabah. However, the larger reserves, such as Bukit Baka/Bukit Raya, Bentuang Karimun, Lanjak Entimau and the Crocker range, are mountainous, largely above an altitude of 500 m, and provide no more than at most 40% of forest suitable for orang-utans. Some 85% of the Danau Sentarum reserve is composed of open lakes; the Gunung Niut area has largely been ransacked and scorched, and has an estimated 2-3 % of habitat remaining. The Gunung Palung Reserve suffered badly from arson and is currently being quarried. The Kutai National Park was burned twice, and is suffering heavily from encroachment and oil exploitation. The largest established conservation area in Borneo, namely the Kayan Mentarang Reserve is not only mountainous and at an altitude above 500 m, but it is also inhabited by tribes living at subsistence level and who appear to have poached the resident apes to extinction.

Because only some 20% of the mountainous (original) Gunung Leuser Wildlife Reserve complex has forest below an altitude of 1000 m, which is suitable for orang-utans to live in, in Sumatra as well it was deemed crucial to enlarge the conservation area so as to protect and restore much of the remaining orang-utan range in the greater Leuser Ecosystem (i.e. 24,000 km^2 at present). The current situation relating to the conservation area network is demonstrated in the tables below.

TABLE XX (A)

Evaluation of the conservation areas in Sumatra relevant for orang-utans; the estimated numbers of apes are the product of the carrying capacity units and an arbitrary standard average density of 3 for Sumatra; the habitat coverage (Hab*) corresponds to Table IX, and refers to the extent of actual forest cover below an altitude of 1,000 m as seen on satellite imagery; the (?) indicates that the presence of orang-utans in this reserve has not been confirmed.

Conservation areas in Sumatra	Total area in km^2	Hab* km^2	Hab %	Cc units	Estim. # of ou
Leuser Ecosystem	24,000	4,513	19	2,029	5,070
SM Siranggas	56	17	30	5	15
SM Dolok Surungan (?)	238	47	20	10	30
CA Dolok Sipirok	69	21	30	19	57
CA Sibual-buali	50	15	30	7	21
Baruman	403	104	26	62	186
CA Rimbo Panti	28	9	32	3	10

Discussion

Traditional swidden agriculture in modern times cannot be tolerated inside reserves, as it destroys disproportional areas of forest (Iban cultivation, Sarawak, Lanjak Entimau).

TABLE XX (B)

Evaluation of the conservation areas in Borneo relevant for orang-utans; the estimated numbers of apes are the product of the carrying capacity units and an arbitrary standard average density of 1.9 for Borneo (see Table V); the cc units derive from the extent of actual forest cover below an altitude of 500 m as seen on satellite imagery. The areas affected by arson during the 1997 drought are indicated in italics, and have been corrected for the estimated loss of habitat.

Conservation areas in Borneo	Total area in km²	Hab* km²	Hab %	Cc units	Estim. # of ou
Gunung Palung (West Kal.)	900	250	28	75	143
Gunung Niut/Penrisen (W. Kal.)	1,800	30	2	9	17
Danau Sentarum (W. Kal.)**	1,250	149	12	52	99
Bentuang Karimun (W. Kal.)	6,000	1310	22	700	1330
Bukit Baka (W. Kal.)	705	142	20	43	81
Bukit Raya (Centr. Kal.)	1,106	165	15	116	220
Tanjung Puting (Centr. Kal.)	3,050	753	25	257	488
L. Entimau/B. Aie (Sarawak)	1,988	217	11	184	349
Kutai (East Kal.)	1,986	486	24	146	277
East Tabin (Sabah)	556	63	11	22	42
Kulamba (Sabah)	207	62	30	52	99
Crocker range (Sabah)	1,269	29	2	15	29
Kinabalu (Sabah)	754	51	7	13	25
Danum Valley (Sabah)	438	131	30	118	224

** the Danau Sentarum Conservation Area was supposedly gazetted in 1982, covering just 800 km² of shallow lakes, but is not shown on the official 1996 map of the conservation areas. An extension of up to approximately 1,250 km² was proposed in 1996 by the ODA Tropical Forest Management Project, to include surrounding peat-swamp forest, and another extension, up to 1,980 km², was proposed by Meijaard and Dennis (1996) in order to link the reserve with the Bentuang Karimun Wildlife Reserve. The size of the ODA extension is used in this report (i.e. 450 km² of suitable swamp forest).

In a comparison of the total extent of conservation area with data on habitat coverage, it will be noted that the latter in several cases deviates considerably from a standard range of 20-35% of the forested area. For instance, the 1,800 km² Gunung Niut-Penrisen Conservation Area has no more than some 3% of habitat left, due to massive forest destruction, in particular along the alluvial sectors. In such cases the observed forest cover (below an altitude of 500 m), as derived from the latest satellite imagery, appears to be considerably smaller than the geographical extent of the conservation area. In other instances it may also be because the conservation area is largely located at an altitude above 500 m, or, in the case of Danau Sentarum, is mainly comprised of lakes. Thus, the habitat coverage ranges between 3 and 30% of the conservation area coverage in Borneo, that is, an average of some 16 %. For the Sumatran reserves this is an average of 26%.

Taking this situation into account, two issues are of prime importance for the survival chances of the orang-utan: (1) representative coverage of the distribution range and (2) protection of the habitat within the established conservation areas.

TABLE XXI (A)

Overview of the formally protected area coverage in the regions related to the total extent of forest in 1996; the term 'Ou forest' stands for the estimated total area of forest cover within the polygons in the region; Ou habitat is the calculated area of habitat within 'Ou forest'; 'Protected' stands for the extent of established relevant conservation areas in the region; and 'P. habitat' refers to the calculated area of habitat within these protected areas. All areas are given in square kilometres. Finally, % stands for the percentage of protected habitat with reference to the total area of remaining habitat in the region.

Region	Ou forest	Ou habitat	Protected	P. habitat	%
West Kalimantan	31,592	8,364	10,655	1,786	21
C. Kalimantan	76,421	19,713	4,156	1,070	5
East Kalimantan	31,476	8,474	1,986	496	6
Sabah	10,675	3,131	3,224	336	11
Sarawak	1,877	510	812	172	34
Sumatra	25,946	8,107	24,844	3,726	45

Thus, the tables relating to protected habitat coverage reveal a remarkable regional imbalance. The overview shows that in West Kalimantan only 21% of the remaining habitat is formally protected in five conservation areas, many of which are of suboptimal – if not outright poor – quality. Central Kalimantan, although having by far the most extensive untouched contiguous swamp-forest habitat complexes in the orang-utan's distribution range, and still the stronghold of the Bornean orang-utan, has a mere 5% of the remaining habitat protected in two reserves. East Kalimantan has also reserved no more than 6% of the remaining habitat, in one single conservation area. In Sabah, 30% of the relict range may be formally reserved, but the six reserves have retained only 11% of the remaining habitat. In Sarawak, 34% of the remaining habitat is to be found in the Lanjak-Entimau / Batang Aie Wildlife Reserve complex.

In Sumatra, 73% of the range falls within the protected area network (of seven reserves), covering some 45% of the remaining habitat. At present the mountainous Leuser Ecosystem in Aceh provides some 4,500 km^2 of lowland forest habitat for orang-utans, discounting the highland plateaus. If its conservation can be established irrespective of the swell of destructive developments, it could become the most important stronghold currently available to maintain the ape and its living conditions. It must be realised, however, that 12 or 13 timber concessions operate within its boundaries, covering not only all the remaining lowland areas, but also many of the steep valleys along the west coast. A major threat to the area is further fragmentation due to road construction, as well as transmigration settlement in all the lowland areas. All of these ill-planned developments are being redressed by the Leuser International Foundation, but often at a considerable financial sacrifice and involving extensive losses of primary forest.

Another, perhaps equally important conservation area of present significance is the Baruman Wildlife Reserve in the Tapanuli Selatan regency (North Sumatra). The distribution of orang-utans in this remote mountainous reserve is still unknown. It covers some 403 km^2 of the eastern mountain chain along the Angkola valley, being part of a forest fragment of some 2,100 km^2, but it is closed in by two major roads (i.e. the trans-Sumatra highway, and the road Padangsidempuan – Pasaribuan).

In geographical terms, no more than an average of 16% of the total forest area in which orang-utans occur in Borneo has a protected status (i.e. 22,009 km^2), with the area divided into what may be considered 14 fragments. Five of these fragments are smaller than 1,000 km^2, of which three (in Sabah) may in any case well be far too small (i.e. < 500 km^2) to conserve their current populations of apes without special management.

However, it will by now be clear that the size of a conservation area reflects neither the extent of remaining forest, nor the extent and quality of remaining habitat. If one considers the lowest number of apes forming a surviving population (i.e. 500 individuals or a limit of 263 cc units for Borneo, and 166 cc units for Sumatra), then altogether no more than three reserve complexes would qualify, namely, the Bentuang-Karimun/Danau Sentarum/Lanjak Entimau complex and Tanjung Puting in Borneo, and the Leuser Ecosystem in Sumatra. In 1997 the Tanjung Puting area suffered a considerable loss of its habitat quality due to the large-scale forest fires which affected at least 30% of the area.

As shown in Table XXI, the total sum of the carrying capacity unit estimates reflecting the assumed actual habitat quality for the protected areas in Borneo amounted to some 1,800 units at the end of 1996. Applying an average of density estimates to the respective patterns of forest types in the areas for Borneo, this implies that some 3,240 apes, or some 17% of the estimated Bornean meta-population, fragmented into 14 populations, live under formally protected conditions.

Application of the same procedure to the Sumatran situation demonstrates that some 5,400 apes, or at most 42% of the remaining meta-population of some 12,700 orang-utans, occur in the seven current conservation areas, while some 40% are supposedly protected in the Leuser Ecosystem alone.

TABLE XXI (B)

The relative distribution of the estimated meta-populations of orang-utans in the regions in 1996, and at the end of 1997, and the proportion of apes in protected areas.

Region	Est. total apes 1996	Est. total apes 1997	% Red	Protected apes	% Prot.
West Kalimantan	6,695	4,194	38	2,028	48
Central Kalimantan	10,158	5,454	46	805	15
East Kalimantan	4,075	3,692	12	230	6
Sabah	1,687	same?	?	419	25
Sarawak	385	same?	?	349	90
Sumatra	12,743	same	-	4,729	37

If, according to a worst-case scenario, the current trend in forest conversion, transmigration settlement, uncontrolled timber exploitation and arson were to continue, and even if the two formal conservation area complexes of significance in Borneo (i.e. the Danau Sentarum/Bentuang Karimun/Lanjak Entimau complex and Tanjung Puting) were to be effectively protected, no more than some 10-15% of Borneo's current total orang-utan population would be conserved. The bulk of surviving apes would then belong to the western type.

The present stronghold population of Central Kalimantan (i.e. the southern population) would be reduced to a mere 8% or less, if the only areas able to be effectively protected were Tanjung Puting and the Bukit Raya – Bukit Baka complex. The northeastern population of Bornean orang-utans could easily be lost altogether. The forest cover of the Kutai Conservation Area and its surroundings was badly scorched and has died. The remnants of habitat in the northern fringe will probably be further obliterated by uncontrolled timber extraction, mining and encroachment. Sabah is also apparently unwilling and unable to establish conservation areas of significant size and habitat quality in the near future. In such a scenario, Sumatra would probably lose no less than some 40% of its current orang-utan inhabitants, because only one of the present seven areas is large enough to contain a significant population.

From this perspective it bears repeating as well, that so far a formal protected status has not prevented all of the conservation areas in Indonesia and East Malaysia from being subjected to authorised timber extraction at the lower altitudes, and, as a consequence, virtually all the prime ape habitat has been degraded by anything

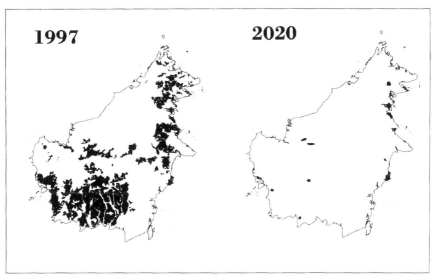

Map illustrating a worst-case scenario involving an ongoing laissez-faire attitude in Borneo; the projected remaining orang-utan distribution pattern in 2020.

between 50% and total obliteration. Indeed, the ape's future is as yet scarcely assured even within the boundaries of established reserves, and even if law enforcement relating to trespassing were effective.

Kutai represents perhaps the most extreme example of an inconsistent, resource-oriented policy, and the insignificance of a protection agency. Nevertheless, the Gunung Niut – Penrisen area, the Bukit Raya – Bukit Baka area, Tanjung Puting, and all the reserves in Sabah, and the (former) designated Gunung Leuser National Park, have also already either lost or suffered severe degradation of their sections of prime quality habitat, i.e. the alluvial valleys, swamp fringes and wet lowlands.

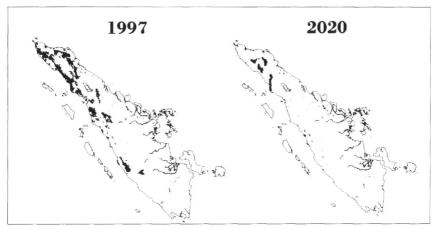

Map illustrating a worst-case scenario involving an ongoing laissez-faire attitude in Sumatra; the projected remaining orang-utan distribution pattern in 2020.

Section III: the decline

It is telling in this respect, for instance, that since the establishment of the Kutai Conservation Area in 1935, its boundaries have been shifted several times in order to accommodate timber extraction, the development of large industrial complexes and the emergence of the large township of Bontang. In the process the supposedly protected area not only floated across the landscape and was seriously degraded, but also was finally reduced to less than 65% of its original coverage, from 3,060 to 2,000 km^2 in 1993, to 1,986 km^2 on the latest official maps of the Directorate-General PHPA conservation areas (1996). Moreover, in 1982, during an exceptional period of drought, illegal squatters in the area set fire to their *ladangs*, and some 1,500 km^2 of the logged-over, and therefore dried-out, forest was burnt (Suzuki, 1989). Proposals for a revision of the boundaries to restore the basic ecological conditions of the Kutai Conservation Area (Wirawan, 1985) and reclaim some 3,200 km^2 of State forest land for conservation were never taken into serious consideration. The occurrence of fossil fuel and coal, and other booming developments, as well as the die-off of trees during the 1997-98 drought in the region, makes the future of this conservation area highly uncertain.

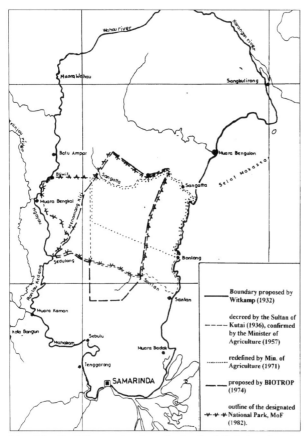

Map of the Kutai Conservation Area, showing the shifts in boundary since its establishment.

Thus, the protection of formal conservation areas leaves much to be desired. A crude evaluation of the management of the significant conservation areas in Kalimantan (Table XXII) reveals so many deficiencies that one wonders whether the existence of an official conservation agency has made any difference at all, or whether, in some cases, its interference has in fact been a liability to conservation. A rough assessment of the situation is outlined in Table XXII.

TABLE XXII

Issues of evaluation	Gunung Palung	Gunung Niut	Bentuang Karimun
Park facilities / housing / office	Deficient	Few	New
Management plan available	Outdated	Outdated	In prep.
Sufficiently motivated staff	No	No	No
Special incentives for good performance	No	No	No
Record of infringements	Few	No	No
Control of staff activities	No	No	No
Recognition legal status of park	Deficient – NP	Poor – SM	? – NP
Budget allocation on time	No	No	Project
Transport means / maintenance	Inadequate	No	Inadequate
Funds for information	No	No	No
Deployment of legal support	No	No	No
Cooperation with army / police	Deficient	No	Poor
Authority in law enforcement	Poor	No	No
Integration in local government	No	No	No
Cooperation by forestry authority	Deficient	No	Insufficient
Warden/guard welfare scheme	No	No	No
Field equipment	Poor	No	Inadequate
Villages in conservation area	No	Yes	Yes
Analysis/ monitoring of threats (GIS)	No	No	No
Natural history knowledge	Insufficient	Fair	Deficient
Licence system for utilisation	No	No	No
English language skills	Fair	No	Insufficient
Issues of evaluation	**Danau Sentarum**	**Bukit Baka**	**Bukit Raya**
Park facilities / housing / office	Adequate	No	No
Management plan available	New	No	No
Sufficiently motivated staff	Insufficient	No	No
Special incentives for good performance	No	No	No
Record of infringements	Fair	No	No
Control of staff activities	No	No	No

Table XXII (continued)

Recognition legal status of park	Poor -?	Poor – SM	Poor – SM
Budget allocation on time	Insufficient	No	No
Transport means / maintenance	Inadequate	No	No
Funds for information	No	No	No
Deployment of legal support	No	No	No
Cooperation with army / police	Insufficient	No	No
Authority in law enforcement	No	No	No
Integration in local government	No	No	No
Cooperation by forestry authority	Poor	No	No
Warden/guard welfare scheme	No	No	No
Field equipment	Fair	No	No
Villages in conservation area	Yes	Yes	Yes
Analysis/ monitoring of threats (GIS)	No	No	No
Natural history knowledge	Insufficient	No	No
Licence system for utilisation	No	No	No
English language skills	No	No	fair

Issues of evaluation	Tanjung Puting	Kutai
Park facilities / housing / office	Adequate	Deficient
Management plan available	Outdated	Outdated
Sufficiently motivated staff	Fair	No
Special incentives for good performance	No	No
Record of infringements	Hardly	No
Control of staff activities	No	No
Recognition legal status of park	Deficient – NP	Deficient – NP
Budget allocation on time	Yes	Insufficient
Transport means / maintenance	Inadequate	Inadequate
Funds for information	No	No
Deployment of legal support	No	No
Cooperation with army / police	Fair	Inadequate
Authority in law enforcement	Yes	Insufficient
Integration in local government	Fair	No
Cooperation by forestry authority	Yes	Deficient
Warden/guard welfare scheme	Inadequate	Inadequate
Field equipment	Fair	Deficient
Villages in conservation area	Removed	Yes
Analysis/ monitoring of threats (GIS)	No	No
Natural history knowledge	Deficient	Poor
Licence system for utilisation	No	No
English language skills	Fair	Insufficient

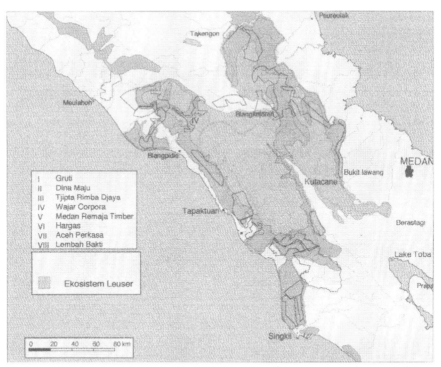

The Leuser Ecosystem: the crude extent of the distribution pattern of the orang-utan is shown; note that 95% of the range was outside what used to be the designated Gunung Leuser National Park (stippled line); timber concessions and conversion projects very precisely overlap with the distribution pattern of the orang-utan and other large mammals, almost as though obliteration of the protected mega-fauna had been planned; the Leuser Development Programme has been designed to reduce these threats.

Evidence of Poaching

Over its entire distribution range the orang-utan is a protected species (*Surat Keputusan Menteri Pertanian* No 421/Kpts/Um/8/1970, and No 327/Kpts/Um/7/1972 in Indonesia, the Fauna Conservation Ordinance 1963 of Sabah, and the Wildlife Protection Ordinance 1957 of Sarawak). Consequently, all hunting of the ape is illegal, and to be formally designated as poaching.

Due to concern in the West about the disappearance of traditional values in tropical countries, people often argue that the hunting of protected wildlife, like orang-utans, by indigenous tribal people may be 'traditional' and 'sustainable' and should therefore be permitted. This is nonsense in view of the fact that several essential traditions like ritual cannibalism (Batak) and head-hunting (all so-called Dayak tribes), which were certainly 'sustainable', can apparently also not be allowed to persist in modern society. Moreover, there is no 'sustainable' hunting possible with respect to the orang-utan; wherever indigenous tribes have hunted the ape for long periods, it has become extinct.

The traditional battle between humans and the ape; illustration from Wallace's famous book The Malay Archipelago (1869) 'drawn on wood' by Wolf; in reality it is always the human who wins.

In the course of the present survey, the poaching of orang-utans in Kalimantan was reported on more than twenty separate occasions, and covered the entire distribution range. The incidence of poaching has attained unimaginable proportions as a result of the 1997-98 fires in Kalimantan (and the following political and economic crises). Apes fleeing the smoke and the burning forest patches were often captured, slaughtered, butchered and eaten by local people who had been waiting at the forest's edge, armed with bush knives, clubs and spears.

Table XXIII

Locations from which poaching was recorded in Kalimantan, and the cultural identity of the poachers:

Kendawangan (West Kalimantan)	Malay, Chinese, urban elite
S. Embaloh (upper Kapuas, West Kalimantan)	Maloh
Gunung Setutuk (West Kalimantan)	Land Dayak, urban elite
Gunung Niut (West Kalimantan)	Land Dayak, urban elite
Nanga Awen (upper Kapuas, West Kalimantan)	Iban/Maloh
Bkt. Menyukung (upper Kapuas, West Kalimantan)	Iban/Kayan/Maloh
Ulu Kapuas (West Kalimantan)	Ot Danum, Chinese
Mabit (Sanggau area, West Kalimantan)	Kantu'/Mualang
Seluwah (Sanggau area, West Kalimantan)	Iban Sebaruk
North of Sanggau (West Kalimantan)	Iban Mualang
Lanjak Entimau Wildlife Sanctuary (Sarawak)	Iban
middle S. Katingan (Central Kalimantan)	Ma'anyan
S. Hanyu (upper S. Kahayan – Central Kalimantan)	Ngadju
S. Arut (Central Kalimantan)	Ngadju
Tapin Bini (S. Lamandau, Central Kalimantan)	Ngadju
Palangkaraya – Kesongan (Central Kalimantan)	Ngadju/Transmigrants
Seluwah hills (East Kalimantan)	Selako
Kutai	Urban elite
Muara Wahau (S. Wahau, East Kalimantan)	Kenyah
Pampang (S. Sebulu, East Kalimantan)	Kenyah
Segah (S. Malinau, East Kalimantan)	Punan Malinau, Basap
S. Tubu (north East Kalimantan)	Penan
Apo Kayan (East Kalimantan)	Kayan
S. Lurah (East Kalimantan)	Kenyah Badeng, Penan

Data on poaching are based mainly on the final results of the illegal activity, notably the detection of illicit trade goods such as young live apes, skulls, pieces of skin, etc., and on second-hand information. Few people openly admit that they have poached orang-utans, because most people in Indonesia are aware of the strictly protected status of the ape. Nevertheless, the circumstantial evidence in the form of infant and juvenile apes (and skulls) in illegal custody and for sale, detected in the period 1990-1997, is so overwhelming that it is justified to state that poaching of orang-utans is shockingly common in both Borneo and Sumatra, especially in readily accessible forest areas. Persecution of the ape, however, is not restricted to any particular social class, such as 'poor' rural forest scavengers or tribal people. On weekends and during holidays, poaching safaris by hunting clubs belonging to both the army- and the industrial elites, are reportedly 'frequent'. Equipped with modern weapons and powerful four-wheel drive vehicles, such safaris enter forest areas along the logging tracks, hunting indiscriminately whatever animal is encountered, including formally protected orang-utans. The survey in Sumatra revealed that within the distribution range of the ape, major poaching locations for such clubs are the timber concessions near Sikundur, the Tamiang area, the Bengkung area and the environs of Natal (Angkola). Some timber companies even offer to such clubs catered hunting trips in their concession areas. The conservation agency, which should be informed of any hunting activity, seems to be wholly unaware of these transgressions.

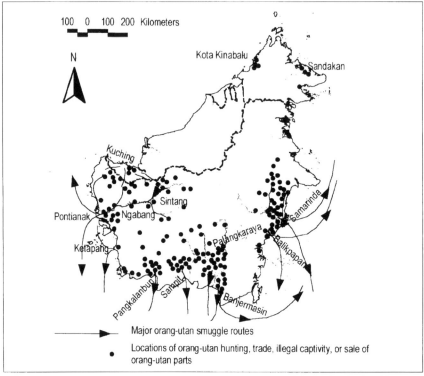

Trade centres featuring orang-utans, and locations where apes were reportedly butchered for consumption and for the production of 'souvenirs' in the 1990s.

In Sarawak as well, an inquiry among the members of several Iban longhouses confirmed the expected trend: Most people denied that orang-utans were ever hunted by clan members, but it was alleged that neighbouring clans commonly did so. However, after becoming intoxicated at welcoming parties, some individuals openly boasted that they frequently hunted the ape, as it was the easiest prey and its meat was delicious (Rijksen, unpubl.), thus corroborating similar, century-old information from all across Borneo. For Sarawak, Bennett and Dahaban (1995) noted, 'any animal seen was usually shot at, including primates ... Night hunting with vehicles and spotlights occurred in any area with road access and was especially heavy in weeks preceding local festivals.' The same is true for Sabah; the fact that fire-arms are more widely available in the East Bornean states has probably resulted in an even more devastating impact on local ape populations than has been possible in Kalimantan, where the possession of firearms is restricted to police and army personnel. In any case, the regular supply of new, young rehabilitant orang-utans to the Sepilok rehabilitation centre in Sabah compounds evidence that the ape is still a victim of poaching in the northern Malaysian state of Borneo.

The surveys have also revealed that a considerable illicit trade in orang-utans occurs across the border of West Kalimantan and Sarawak. This international smuggling was reported by local people to occur at all border crossings. A check at one of the border crossings showed it to be unguarded, with people and their assorted trade goods passing through uncontrolled. It is probable that most of the orang-utans confiscated in Sarawak to be sent to the Semengok centre are of Kalimantan origin.

Furthermore, intensive illegal trade was reported to occur via the Apo Kayan and upper Kapuas areas from Indonesia into Sarawak, from Sampit and Banjarmasin in Central Kalimantan. Also from Pontianak in West Kalimantan, from Sampit and Banjarmasin in Central Kalimantan, and from Samarinda and Sangkulirang on the east coast, regular trade in apes was reported to ports in Java and to Singapore, Bangkok, Hong Kong and Taiwan, and other destinations abroad (Smits, pers. comm.). News reports of the early 1980s had already made mention of the illegal trade in orang-utans from Samarinda and Pontianak (e.g. *The Jakarta Post,* October 24th. 1980).

> During the surveys in Kalimantan, an old and persistent Asian tale of superstitious barbarism was repeated by several informants. During the last decade shipping crews from Taiwan, Korea, the Philippines and China purchased orang-utans, not only to sell as exotic pets at home, but more often to devour the brain of the creature while it was still alive. One informant in Samarinda Harbour even claimed to have witnessed such substitution cannibalism on a number of occasions. Amidst a crowd of drunken men, the orang-utan was tied, and its head inserted through a hole in a board. The owner then sliced open its scalp to expose the ape's cranium, and the brain of the dying victim was spooned out and eaten by many of the participants, allegedly to increase their sexual potency.

Numerous international smuggling routes also reportedly exist from Medan and other major ports in northern Sumatra to Singapore, Thailand, and towards further Asian destinations. Notably the large fishing ships and the cargo vessels carrying timber and shrimps often transport illegal shipments of protected wildlife. Law enforcement is virtually absent, and several informants mentioned that the detection of an illegally transported protected species is usually an occasion for serious graft.

Until recently orang-utan skulls were openly traded in all the main tourist cities in Kalimantan, notably Pontianak, Muara Teweh, Balikpapan, Banjarmasin and Samarinda. Prices ranged between thirty and one-hundred thousand Rp (US$ 15 – 50). The more expensive skulls were those decorated with so-called traditional carving. It is perhaps relevant to note that there is nothing traditional at all in such decoration; the human and ape skulls acquired, venerated and maintained in the religious head-hunting tradition of Borneo could be dressed but were never carved. The skulls sold as 'antique' artifacts were in fact obtained from recently slain orang-utans, decorated with carvings in small cottage industries and dyed with shoe polish to make them look antique, as though they had been smoked over the hearth.

Skulls were on sale in numbers varying between one and ten in any souvenir shop, and usually were offered for sale along with the supposedly magical parts of other protected species, notably bear and tiger, as well as the skulls, teeth and claws of the clouded leopard, and the antlers of deer. Orang-utan skulls were also on sale in Jakarta, Medan and Bali.

In 1991, the Ministry of Forestry issued a special instruction to the police in Kalimantan to put a stop to any further trade in orang-utan skulls. Since that time, skulls are sold from under the counter, and their prices have increased.

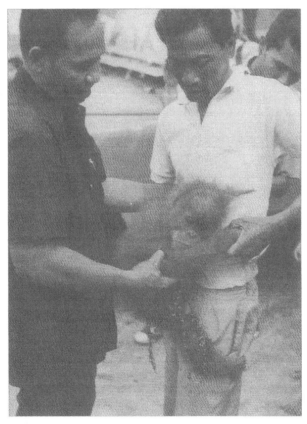

Every captured orang-utan has been the victim of an unbearable tragedy: a ransacked habitat, often a butchered mother, imprisonment in a stinking cage, manhandling by noisy, aggressive humans, inadequate and inappropriate food, and loneliness and torture in custody or while being traded and transported.

Diagnosis of the underlying (ultimate) causes of decline

General

The surveys have revealed that the survival of the orang-utan is critically endangered because of poaching, habitat loss (forest conversion and encroachment), habitat degradation (timber exploitation, arson) and its shock-wave effect on the orang-utan's socio-ecology, and habitat and population fragmentation. These processes are a manifestation of a complex of more fundamental underlying causes, of which (1) the global market forces seeking resources and commodities (2) greed for land, and (3) deficient conservation and protection are the most prominent issues. Poaching is fuelled by a primitive predatory attitude, lack of education and inadequate economic conditions. These factors will be examined in this chapter in detail. The manifestation of these underlying regional causes may in fact differ somewhat per political unit, but since almost 95% of the orang-utan population occurs in Indonesia, emphasis is given to the Indonesian situation.

In its report entitled *Corporate Power, Corruption and the Destruction of the World's Forests*, the Environmental Investigation Agency (EIA) in 1997 presented the results of a study on the impact of transnational timber trading corporations. Its major conclusions were:
- The global timber trade, dominated by transnational companies (TNCs), plays a central role in the destruction and degradation of the world's diminishing natural forests and the ensuing loss of environmental, social and economic functions. This trade remains outside of any coherent global regulation.
- While forests are national assets, they are of immense international importance, particularly as storehouses of biodiversity and in their role in climate regulation. Similarly, both the causes and effects of forest loss are frequently international in nature and extent.
- Increasing trade liberation and the globalization of trade have led to a shift of power away from national governments to corporate leaders and have undermined the ability of national governments to control and regulate effectively the international timber trade.
- Transnational companies often have the economic and political power to undermine national management of forest resources by gaining unrestricted access to forests.
- Excessive, increasing and unsustainable international demand by the developed world for timber and other forest products (e.g. *rotan, gaharu*) is having a devastating impact on forests worldwide.

To this last conclusion can be added (a) the world-wide demand for all sorts of industrial cash-crops, such as palm-oil, corn and cassava, for which thousands of square kilometres of forest are being destroyed each year, until all accessible land will have been appropriated. And (b) a global policy of international organisations to frustrate conservation through euphemistic concepts, such as 'sustainable development' and 'participatory management.'

> In 1997, the total area of oil-palm plantations in Indonesia covered some 25,000 km². The annual increase in plantation size was close to 2,000 km² during the 1990s, and is expected to reach well over 50,000 km² soon after the turn of this century. The forest area has already been allocated in the latest TGH procedure, and in January 1998 the Minister of Forestry formally released some 40,000 km² of State forest land for this purpose. Ninety-two percent of Indonesia's palm-oil production is in Sumatra, and some 5% in Kalimantan. Malaysia, including Sabah and Sarawak, has some 28,000 km² at present, which is to be extended to a maximum possible coverage of 36,000 km² by 2008.
>
> As with other 'natural resource industries' in Indonesia, the production and sale of palm-oil is controlled by a limited number of powerful companies in which family members of the former President Suharto played major roles until June 1998. Four companies, namely Sinar Mas, Astra, Salim and Raja Garuda Mas, control 68% of all privately owned estates; they are also the main players in control of Indonesia's timber industry. Palm-oil is used in many household and luxury products (e.g. soap) worldwide. Indonesia produces palm-oil cheaper than anywhere else in the world, out-competing Malaysia fivefold in labour costs. Malaysia and Indonesia cater to some 80% of the world's steadily increasing demand. The main industrial buyer of raw palm-oil and crude derivates is the Anglo-Dutch multi-national Unilever; virtually all Indonesian palm-oil passes through the Netherlands. Main investment loans to boost conversion of forest and production of palm-oil have been provided by the World Bank, two large Dutch banks and a Japanese bank (Wakker, 1998).
>
> After the economic crash in 1998, Indonesia relaxed its policy for foreign investment in oil-palm plantations, and allocated one million hectares for this purpose in eastern Indonesia. It also transferred governmental control of plantations from the Ministry of Agriculture to the Ministry of Forestry. In March 1998 North Sumatra was officially declared 'closed' for further forest conversion.

A central issue in this complex of problems on the regional level is the competitive conflict between the orang-utan and humans for space and living conditions. The orang-utan has specific habitat requirements, and unfortunately humans have a preference for the places where these requirements are filled, notably in alluvial valleys. As a matter of principle the legal framework should allow a government to regulate this fundamental conflict. However, for several reasons this is not so simple in practice: First, because the orang-utan, which is considered 'just' an animal, is an incompatible party in the conflict; second, because a hugely powerful international market for consumer products (wood, pulp, palm-oil, *kemiri* or lumbang oil, patchouli oil, incense, coffee, beef, animal feed, sugar, etc.) entices both local people and their government structure to clear and convert ever larger areas of wildland forest, irrespective of its ecological value, status and the presence of formally protected species; third, the still considerable human population growth in some regions is spilling, both unscheduled and by means of transmigration, into the forest estate. Both these latter factors render land-use planning and other regulatory measures out of date by the time the actual field situation is assessed.

Finally, the allocation of State forest land, without a firm enforcement of the territorial status, leads to an image of 'free land'. After all, in 1954, the economist Scott Gordon had already noted that 'everybody's property is nobody's property', and that 'wealth that is free for all is valued by none, because he who is foolhardy enough to wait for its proper time of use will only find that it has been taken by another.' Hence, State forest land is pillaged ultimately because entrepreneurs are able to own the timber – and often even acquire a title to the land – in a way that they cannot own the forest as a living and productive system (De Soto, 1993). Thus, surprisingly enough, by downright ignoring, or being unable to enforce the territorial right in the hope that regulation alone will be enough, the government inadvertently becomes the prime cause of the exponential tragedy of the commons concerning the rainforests and their biological diversity (Ridley, 1996).

Tragedies of the commons characterised the evolution of European states during the late Middle Ages, and resulted in the total annihilation of the once virtually uninterrupted expanse of wildland forests spanning the continent, as well as in the extinction of all the larger mammals. Indeed, the disastrous and amoral conduct of a few increasingly powerful 'free-riders,' exploiting and laying to waste every single marketable resource, seems to have characterised all the so-called developing societies in their apparent striving for an ideal of civilisation. Ironically, this factor as the driving force of industrial and trans- or multinational neo-colonialism is still clearly recognisable.

Swamp-forest areas in the southern half of Borneo became the frontiers of 'development' in the 1990s; draining and levelling the area is easy, growing crops is almost impossible.

The sociologist Elinor Ostrom has demonstrated theoretically that the best way to avoid such tragedies, and to evoke the required general constraints as regards sustainable utilisation of a resource, is to facilitate free and open communication (Ostrom, *et al*, 1992). After all, at the basis of all tragedies of commons lies a

Section III: the decline

> The natural forests in Southeast Asia have survived into the twentieth century largely by default. That is, until the end of the nineteenth century the low human population density and simple technology, in combination with tribal territoriality and personal *adat* claims on certain resources, meant that indigenous tribal societies (outside Java) had an almost insignificant impact on the natural structure of biological diversity, with the exception of rhinos, and, in some areas, orang-utans. Moreover, the international market had scant interest in the timber resources until the late 1960s.
>
> With the formation of a State during the first half of the twentieth century, a central government assumed full responsibility for the whole territory, abrogating tribal and regional territorial claims. Indeed, the State philosophy in Indonesia (i.e. *Pancasila*) explicitly rejects cultural or tribal segregation and does not recognise 'indigenous communities' (Sumardja, 1997). In effect, all land beyond the cultivated estates for which a right of use could be claimed became 'State land', and the government, like any government at a similar stage of development, hoped to control the use of such (forested or wild) land by means of regulations. In reality, however, the originally sound principles have been corrupted to yield an ambiguous framework in which opportunities abound for piracy and land grab by enterprising individuals wielding some authoritative power.
>
> Thus, notwithstanding the best intentions of the authorities (e.g. Suryohadikusumo, 1992), in reality the regulations are designed primarily to control profit of crop production relating to the land (Ostrom, *et al.* 1992), rather than to induce sustainability, or protect possible long-term life-support functions in a given area.
>
> Under such social transition conditions as emerge with the formation of a State, by claiming the territory in the common interest, yet failing to enforce the claim, a government in fact creates a free-for-all resource. Under these circumstances, the winner is the person(s) who – when everyone else is exercising restraint on the moral conviction that such a strategy is for the 'common good' – has no compunction whatsoever about plundering environmental resources. A policing force is commonly intended to guarantee that the proper control is consistently applied, lest a 'free-riding' opportunist bring about a disaster for all, enriching himself at the expense of others, i.e. in a tragedy of the commons (Gordon, 1954, Hardin, 1968).

'prisoners dilemma', a game-theory model in which two egoists must be made to cooperate for the common good and resist the temptation to profit at the other's expense. Such mutually beneficial cooperation is most readily achieved if the two egoists communicate openly (Ridley, 1996). However, for conservation of wildlands and delicate organisms beyond the scope of (sustainable) utilisation, the theoretical solution involving open communication must still be supplemented with effective law enforcement.

Land-use conflict

At the close of the twentieth century most people worldwide do not seem to have yet reached the developmental stage at which any deep concern about evolution plays a role, or at which a serious kinship can be considered with anybody outside of their primary ethnic sphere, to say nothing of a separate species altogether. Southeast Asian people of today are no exception: By people in Borneo and Sumatra, the orang-utan is commonly considered 'just another (protected) animal', with no consideration given to its existential requirements. The conflict which arises from the prevailing view is that in the surge of human demands for land, resources, wealth and welfare, the formally protected ape is denied, directly or indirectly, its habitat and living space. In fact, the reports of rural people poaching and slaughtering orang-utans, for whatever reason, indicate that the creature is denied its very life when and wherever it is encountered. One wonders how such an abysmally low ethical standard could develop, much less be retained, in the twentieth century.

Be that as it may, the conflict over land use by humans and by the ape arises from a complex of issues, which can be roughly differentiated into the following:
- exponential human population growth, and correspondingly expanding consumer demands;
- a world market and its canonized ideology demanding socio-economic growth.

When fed into a computer model, data on population dynamics and expanding land use can readily demonstrate that if Indonesia were to succeed in reducing its annual population growth rate from the currently already impressive 1.6% to 1%, the demand for farmland would drop from a current 2% expansion to 1.4%, and the projection of related forest destruction up to the year 2030 would be almost halved, from a foreseen 450,000 km^2 to some 250,000 km^2 (Jepma and Blom, 1990) if all other factors affecting the despoliation of forest remained the same.

The world market is at present undoubtedly the most formidable destructive force with regard to the forests in Southeast Asia because of its voracious demand for timber, coal, gold, charcoal, and minor forest produce (e.g. rattan, resins, copal, incense, ornamental plants, animals), as well as for land for the production of vegetable oil (oil-palm), rubber, animal feed (e.g. tapioca, maize) and luxury consumers products and drugs (coffee, tobacco, spices, tea, patchouli oil, etc.). Since the 1990s, Indonesia's annual log harvest has reached its peak, exceeding 30 million m^3, and with more than 17 million m^3 annual input into the international market it has surpassed Malaysia as the world's largest exporter of tropical timber, much of it in the form of disposable plywood (Hurst, 1990). Indonesia is also striving to surpass its neighbours in the production of palm-oil and consumer goods, by keeping labour costs as low as possible.

Considering these forces of 'development', it is easy to ignore the fact that the population is still increasing and expanding into the forest estate at an alarming rate. Furthermore, it is never taken into account that if the entire present human population of the world, i.e. over 5.400 million, were to seek to enjoy the same standard of living and consumer opportunities as those prevailing in northern Europe, Japan or the USA, it would require four times the surface area of the earth to produce the desired goods and ecological services. In other words, if the current state of the world market reflects the desired consumption level, and this would be steady, while acknowledging that the earth is presently at the very limit of its resources, the human world population would need to be reduced by at least 75% in order to allow uniform standards of living and welfare. In such a scenario, it would not be possible to reserve habitat for wild orang-utans to survive.

> For example, the almost 16 million inhabitants of the over-populated industrial city-state of The Netherlands require some 114,000 km^2 of land area for agricultural production, of which some 60,000 km^2 is used for livestock feed only. With a national surface area of only 24,000 km^2 allocated for agricultural land, this implies that some 90,000 km^2 of fertile land **outside** the territory of the country is in use to support the Dutch economy, of which some 20,000 km^2 is situated in tropical countries (mainly Asia and Latin America). In terms of wood consumption and trade, the Dutch economy makes use of the annual harvest of timber from some 150 km^2 of tropical rainforest in Indonesia, and 102 km^2 in Malaysia (data from *The Netherlands and the World Ecology*, Netherlands Committee for IUCN, 1994).

The islands of Borneo and Sumatra are no exception to this global model; the current density of consumers already overshoots the meagre carrying capacity of the few fertile productive regions in these islands. Not only consumer goods, but even food (i.e. rice) has to be imported from other regions. A prolonged drought coinciding with a drop in international economic exchange, such as happened in 1997-98, immediately resulted in a serious famine in what was once considered a 'paradise' in the early 1990s. At present, people are moving into the marginal areas, destroying the vestiges of the ecological assets all throughout Borneo and Sumatra.

In order to give the wild orang-utan – as a representative of ecological integrity – a fair chance of survival under natural conditions, not only the governments of Indonesia and Malaysia, but also of all the countries with which these nations maintain major trade relations, must, with considerably greater zeal than hitherto, address their demographic problems concerning human population growth and land allocation. Concerns and talks about a desire for sustainability, ecological integrity and the conservation of biodiversity are absolutely meaningless if these fundamental factors are not taken into very serious account. Indeed, rather than for anything else, such terminology is more often used to conceal perpetuation of the *'laissez-faire'* attitude resulting in ongoing exploitation and to open up previously inaccessible forest areas.

Thus, a person with any interest or authority whatsoever in relation to the rainforest areas in which orang-utans occur, be he (or she) an illiterate gatherer, a subsistence farmer, a timber tycoon, a tourist, a scientist, a bureaucrat, or a minister, is little more than a *wayang* puppet on the screen of a mighty international market directed by a powerful meta-population of technologically advanced and over-populated nations, notably Japan, Taiwan, northern America, China and Europe, or of regions, especially Java, Selangor, Singapore and Hongkong. Every individual in his or her legitimate desire for economic security and welfare will make rational and expedient decisions with reference to forest and land use on the basis of the conditions which are, presumably, set by national authority and power structures, but ultimately are dictated by this interlinked and voracious market system. It is doubtful whether room exists at all for the survival of any wild orang-utan into the twenty-first century, unless at least Indonesia can reassert a more independent position and determine its own agenda for a future in which conservation of its unique natural resources is a priority issue.

The legal framework: asset or liability?

Ultimately, the legal framework for the protection of biological diversity (i.e. fauna and flora) is the basis for a possible resolution of the human *versus* ape land-use conflict. In Indonesia the legal framework of relevance for land use with respect to forests comprises (1) the National Constitution of 1945, (2) the Basic Forestry Law (1967), (3) the Act concerning Basic Provisions for the Management of the Living Environment (1982), and (4) the Act concerning the Conservation of Living Resources and their Ecosystems (1990). The whole framework is explicitly 'based on the principle of harmonious and balanced sustainable utilisation of living resources and their ecosystems' (General Provisions, Ch. 1; Art. 3), referring back to the National Constitution of 1945.

The Constitution states: 'Land and water and the natural riches therein shall be controlled by the State and shall be made use of for the greatest welfare of the people'. While the legal framework contains a strong element of ambiguity, between an *obligation to use* natural resources, and a *desire to conserve* (some) valuable landscapes, animals and plants, it does not provide prescriptions to safeguard the habitat of protected species, despite the title of Act No 5 concerning the protection of organisms, which includes the phrase 'and *their* Ecosystems.' With regard to the conditions for survival of species, the Act refers to 'sanctuary reserves' (Art. 15, 16, and 19), and Articles 8 and 9 concerning 'life support systems' may also have some bearing on this issue[78], although by life-support system is meant 'a natural process of various elements of both living and non-living resources which ensures the continued existence of living organisms' (Art. 6), 'for enhancing human welfare and the quality of human life' (Art. 7).

> Art. 13 states: '(1) The preservation of plant and animal species shall be implemented both inside and outside natural sanctuary reserves'; and '(3) The preservation of plant and animal species outside the sanctuary reserves shall be conducted by protection and promoting breeding efforts of the species to avoid their extinction.' Chapter V of the Act on the preservation of plant and animal species, however, fails to make any reference to habitat protection.

Further, the legal basis for land allocation other than exploitation is extremely weak while ownership of land is legally impossible. Most important, however, is that the representation of the State in potential conflicts pertaining to land use is often ambiguous, seeking paths of least resistance and maximum economic gain, rather than long-term common-interest strategies. Hence, in cases concerning forests, the authorities often yield to claims by individuals, with reference to the constitutional issue of 'greatest welfare of the people' or traditional rights (i.e. *adat*), as long as the

---

[78] It is remarkable, however, that in Article 10 a sense of realistic caution is detected with regard to law enforcement as outlined in the preceding Article (9), which states: 'degradation within a life support system area due to natural processes or unwise utilisation or other causes, shall be followed by planned and continuous rehabilitation efforts.'

claims do not conflict with any immediate economic interest on the part of the State. In any event, some private sharing in the profits ensuing from the land use may be involved.

Part of the legal framework ambiguity undoubtedly stems from the general ideology underlying the Indo-Malay social order which is rooted in consensus, i.e. the *musyawarah* principle. It is a system of joint agreement and compromise to be achieved under the guidance of a respected and elderly member of the respective society (Koesnoe, 1969). In its report concerning *Forest Policies in Indonesia* (1985), The International Institute for Environment and Development (IIED) has also credited this exalted principle to justify the premise that 'policing actions to prevent illegal activities is not the preferred method of protecting natural resources. If it is deemed necessary to take certain action such as protecting forests, then a management system which deters or deflects destructive elements is much the preferred path.' Unfortunately this standpoint appears to be applied almost exclusively in instances **when conservation or protection** of natural resources is required. In cases where exploitation of the resources is the issue, however, such liberal idealism is unheard of, and a simple power-based system aimed at consumption readily overrules all social and traditional considerations. It is remarkable that, in all of the present world cultures, ethical and humanitarian purism is indicated most readily and almost exclusively where conservation of nature is the issue.

Quite naturally, the human communities in and around forest areas are always on the lookout for any opportunity for gain. This may apply in particular to transmigrants and pioneer settlers, but long-term residents having an indigenous background usually display a similar opportunism (Vayda, *et al.* 1980). In this all people are evidently guided by a rational risk assessment of the chances involving the least resistance and fewest constraints. With respect to land, the most fruitful opportunities in this respect concern State forest land.

In Indonesia, virtually all the forested areas in which orang-utans occur comprise State forest land (*Hutan Negara*), in the custody of the Ministry of Forestry, and according to the legal framework, 'all land within two kilometres of a river is available for use by local inhabitants.' As a consequence, the alluvial valleys and well-watered lowlands, indeed the prime habitat of the orang-utan, have in fact been designated for deforestation, and are falling prey most readily to conversion and settlement. Earlier it was demonstrated that even a formal conservation status can do little to prevent this destructive development process, because the organisation assigned to protect a given area has neither sufficient power nor the uncompromising commitment to do so (Davis and Ackermann, 1988). It is therefore perhaps not surprising that many people have a misconception about 'empty' land (*tanah kosong*) with regard to State forest land, including the protected areas.

In order to regulate land use, in the 1980s the government established a new, integrated land-allocation procedure with respect to State (forest) land. Special interdepartmental committees in every province were charged with implementation of the procedure by consensus under the guidance of the State Planning Agency (Bappenas/Bappeda), and in 1986 the first formal land-use planning (TGHK – *Tata guna hutan kesepakatan*) maps were issued at the regional level. Since that time all State forest land has been accorded an official status with reference to its present and future fate regarding land use. The procedure placed 75% of the country within State forest boundaries, of which some 60% was originally (1983) intended for permanent forest cover (Davis and Ackermann, 1988).

The status categories devised according to the TGHK procedure are formally laid down in the legal framework, and concern:
– Conservation areas (*kawasan pelestarian alam*),
– Protection forests (*hutan lindung*),
– Common production forest (*hutan produksi biasa*)
– Limited production forest (*hutan produksi terbatas*)
– Conversion forest (*hutan konversi*), and
– Forest plantations (*hutan tanaman industri*).

> The government's land-allocation procedure was a step towards decentralisation and is soundly based on consensus from the regional level all the way up to the central government. It was primarily intended to better regulate the authority of the Ministry with respect to the forest estate, but also served to regulate exploitation of the vast forest resource and call a halt to the devastating slash-and-burn cultivation. Brought into being on the recommendations of FAO, the procedure was intended to bring the unruly mass of subsistence farmers (*perambah hutan*) under control of the Forestry Department for encouragement to adopt a more sedentary lifestyle. However, the effect was that slash-and-burn cultivation was further encouraged, as the areas of the local communities were not demarcated and the intended control remained virtually absent.

In the process, a number of local communities were incorporated into what had now become State forest land. Their presumed customary rights to previously cultivated, and henceforth abandoned (or fallow) lands were the core of an ambiguity in legislation which has facilitated the rape of the wildland forests as well as collusion. The abrogation of supposed customary territorial 'rights' is evidently one of the prices to pay to achieve a modern civilised State form, but it does not come without resistance. With Indonesia's independence, when a scatter of colonially dominated sultanates, fiefdoms and tribes were forged into one nation, it was apparently unacceptable to enforce such abrogation altogether. Thus, the issue of traditional (*adat*) rights simmers on under the surface of unified nationalism, making well-regulated land use in general, and conservation in particular, virtually impossible, in the light of corruption and the desire to wield power. Nowadays, all but the conservation areas on State forest land are formally given in concession by the Minister to a third party, either the private sector or a semi-government agency (*Perum*) for exploitation of the timber resource, and, in some cases, the land, but an uneasy feeling about possibly forsaken rights cripples both the acceptance and the enforcement of long-term responsibility for a sustainable harvest or the conservation of forest functions.

> The Basic Forestry Law (1967) of Indonesia classified all State forest land as Production- and Protection Forests, and as wildlife and nature sanctuaries. Production Forest (*Hutan produksi*) is subdivided into the permanent forest estate (*Hutan produksi tetap*), and into Conversion Forest (*Hutan konversi*). The former concerns land that should remain under a continuous tree canopy for the sustainable harvest of timber, the latter is destined to have the forest cover removed and the land cultivated for horti- or agriculture. Protection Forest (*Hutan lindung*) should remain untouched in order to conserve the general environmental services, notably soil retention, erosion control and flood alleviation. The sanctuaries are meant to protect wildlife or a particular ecological or scenic value. The nature sanctuary (*Cagar Alam*) is, in principle, the most strictly protected form of reserve; the regulations concerning the wildlife sanctuary (*Suaka margasatwa*) are particularly deficient in effective habitat protection. However, in practice there is no difference in the effectiveness of protection. And where nature sanctuaries are usually too small to be of exploitative interest in any case, in reality the wildlife reserves as well as the Protection Forests virtually offer a carte blanche for officially sanctioned timber exploitation.

Timber-concession holders are formally bound by a complex set of rules and regulations, which accord closely with the ITTO guidelines. In Indonesia a concession holder must produce annual (*Rencana Kerja Tahunan*) and five-year plans (RKL) as well as an overall operation strategy, including an assessment of the timber stock which covers the entire period of the concession right, i.e. 20 years. The concession holder is obliged to have an independent Environmental Impact Assessment (AMDAL) conducted and must provide the Ministry of Forestry with aerial or satellite photographs of the entire concession area. The concession holder 'shall not carry out any logging activity in areas designated as protection forest, if such areas overlap with his concession' (as the Minister emphasised publicly in the *Jakarta Post*, March 22nd. 1993). The concession holder is also required to establish a wood-processing plant and must make provisions for reforestation. Finally, he is obliged to protect the estate against encroachment and (timber and animal) poaching by deploying one forest guard (*jaga wana*) for every 3,000–6,000 ha, although a formal mandate for enforcement is pathetically weak in the face of the ambiguous legal framework.

Failure to comply with any of these regulations, when detected, is supposed to lead to a considerable fine, confiscation of equipment and harvested produce or, in the case of recidivism, revoking of the concession contract. But the court, under the ambiguous legal framework, and faced with financially powerful concession holders, may not invariably be on the side of the Ministry. Moreover, it will be evident that all these regulations stand or fall with the effectiveness of control, and although the Ministry has put great effort into organising the controlling system, the extraordinary economics of timber extraction and other major interests concerned with land and natural resources frequently overshadow every good intention. Indeed, other government departments often override the authority of the Ministry with respect to the classified forest estate for their own 'development' ends (Hurst, 1990), in spite of an increasingly organised land-allocation procedure (TGHK) being implemented since the late 1980s.

Furthermore, the Indonesian legal framework pertaining to State forest land has recently issued a special permit for tree harvesting (IPK or *izin potong kayu*). Meant to facilitate 'traditional' harvesting of trees for the building of a house or dugout canoe, this permit is issued at the local level. In many areas it has proved to be a major opportunity for the quasi-legal exploitation of logged-over forests, resulting in the utter demolition of the remaining forest cover. Powerful entrepreneurs as well as police and army personnel are frequently behind this very lucrative form of 'traditional' exploitation.

Thus, notwithstanding the legal framework regulating the custodianship of forest areas, all natural (or wildland) forest in Indonesia and the east Malaysian States is, in the minds of most inhabitants, free or empty land (*tanah kosong*). It is seen as an open-access resource to be exploited or converted, precisely because it is covered with 'wild' forest, and hence unprotected. After all, ancient tradition held that clearing an area of wildland forest constituted a publicly sanctioned claim regarding the use of the land. Only if, at the onset of clearing, some augur, sign or adverse event (e.g. a disease, or a raid by a neighbouring tribe) occurred, would the claimant feel obliged to cease activities and move on to another location.

> In contrast to what might be expected, as regards modern subsistence, abandoned, derelict land, or regenerating *ladang* is a far less attractive option for occupation and settlement than virgin forest on State forest land. This is in spite of the fact that reclamation of abandoned land often requires considerably less investment. The main reason is that abandoned land of secondary forest almost invariably has a former user who will demand compensation from the newcomer after major investments have been made.
>
> Superficially this seems to be quite different from the traditional practice of shifting cultivation; shifting cultivators used to prefer secondary forest for clearing rather than expanding further into the primary wildland forest. There is, however, no need to suppose that a sense of 'ecologically sound' farming or 'sustainable forest use' existed in the practice of traditional shifting cultivation, as many romantic anthropologists have been led to believe. That subsistence tribes did rotate their agricultural use of the forest land within areas of secondary forest already claimed is mainly because of (1) increasing insecurity in forest areas further away from the settlement, with reference to possible territorial claims and raids by neighbouring tribes, and the impact of nuisance animals (2) higher input in labour to clear primary forest, and perhaps (3) fear of upsetting the spirits of the primary forest, and the high price of propitiating these spirits (see e.g. Ling Roth, 1896). Hence, the decision to engage in shifting cultivation was based primarily on fear and economic constraints, rather than on some exalted insight. There is no environmental ethic in tribal people. Nicanor Gonzales, the leader of the indigenous peoples' movement in Bolivia, has stated, 'at no time have indigenous groups included the concepts of conservation and ecology in their traditional vocabulary' (Stearman, 1994).
>
> Apparently the same rational decision making underlies the modern frontier practice of slash-and-burn cultivation. Now that the spirit world and intertribal strife have been abolished, and the chainsaw has removed several physical constraints, the rational frontier-mentality decision-making process will not change until fear (of consistent law enforcement) can be re-instilled to replace the current *laissez-faire* attitude.

Natural forest is 'wild'. It bears no indication of being a private possession, and thus no threat of potential retaliation against private territorial claim violations is perceived. Moreover, any traditional desire not to upset the harmony of the spirit world, or to call down its wrath – if it was ever a major inhibiting factor – is virtually gone, now that the animistic view of life has been banished by modern ideologies and montheistic religious teachings. It is ironical, however, that in an area of State forest land which has been given out under a concession right for the purpose of exploitation, either for oil and coal, or for timber (*Hak Pengusaha Hutan*), the opportunities for protection against encroachment seem to be better if the concession owner is prepared to assert 'ownership', acknowledge the regulations and to implement law enforcement. The same applies when research is being conducted and the regular presence of researchers in the area is guaranteed. In both instances the psychological effect of a clearly distinguished territorial right (*hak*) is the key to inhibit further encroachment upon the area.

Land: between subsistence and speculation

With the current human population expansion and the development of an industrial infrastructure, land is becoming relatively scarce, and people with any foresight, power and financial resources are seeking to acquire land for speculation purposes. Despite the conditions inherent in the land-allocation process, the only land to be readily usurped is State forest land with its public image of open access (*tanah kosong*). In this, the current development of human society in Southeast Asia appears to follow very closely the early medieval European model (see e.g. Thomas, 1983).

Two powerful means can be deployed by an enterprising modern landlord, either separately or simultaneously. First, one can obtain a concession contract for timber exploitation which in any case yields disproportional profits. If a State forest area holding a Permanent forest status appears to have less than 21% of tree cover (remaining) the entrepreneur can apply through the local government for a change in status to Conversion Forest. Since control of logging operations in accordance with the regulations for forest exploitation is extremely weak, it is not surprising that such a formal opportunity strongly facilitates forest destruction. Second, once much of the forest cover has been removed, an entrepreneur may deploy local poor people or transmigrants as pioneers to further clear and settle forest land. Clearance can even be legitimized by a wood-cutting permit (IPK) for 'local people' to cut remaining trees for 'traditional use', or by a 'prospective permit' (*izin sementara*) issued by a local authority. In addition, as a quick ultimate solution, arson may even be resorted to.

> Notwithstanding the fact that all forest land has been formally in the custody of the State since the mid-1980s, Indonesia has a standard procedure for obtaining title to what is, rightly or wrongly, considered to be communal or *adat* land (MacAndrews, 1986). Officially this involves checks and the consent of authorities at various levels, which, in the event the claim concerns State land, must be referred to the central authorities in Jakarta. However, clearing of the forest and occupation of the land usually takes place before the procedure even begins, and in order to influence decisions, negotiations are often initiated to offer authorities a share in the profits.

Wherever supposedly landless people encroach upon State forest land, they are often instruments in this process of deliberate despoliation: cheap labour able to readily evade the modern land allocation procedures and regulations by means of an appeal to their poverty, supposed traditional rights and an acknowledged paucity of economic alternatives. They remove the forest, cultivate and occupy the land, claiming subsistence, and when productivity appears to be sufficient, are either bought out, evicted or coerced into plantation labour by the emerging landlord entrepreneur, at precisely the time when the government has facilitated the infrastructure for accessibility. Such people have been variously designated 'truck farmers' (Dove, 1993), 'urban-based entrepreneurial shifting cultivators' and 'transitional or opportunistic shifting cultivators' (Ohlsson, 1990).

The clearing of virgin forest requires some relatively modest investments with reference to the ultimate gain: For the concession holder it implies no more than some minor additional deployment of labour, and even for the impoverished, landless cultivator moving in the wake of the timber extraction, it constitutes a relatively minor financial sacrifice for the 'permit' and very little physical input. Furthermore, the authorities may eventually be persuaded to ignore encroachment or extraction beyond the formally allowed cut (e.g. Ohlsson, 1990; Brown, 1991). Indeed, any such investments as may be required for forest clearing are usually more than adequately compensated for by the yield in timber which is harvested illegally from the area (Hurst, 1990).

The extra income generated from illegal timber felling, poaching and illicit trade appears to be so lucrative that a substantial proportion (i.e. > 15%) of the rural subsistence farmers near wildland forest areas neglect their highly productive wet-rice cultivation to devote their efforts to this form of poaching and land appropriation (Rijksen and Griffiths, 1995). As a matter of fact, a team of only three persons, equiped with a chainsaw and one or two buffaloes for traction, can earn anything up to 15 times more than if they work their own rice field or seek regular employment (see also Vayda, *et al.* 1980).

It is therefore not surprising that enquiries among transmigrants have revealed that over 65% of their income usually derives from extracurricular activities in the surrounding forest, and some 30% derives from their labour in regular timber concessions and nearby wet-rice fields belonging to indigenous landowners. Less than 5% derives from cultivation of the (2.5 ha) land which was allocated to them (Rijksen and Griffiths, *op. cit.*).

In addition to this complex of forces, concealed behind a facade of development that fuels encroachment upon and the despoliation of forest areas, one may discern other incentives, including of course the genuine expansion of a growing body of (landless) subsistence farmers with scarcely any other livelihood alternative (Hong, 1987; Hurst, *op cit.*). Moreover, in some regions, the local culture reinforces illicit forest clearance, arson and poaching as a challenge to central authority (e.g. *Jakarta Post*

December 2nd. 1989), yielding status within the community upon those members who carry it out. Be that as it may, during the late 1980s, the World Bank estimated that smallholder conversion was responsible for up to 55% of the annual deforestation, development projects contributed almost 30% and logging approximately 10% (Davis and Ackermann, 1988).

High-level forestry authorities in Indonesia appear to be well aware of the major constraints on regulated sustainable forest management (Suryohadikusumo, 1992), but are not in a position to improve the situation unilaterally. Perhaps in its 1991 policy document the World Bank has most succinctly summarised the forces which maintain the 'moving frontier' of forest destruction in all tropical countries. These forces are largely extraneous with respect to the people comprising that frontier, and concern deficient government. They can be differentiated into:
– failure to decrease population growth, especially in the frontier communities, due to cryptic ideological or religious support in favour of, rather than against, reproduction, and discrimination against women;
– provision of direct or indirect subsidies and incentives for transmigration and translocation in State forest land;
– failure to secure sustainability of resource use, through effective law enforcement, education, extension, etc.;
– failure to provide for appropriate agrarian and social reforms on productive land;
– improving rather than blocking access to forested areas.

To this already extensive list, however, should be added:
– failure to facilitate alternative jobs in industry and agro-industry (Alvim and Alger, 1993), with a clear functional vector away from the forest (Rijksen and Griffiths, 1995).

Considering the possible internal forces opposed to such wanton forest destruction and land grab, it is remarkable that the cultural atmosphere in Southeast Asia rarely, if ever, fosters a sense of responsibility in people who actually cause serious damage to the community by means of controversial or illicit land use (see e.g. Suryadiputra, I.N. *Whole Society Responsible for Damage to Environment*, in *Jakarta Post* February 10th. 1992). Consensus is apparently concerned primarily with regulating the potential conflicts over utilisation, rather than with the conservation of ecological conditions in the interest of the community.

It should, however, be of major concern that reckless forest destruction, arson and poaching are frequently carried out as acts of subversion against the centralised State government. Since the land and resources belonging to the State are leased out to third parties without the regulation of any returns to local communities, the forest and its organisms are being sacrificed as tokens of dissent. Such challenges can only be contained when the government at all levels demonstrates that public revenue and taxation of income resulting from the land products is to some considerable extent

> An illustration of deficient public responsibility is the case in what was locally claimed to be a natural disaster occurring in the Alas valley (Aceh, northern Sumatra) in 1981: A handful of local farmers removed much of the forest cover on a slope of the water catchment of the river Mengkudu, first to extract the valuable timber and subsequently to add to their subsistence plot by traditionally claiming some supposedly suitable sectors of the slope as a cleared *ladang*, with the intention of eventually planting a cash crop of 'kemiri' trees (*Aleuritis moluccana*) for the production of lumbang oil from the seeds.
>
> Within one year their clearing operations had caused the erosion of some 2,325 ha of slope (45°-60°) which, during a heavy downpour of rain, led to a devastating flash flood. It washed away 17 houses in the village downstream, killing 13 people, covered some 80 ha of sawah land with stones and debris and destroyed about 100 m of the hard-surface road, including a bridge (Robertson and Soetrisno, 1982).
>
> For the community alone, the material damage resulting from this illegal activity could be assessed to exceed the average annual income of over two hundred households. The timber removed illegally from the slope may have yielded up to 10 times as much, but the cleared 10 ha *ladang* would never yield more than 2% of that amount in the years in which the soil would support a harvestable crop. Although the cause-effect chain was readily recognised, it was never interpreted as greedy piracy on the part of a few individuals culminating in a great loss to the community. Indeed, the issue was never even raised at that level. Instead, help was expected from the government to compensate for this allegedly natural disaster. A year later the same thing happened near the river Penanggalan, some 5 km to the north.

returned and channelled back into meaningful and acknowledgeable support of development in the region; this is, however, a process of fair government far beyond the scope of nature conservation. The current move towards forcing the management of a wildland forest, whether concession holder or conservation area manager, to accept financial responsibility for directly supporting the development of nearby rural populations is unwise and probably unconstitutional, as it dilutes the responsibility of the government.

Law enforcement?

Although illegal hunting is the most acute threat to the survival of the orang-utan, massive habitat degradation, forest conversion and fragmentation have an equal, albeit often concealed and indirect, impact. Indeed, as was elaborated earlier, both factors amplify each other's impact. However, the whole sequence of destruction and (local) extinction is invariably initiated by poorly controlled timber exploitation and exploration for oil, gold and marketable forest produce. It is essential that the authorities become fully aware that 'excessive and irrational utilisation of the forest resources will cause problems in environmental balance which in turn may disturb political stability. Inadequate quality of government employees to oversee and enforce control will cause uncontrollable utilisation of the forest resources' (Suryohadikusumo, 1992).

> **Hunter's fatal mistake**
>
> KUALA LUMPUR (Reuter): A Malaysian businessman out hunting shot dead a tribesman sitting high up in a tree yesterday, mistaking him for a monkey, the national news agency *Bernama* reported.
>
> Police and firemen, using ropes, took several hours to bring down the man from an 18-meter (60-foot) tree in the Petaseh forest reserve, about 120 km (75 miles) south of Kuala Lumpur.
>
> The businessman fired his shotgun at a figure on top of the tree, thinking it was a monkey which his dogs had been chasing, *Bernama* quoted police officials as saying.
>
> The man reported the incident to the police when he realized his mistake. They classified the case as murder and detained him, *Bernama* said.

An almost insignificant newspaper article (Jakarta Post, October. 8th, 1993) reveals the malevolent force behind the destruction of our relatives. Here one of the hunter's own kind is accidentally bagged.

Timber exploitation and the scouring for forest produce extraction and gold opens up access to the most remote former refuges of the ape, and greatly facilitates poaching. The local labour in timber concessions is often recruited from among the subsistence farmers and hunter-gatherers in the surroundings. The labourers in the timber industry as well as outsiders gaining access to the forest add to their income by gathering minor forest produce (rattan, incense or *gaharu* wood, illipe nuts, copal, wild fruits, medicinal plants, orchids, fruit bats, birds, snakes, monitor lizards, terrapins, fish, etc.) and, if they can lease a firearm, large mammals as well. All labourers are commonly obliged to rely upon the forest for at least part of their subsistence. Many of them are highly professional in terms of bush skills and scraping along, indiscriminately pillaging anything of consumptive or commercial value from the forests they scour (Tillema, 1990). They will not exclude the orang-utan, which is by far the easiest prey item in the Malesian rainforest.

Moreover, forest conversion, both planned and unscheduled, has increased immensely the risk of direct human-ape conflicts and in fact forces displaced orang-utans to wander about and stumble into confrontations in people's domestic gardens or in freshly sown plantations belonging to commercial enterprises. This implies that even if the majority of people had, until recently, felt no reason to confront an orang-utan, modern developments in land use have rendered confrontations unavoidable. And since law enforcement for wildlife protection has been lax, any confrontation is likely to result in a dead ape. In some areas this has probably reinforced the trend in active persecution when the reward for the sale of an orang-utan skull or a captured juvenile is in cash.

All these factors driving the orang-utan to extinction could be either avoided or contained, if only regulations and laws were designed with a clear objective in mind and were enforced consistently. The governments of those areas in which orang-utans occur have a legal framework in place which should not only protect the ape and the gazetted conservation areas in which it occurs, but also the major environmental functions of virtually all forest on State forest land. However, the law is ambiguous, and enforcement is virtually absent. The question is why is law enforcement with respect to the ape and such areas so deficient that trespassing, encroachment and poaching are common?

If people in Southeast Asia appear to have a sound respect for private property, why then do many people fail to appreciate the concept of common interest where State forest land, conservation areas and protected wildlife is at stake? Why is the legislation so ambiguous with reference to land rights? Could it be that a collective public perception of the value of protected areas and wildlife has never been called into existence because people had never heard of, or clearly understood, the functions of nature? Do people accept the State concept and its laws only out of fear of repression? Or is it because they are constantly confronted with ambiguous conduct and collusion on the part of authorities, and hence experience no, or at best inconsistent, law enforcement in this respect? Or all of the above?

Ill-informed politicians and bureaucrats often challenge nature conservation actions with the question, 'What is more important, a human or a beast?' This should of course refer to human welfare versus the needs of an animal. However, if applied to the existential conflict between humans and the ape, such a question is invalid because the orang-utan is a legally protected species. The issue here is not one of choice between humans and orang-utans, but between law and opinion, or between the State and individual peripheral interests. That the ape is also of extremely high, internationally acknowledged, kinship value to our own species, is here a subsidiary moral issue that, in a world based on logic, rather than mythical humbug, would even justify protection on a par with humans.

The ambiguity of many authorities with respect to the protection of wildland forest areas and wildlife is not exclusive to Indonesia and Malaysia. For most human cultures, the integrity of nature and the basic interests of wildlife are subordinate to peripheral human interests. In other words, conservation of nature is all right only as long as it does not stand in the way of human desires and aspirations. Indeed, this ambiguity has even become prevalent in the international conservation organisations, and is reflected in the internationally trendy concepts of sustainable development and

> If one were to address the ethics of pitting a human against his closest living relative, it would be appropriate to consider that it is a controversy between what Maslow (1970) has identified as a *basic interest* in the case of the ape (i.e. survival), against a *secondary* or *peripheral interest* for humans (i.e. an interpretation of welfare). A basic interest cannot be sacrificed for long without either loss of life or general loss of wellbeing; peripheral interests can be foregone forever without serious consequences to one's wellbeing. In a civilisation built upon ethical foundations one should not subordinate a basic interest of another sentient being for the sake of promoting a peripheral human interest (van de Veer, 1979). The original question can now be transcribed in more mundane terms into one concerning the ape's extinction as opposed to human greed.
>
> Unfortunately, for many people in the late twentieth century, greed prevails over ethics: Against an annual revenue of several billions of dollars from the exploitation of wild-grown timber and other natural resources, the fate of what is commonly seen as an essentially 'worthless' ape may seem inconsequential for a simple materialist, especially when the euphemistic term 'basic needs for poor people' to justify demolition of a natural forest area is readily accepted by an ignorant public.
>
> Considering that in a civilised, enlightened world both the ape and humans should be accorded equal ethical value, it is obvious that the **survival** of fewer than thirty thousand orang-utans on earth requires at least as much attention as the problem of the **distribution of welfare** of well over 5,000 million humans worldwide, or even of the approximately 230 million inhabitants of Indonesia and Malaysia.

participatory management (e.g. *World Conservation Strategy*, 1980). Promoted alongside financial support, these concepts quickly became the guidelines for conservation authorities in many developing countries during the 1980s. Ironically, any motivation for protection was effectively replaced by an attitude of *laissez-faire* resulting from the impact of international support for nature conservation under this new resourcist banner. The new trend may even have spurred ecological degradation and the extinction of the most sensitive species (e.g. the Sumatran rhino in Borneo and in much of its former distribution range in Sumatra).

International support?

In the 1980s, several influential international conservation organisations established official representations in Indonesia. However, coordination has been weak and an integrated differentiation of activities in support of the Indonesian nature conservation structure has barely been attempted. One may well wonder why the organisations have chosen to operate like proselytisers, claiming a similar general objective, yet functioning under different ensigns and displaying competitive territoriality. If their alleged objective is to conserve nature, why then have they not joined strategically and differentiated their support into specific fields, such as conservation policy, protection, species conservation, education, awareness, and NGO-development, in accordance with a joint strategic plan?

For many people in the international circuit of conferences, symposia and workshops, this last question may cause indignation. For did not the international conservation community unconditionally adopt a unified conceptual approach, known as the *World Conservation Strategy*? Indeed, the many participating organisations may abide almost dogmatically by this strategy, but the problem is that it deals primarily with the interests of people rather than with the conservation of nature. It rejected the traditional focus on endangered species, and installed the concepts of 'sustainable development', 'local people's participation' and 'ecosystem conservation', which cannot be interpreted in practical terms.

As a consequence, since the early 1980s most international and bilateral support for nature conservation has focused on rural development in the guise of conservation, frequently with teams of untrained, expatriate volunteers and no conceptual guidance whatsoever. It has been fully dedicated to the economical development of the supposedly 'poor stakeholders', to make it possible for them to participate in a consumer society, while hoping to save some wildlife on the side. Thus, during the past decades it has looked as though international conservation organisations have carefully avoided addressing the real conservation issues (Rabinowitz, 1994), and with disastrous consequences.

The *World Conservation Strategy*, launched in the early 1980s at the cost of some US$ 60 million, proffered the perfect excuse for all-out ignorance of the real issues, although it was probably meant to achieve the opposite. The essence of the strategy was to seek universal adoption of the fundamentalist concept, embedded in the US legal system since the beginning of the twentieth century[79], that the conservation of nature, rather than the protection of organisms and natural landscapes against human exploitation, should mean 'wise use' or 'sustainable utilisation' of natural resources in the service of a 'sustainable development.' That sustainable development is unattainable in an ecological sense was apparently not contemplated. The new strategy was a desktop solution for bureaucrats to redress an uneasy feeling that the establishment and protection of nature reserves could be considered some kind of neo-colonial interference in the supposed territorial 'rights' of impoverished rural people. An intellectual juggling act with new terms gave the illusion that finally one could run with the hare while hunting with the hounds.

That new concepts of 'sustainable utilisation' and 'participatory management' for established conservation areas found a remarkably ready acceptance in many developing nations is perhaps not surprising in a world governed by euphemisms and voracious industrial market forces. First, such concepts were, and are, promoted by allegedly respectable and powerful organisations, influencing even the major international development banks. Second, they comfortably justified a *status quo* and the avoidance of moral and political problems with land allocation at the grass-roots level, which inevitably arise during the evolution of a society from a traditional to a

---

[79] The official US definition, stating that conservation is 'the use of natural resources for the greatest good to the greatest number of people', (or, 'wise use of natural resources') was designed by the State Forester Gifford Pinchot on the basis of his fundamentalist Christian conviction that the earth was created for humans, during an ideological dispute with the naturalist John Muir concerning the inundation of the unique Hetch-Hetchy valley at Yosemite National Park in California.

> During the 1980s, under the mounting pressures that were generated by the expanding international market and growing populations to exploit more land and resources, the traditional American illusion of 'wise use' or 'sustainable utilisation' came to supplant entirely the original concept of conservation in the international circuit of professional conservation diplomats. Since then it has been advocated that improved environmental awareness and giving management responsibility to the local people for utilisation of a wildland area and its natural resources would result in the conservation of biological diversity (e.g. McNeely, 1989). It is hard to understand how such a neo-romantic desktop revival of the 'noble savage' image can be maintained while wildlands are being ransacked and exploited by market-driven opportunists, including such aforementioned local people[80] who usually play major roles, either operating on their own behalf or in the interests of a trader, an industry or a prospective landlord (Dove, 1993). Moreover, there is absolutely no evidence that a policy of so-called sustainable land use under the impact of market forces is effective in the conservation of biological diversity (Wells, 1997; Brandon, 1997). Indeed, there is ample historical and empirical proof to demonstrate that increased welfare has no positive effect whatever for nature conservation. Further, on the basis of some insight into human sociology, there is no doubt that, under the given conditions of chaotic development in a developing country, a 'participatory approach' will only boost resource depletion rather than serve nature conservation (Murphree, 1994). The concept of 'participation' was a disastrous policy for the forests and wildlife of Europe (see e.g. Thomas, 1983; Buis, 1993), and there is no reason to believe that it would work for the conservation of rainforest and orang-utans in Southeast Asia (see also Hannah, 1992; and Murphree, op. cit). To allow people, in transition from tribal subsistence to a consumerist state-society, access to what has in effect become an open-access resource or a commons, leads inevitably to an environmental tragedy (Gordon, 1954; Hardin, 1968).

civilised state. The latter cannot accept the concept of tribal territories or traditional commons in land rights, but its government can expect serious political problems if it enforces the conceptual change required, especially when it concerns land use. Moreover, it must be borne in mind that even for government authorities an immediate financial interest in the yields of land *use* and resource exploitation can readily overrule any desire for nature **conservation**.

> The participatory approach to local claims regarding land use in conservation areas is based upon ignorance of the traditional mind and upon inconsistent reasoning. Tribal units never 'owned' the land they occupied or considered to be their territorial hunting grounds; the traditional animistic concept meant that people felt they 'belonged to the land.' From a traditional perspective, an autochthonous person can therefore not claim 'ownership' of land; the occupation and use of land under tribal conditions was related to crude territorial power over other tribal claimants, rather than a commonly accepted 'right.' Now that tribalism (and animism) has been abolished, and a government installed to safeguard the interests of its citizens, many nations have adopted the concept of State land, i.e. land to be managed or protected in the common interest. No evidence exists to support the contention that local stakeholders with a vested interest in the land are able to conserve the ecosystem and wildlife on such land, although there is evidence that wealthy private owners under certain circumstances did and do protect and conserve the wildlife and natural conditions on their estates. It is important to realise that behind all claims for 'traditional rights' and participation there is usually a hidden agenda of private or community interest related to exploitation, power and greed, but never an unadulterated wish for preservation of the ecosystem.

---

[80] For some recent empirical evidence exposing the fallacy of this concept see e.g. Browder, (1992), Southgate *et al.* (1996); van Schaik *et al.* (1996).

There is no doubt that where the concept of sustainabilty and participation is directed at types of land use and at organisms which are formally allocated for use, it could make theoretical sense. But why should conservation areas and the last wilderness on earth be sacrificed[81] to such an anarchistic experiment with dubious results for many extinction-prone, sensitive plants and animals?

After all, protection for the conservation of wild organisms is essentially different from a supposedly 'wise' or 'sustainable' use of species which inevitably become domesticated in the process (Rijksen, 1984). Any form of use or consumption tends to select or domesticate the exploited organisms, undermining the essence of conservation, i.e. the acceptance of wildness.

Where the integrity of the few conservation areas is at stake under mounting pressures of forest exploitation, encroachment and poaching, one should seriously wonder about the meaning of statements such as 'conservation is sustainable use' and 'local people must be involved in solutions to the conservation problem' (Golder, 1996)[82]. What is the manager of a sensitive conservation area to make of international support that coerces 'strategies which help protect biological diversity **by promoting** sustainable **use of fragile and threatened natural areas**', ... 'helping the **poorest** to conserve **their** natural capital', ... 'on recognition that conservation goals can best be achieved by meeting human needs', as the 1995 WWF International folder entitled *Conservation For People* stated?

Is it surprising that under such expatriate guidance and powerful financial support, activities have shifted from unpopular law enforcement and other protective measures, to the implementation of *laissez-faire* strategies involving so-called 'sustainable' use of natural resources and popular rural development aid to poachers and ravagers of conservation areas?

One may well wonder why nations would consent to become the experimental playground for the revolutionary ideological concepts of international conservation organisations (see e.g. McNeely, 1989; Golder, 1996), if such concepts entice local anarchy and, in practice, make protective management virtually impossible (Rijksen, 1984). Could it perhaps be a reflection of indifference to the natural heritage and its unique organisms in the swell of modern economic and ideological developments? One would hope that to be unlikely. But why then is active protection of State forest land in Borneo and Sumatra even rarer nowadays than several decades ago? And why has there been hardly any development regarding better control of the utilisation of

---

[81] During the IUCN World National Parks Congress in Bali (1982), the then chairman of the Commission on National Parks and Protected Areas (CNPPA), Kenton Miller, promoted in a special address the possibilities of 'sustainably exploiting' medicinal plants, wildlife, timber, fossil-energy reserves and all other natural resources *from conservation areas*, while stressing the moral obligation to invite local people to partake of the natural system.

[82] With such propaganda a professional advertising corporation can certainly tap the resources for development aid, but, conversely, it effectively degrades the profession of conservation management. Indeed, humanitarian 'aid' agencies came to believe that conservation was an opportunistic side-issue of their own objective and also claimed resources and responsibility for conservation areas under the false ideology of 'sustainable use of natural resources', with disastrous consequences.

Section III: the decline

> This is not to say that every form of 'participation' is inevitably counter-effective for conservation; after all, a civilisation cannot function without the active support of many of its citizens. But the application of an undefined concept of participation, as advocated when conservation areas are at stake, can result in the erroneous giving of unwarranted responsibilities to interest groups, while the sociological conditions and political backgrounds of such groups are ignored (see e.g. Dove, 1993), or entirely misunderstood.
>
> The concept of participation may be interpreted appropriately as a way to establish communication and seek consent for decisions during a planning stage in order to deflate a tragedy of the commons (Ridley, 1996). It is insufficiently realised, however, that participation alone cannot solve the fundamental problems of economic imperatives destroying forest and squandering biological diversity. It addresses neither the essential misconception of territorial ('property') rights at the appropriate levels of society, nor the slack law enforcement whenever the adjective 'wild' is involved. Contrary to the ideological illusion, a utilitarian interest cannot be conciliated with conservation: Something to be conserved cannot be used.

land and natural resources? Any control which has been exerted was primarily to minimise a possible tax loss on the profits made from forest land use, but it is questionable whether it concerned the sustainability of the use itself, and certain that it rarely if ever challenged encroachment and forest destruction. Is the answer perhaps that developing nations, in addition to nurturing some opportunism, put their trust in a concept which is promoted so powerfully by international conservation organisations, and is adopted as a guideline even by the major international and bilateral donors? It is certainly difficult to challenge such ideological power when large project grants (or even loans) are at stake – after all, in Southeast Asia too, one does not look a gift horse in the mouth.

Further, one may also ask why international organisations claiming to promote nature conservation shifted focus. From being the professional advocates for the integrity of the last fragments of wild nature in the world, they became dabbling champions of 'sustainable development' of 'the poorest', and in that respect joined a host of other adequately specialised international aid organisations.

There is no doubt that in order to be effective, conservation, in the proper, protective sense of the word, must be integrated with all other aspects of development under skilled governmental guidance. However, should one not expect the powerful international conservation agencies to have sought to apply their revolutionary concepts to all government land destined for use **outside** the few established conservation areas, notably the production forests? And then, should one not expect them to have sought effective cooperation with the specialised humanitarian aid organisations, e.g. UNDP, FAO, national development aid agencies and NGOs? These latter should have carried out their specific tasks outside the field of nature conversation, while the international conservation organisations should have consistently helped and backed the national conservation agency to be recognised and acknowledged by other, more powerful departments, and stood in defence of the integrity of nature on the fringes of conservation areas.

> One can only guess at the motives of international conservation organisations that deviate from a policy of consistent protection for conservation areas. They may have been rooted in frustration over the ineffectiveness of conservation, in a romantic illusion concerning the vanishing noble savage image and in sympathy for rural people who could be forcibly wrenched from their customary exploitation of natural resources, or they may have sprung from political or economic opportunism. It is certain, however, that there is no empirical evidence for the possible efficaciousness of the advocated change from the well-proven, albeit unpopular, protection into a 'sustainable utilisation'- strategy. Improving people's welfare has a negative, rather than positive, effect for the conservation of nature. It does not reduce the desire for consumer goods, but stimulates it. Thus, the change of policy seems to be rooted in false sentimental ideology and financial opportunism, rather than in wisdom. For under the surface of 'participation' and 'collaborative management' of sustainable utilisation one may certainly recognise the romantic ideology of anarchy. The so-called technical assistance for many post-1980 'conservation' projects is characterised by a combination of subversion against a centralised government with its regulations and corrupt structures, and nostalgia for self-governing of 'the local people', a belated echo of exalted ideals contained in the famous book *Mutual Aid, A Factor of Evolution* by the Russian prince Petr Kropotkin (1902).
>
> In any case, a desire to be accepted among the wealthy international community of the UN and other diplomats, and to partake in the mounting financial resources for alleviating global misery, poverty, and resource shortage, may well have preconditioned the ideological shift to a considerable extent. It would not have been for the first time that international nature conservation was corrupted by politics and the availability of financial resources.

After more than a decade it must be realised that in terms of conservation effect some serious mistakes have been made regarding the conceptual 'revolution' that exchanged conservation for sustainable utilisation and enforced misplaced 'participation' by local people with respect to 'management' of natural forests in general, and conservation areas in particular (Brandon, 1997). The devastated conservation areas which have suffered such 'management' are convincing evidence that conservation and consumer-driven use are essentially different issues, based on fundamentally opposed objectives (Rijksen, 1982). It is possible that this was finally, and rather half-heartedly, recognised by the IUCN in the early 1990s, a decade after the launching of the *World Conservation Strategy,* in a new document called *Caring for the Earth* (1992). The message, however, has unfortunately not yet reached the field. It is deplorable that sensitive conservation areas under the assertive guidance of international organisations have been, and still are, sacrificed to naive experimentation with *laissez-faire* resource utilisation by local people as though it were a form of management, but there are also important far-reaching socio-political effects. In a final evaluation, an ignorant international conservation-minded constituency will simply blame the national conservation agency for the expanding forest degradation and the loss of unique species. In psychological terms such unwarranted censure does not heighten commitment, but instead tends to induce in the national agencies a mixed, paralysing feeling of inferiority and arrogant xenophobia. One can already discern such a trend in Malaysia and Indonesia.

Seen in a critical perspective, the result of three decades of international assistance has been the functional alienation of the national conservation structures from international nature conservation, great confusion in the directions of management at the field level, and a steadily deteriorating protective structure. Despite the good intentions of various individuals, it seems that the scatter of so-called support for nature conservation, guided by an ambiguous objective, has prevented rather than boosted the development of full national commitment and a commonly accepted responsibility.

> Conservation programmes with ambiguous objectives, unguided or misguided experts, no feedback, and hidden agendas prevent the development of trust on the receiving end and preclude effective cooperation. One may well conclude that under the selection pressures of corporative growth, socio-economic geo-politics and inconsistent ideology, during the last decades the commitment and mutual trust necessary for successful cooperation has barely had a chance to develop in the dealings with powerful international conservation agencies – with dire consequences for the ecological integrity of wildland forest areas and wildlife in Southeast Asia.

The key question is: How can international support be coordinated and deployed to make national nature conservation structures in developing nations more effective? One answer may be that international donors must be possessed of a realistic, unambiguous objective, based on knowledge rather than romantic myths and political ideology. Another may be that their actions should not be dominated by self-aggrandizement. Instead, they should be eager to learn from the history of conservation, including their own mistakes, and to seek result-oriented cooperation rather than provide 'aid', while incorporating standard monitoring and feedback mechanisms into their support system. Finally, an international conservation community should be able to support any serious plan for conservation readily and conditionally, with financial and political means, invited expert assistance and advanced information processing.

However, it requires in the first place an unwavering commitment to the cause of nature protection on the part of the donors, in order to facilitate development of the necessary mutual trust, and to achieve effective cooperation. It is probable that focusing on an attractive, endangered 'umbrella' species, such as the orang-utan, will facilitate a clear, unambiguous project formulation, and a simple, straightforward implementation.

Unfortunately, international financial support is most readily generated through media coverage of misery. Consequently, for orang-utan conservation such coverage has been most successful in raising funds for the rehabilitation of (immensely appealing) orphaned apes. A single WWF campaign in the Netherlands in 1997, including television coverages yielded half a million US$. The TV programs featured cages full of confiscated, miserable ape infants, a surgical operation on a juvenile male ape and a good interview emphasizing the supposed financial inability of a developing country to cover such conservation expenses. It may be expected that sensationalistic media coverage of the devastating fires and a few of the abused ape victims will have raised much more compassion and substantial, albeit misguided, 'aid' resources.

> There should be no doubt whatsoever that the misery undergone by several displaced and persecuted orang-utan populations in Borneo is beyond imagination, and that a large number of surviving apes are suffering badly. Nor is there any doubt that, in a civilised world, care and appropriate treatment should be accorded to the few surviving ape victims who can be found and saved from the butcher's knife or from a contracted, miserable imprisonment as a 'pet.' However, the emphasis on such *ad hoc* addressing of emergencies gives an exceedingly false image of what conservation is about, and will not in the least diminish the risk of extinction of the species. Indeed, it may devour an extraordinary percentage of the financial resources which could better be applied to actions which do give the remnant wild orang-utan populations a realistic chance of survival[83].

In the real socio-economic context of conservation, support for emergencies could well be transcribed as 'banking on misery with an opposite effect.' What happens is that affluent outsiders rush to the supposed rescue with the best of intentions but wrong assumptions about 'poverty', in order to provide what in effect boils down to a financial 'reward' for poor conservation performance on the part of the country in question. After all, even with international assistance the local organisation failed to prevent the disaster, and is reluctant or unwilling to acknowledge due responsibility. This type of emergency support cannot be expected to promote future commitment to conservation; on the contrary, it is as likely to facilitate (a desire for) further even more such deplorable situations as to generate financial support.

Conversely, a well-publicised campaign exposing the extent of the orang-utan's peril may generate an international shock-wave which has political reverberations in areas where it is most needed; perhaps collective shame can induce effective commitment. Under such conditions it may be possible that the resources raised through well-focused media coverage can be deployed in professional negotiations for better conservation of the surviving orang-utan populations. If the rich nations of the world are willing to pay for the survival of the orang-utan, it should expect effective conservation to that effect: Only a hard *quid pro quo* attitude can force Indonesia and Malaysia to take the required measures to conserve the wildlife and ecosystems in *their* custody. International financial support is necessary to command negotiations for joint professional actions such that lead to local commitment and sustained success. The key question is to what extent Indonesia and Malaysia are willing to contribute, so that the price for survival of the ape is affordable in time. It is regrettable that in 1998 Indonesia's economy collapsed at the same time that this century's greatest ecological calamity occurred. As a result, the country at present has reverted to such public poverty that politically it is unable to place on the priority list measures to combat the imminent loss of its forests and apes.

---

[83] This seems to have been realised gradually by at least two major international organisations, notably WWF for Nature and IFAW, but it is not easy to adjust policy and re-allocate funds which were generated by media-driven sentiments.

The value of law enforcement

When an area or species has been legally designated as protected, by definition it has been done so in the common interest. In other words, an authority representing the entire community has decided that protection of a given section of land or of a species is in the interest of all members of the community, even if thereby the interests of certain individuals may be violated. It is considered subversive to challenge such a decision on the basis of short-term peripheral interests, or by a reference to outdated 'traditional rights'.

This is not to say that under every circumstance people are obliged to refrain from interference with a protected species, and cannot defend themselves against physical attack. 'Under common ethical rules, it is permissible so to act that an interest of another sentient being is subordinated for the sake of promoting a like interest (or a more basic one) of a person' (van de Veer, 1979, p. 183), as in a confrontation between a human and a charging elephant or stalking tiger.

However, the doctrines of sustainable use and participatory management with reference to protected wildlands and wildlife confuse long-term basic interests with short-term peripheral interests. In effect they subordinate the ecological integrity of regions of land and the basic interests of its wildlife to any kind of short-term human interest. This is the result of a materialist ideology which turns ecological wisdom on its head, claiming that natural resources were created solely *for* human use.

Such an approach causes uncertainty in any executive agent, inviting a double standard in the execution of law enforcement and making possible uncontrollable graft, ultimately resulting in the illegitimate destruction of wildlands and the extinction of wildlife. Equally important, however, is that this approach exposes the low value which an authority apportions to the role of the State in guarding the long-term basic interests of the community. Under such conditions the average

> Usually the protection of private property is fairly well enforced by governments, but the protection of State property with reference to wildlands is not. If an offence involving private property or against a person is effectively dealt with, why then is an offence against the interests of all so readily considered almost insignificant? In instances of poaching, trespassing and encroachment upon State land, why does ambiguity on the part of the authority often appear to transform a risk of law enforcement into complacency, if not ill-placed sympathy for the perpetrator? If the reason is simply a lack of functional commitment in the authority, training and consistent discipline could improve the situation considerably.
>
> The point is that any person who trespasses into a publicly protected area demonstrates an outlaw mentality, because he (or she) challenges the authority (which established the area as protected) and hence violates the foundations of a civilised community. If it cannot be accepted that precedence must be given to the protection of anything of common interest against the natural desire of any individual to pursue his (or her) own peripheral interest, then the concept of a State and its authority is meaningless. It is irrelevant whether such a trespasser is driven by poverty, tradition or greed to go after resources now protected in the common interest.

person will never begin to understand the inherent values of civilisation and nature, and a conservation authority has an impossible task.

For several decades the governments in Borneo and Sumatra have shown token concern about soil erosion and other environmental losses due to the devastation of forest land. Such accountable losses are, however, never recovered from those who caused them. On the contrary, substantial financial resources are applied each year to resettle and compensate illegal squatters, and more public resources are wasted in attempts to undo the terrible environmental damage caused by bad logging practices, slash-and-burn cultivation, arson and failed transmigration projects. If an illegal squatter is evicted from a conservation area he can count on considerable sympathy and even compensation, whereas the aboriginal apes and other wildlife displaced and probably killed during the 'development' process were, and continue to be, entirely ignored. After all, the emphasis is exclusively on *peripheral human* interest.

It is realistic to expect rational people to seize any opportunity for personal enrichment if the conditions allow, that is, when traditional social mores and constraints have been lifted and modern regulations are not enforced. A supposed environmental ethic is neither innate, nor can it simply be instilled by means of education, without consistent social or institutional enforcement. This principle lies at the basis of encroachment upon State forest land, including conservation areas; it causes concessionaires to ransack the forest estate in their custody, and it applies to the poaching of, for instance, the orang-utan. After all, no challenge of interest can be envisaged while law enforcement is deficient, if not virtually absent at the level of the poacher.

In Borneo and northern Sumatra, most of the human population has lost its traditional behavioural constraints (i.e. adat) as well as any traditional sensitivity, awe or respect for the spiritual harmony of wildland forest (see also Redford, 1990), while having ready access to modern technology (chainsaws, motorised transport, radio, TV, etc.). The people seem to be insensitive to the consequences of forest destruction even if they are or will be affected. For them the importance of and need for long-term conservation appears utterly irrelevant in the light of short-term incentives (e.g Stearman, 1994). The reason is that they have been conditioned to consumerism, and it would be unrealistic and unfair to expect that so-called indigenous or native peoples with a roughly distant tribal background are insusceptible to the addictive lure of material luxuries (Vayda et al. 1980; Redford, op cit.; Persoon, 1994; also van Schaik, et al., 1997). That more or less serious conflicts of interest may arise, for instance, when the State has issued a timber concession on State forest land, or where tribal communities claim traditional rights over the area of forest[84], is due primarily to reasons involving short-term peripheral (economic) interest, rather than any exalted sense of obligation regarding environmental and cultural protection. In western Indonesia such potential conflicts have usually been prevented by the deployment of extensive local participation in the exploitation, both by making use of locally recruited labour, and because the poor supervision of operations often made illicit subcontracting and fencing of poached timber possible (Hurst, 1990).

---

[84] The cases in which such conflicts were exposed in the media were usually instigated by expatriates and acquired an ideologic dimension (e.g. the Penan conflict in Sarawak; Davis and Henley, 1990; Manser, 1992); more commonly, however, the basic incentive for the conflict is socio-economical (i.e. no financial compensation for the loss of traditionally claimed resources) and/or a conservative fear of, or resistance against, enforced acculturation (Hong, 1987; INSAN, 1989; see also Persoon, 1994).

The poaching of orang-utans is by far the easiest way to extract income from the forest, although the profits to be made at the poacher level may seem relatively low (i.e. up to US$ 50). It requires no more than some basic bush skills and the simplest of hunting equipment. Hence even fishermen and gatherers of forest produce will not hesitate to track down and kill an orang-utan whenever they detect one in the forest.

Thus, as long as authorities are unable or unwilling to apply consistent law enforcement and to exert effective control over the use of State forest land, the false image of an open-access resource will be preserved, until all the forest is degraded or occupied by 'participating' squatters, and all the spectacular wildlife is exterminated.

It is often claimed that policing or active enforcement of the law would not work for conservation, because no force would be large and effective enough to control the numbers of people seeking to trespass. However, although conservation invariably fails due to poor enforcement, there is no evidence whatsoever justifying such a negative statement. Often such claims are accompanied by arguments which reveal that, rather than involving unsuccessful conservation, the issue is a moral one, having to do with an unwillingness to prevent by force people's access to wildland resources, especially if these people are considered to be poor or claim a tribal background.

However, many examples indicate that consistent law enforcement has an educational effect which goes far beyond the few exemplary cases of assertive action for conservation of wildlands and wildlife. For instance, a consistent army-supported law-enforcement strategy has effectively saved the Indian rhino from extinction in Nepal (Martin and Vigne, 1995). Indeed, there is a wealth of empirical evidence to show that where conservation is successful, intelligent, consistent policing in combination with propaganda is the key factor.

This is not to deny that the prevention or control of incidental poaching of orang-utans in the remote and fairly inaccessible areas in Borneo and Sumatra is scarcely feasible. While it is conceivable that consistent patrolling and surveillance can prevent encroachment upon conservation areas, the poaching of wildlife, being a hit-and-run action, may seem virtually impossible to prevent. Indeed, law enforcement for the protection of orang-utans will be mostly reactive and delayed. The poacher is rarely apprehended. Even if it is evident that the ape has been poached from a conservation area, any possible legal action usually cannot extend beyond confiscation of the illegal remains of the hunt.

Such remains, however, e.g. a young ape or carcass, a skull or a piece of skin, are usually not difficult to detect, and provide good evidence for a legal case. It is important to realise that consistent detection, confiscation and subsequent indictment somewhere at the level of the middleman could crush the market, and hence remove the incentive to acquire young apes[85].

---

[85] Note, however, that this will not prevent people from killing an orang-utan for its meat, or just for 'sport'.

When indicted, a middleman would certainly suffer under the impact of a US$ 50,000 fine, since his retail price for an ape rarely exceeds US$ 100, while the final trader cannot make more than at most US$ 1,000 on the pet market in a large city in Indonesia[86]. Therefore, it does not matter whether poaching is detected after a whole sequence of illegal local sales, anywhere in a train of transactions between middlemen and traders down to the eventual 'owner'. When detection is followed up by consistent prosecution and subsequent publicity of the indictment it will have a deterrent effect on further poaching. When such legal cases are widely publicised, and properly interpreted, they create a powerful educational momentum among people, which has a tremendous preventive effect demonstrating the non-monetary 'replacement value' of the ape. Such an effect was clearly discernible in the early 1970s when rehabilitation of orang-utans facilitated the confiscation of illegally kept apes. It is highly unfortunate that the renewed interest in better re-introduction has not been accompanied by consistency in a confiscation and prosecution policy. Indeed, the authority for protection (Directorate-General PHPA) has taken barely any initiative or uninvited action in this important field. The result is that rehabilitation/re-introduction in the 1990s appears to be much less effective than it could have been. It is important to realise that if rehabilitation/re-introduction fails to have a powerful and effective deterrent effect on poaching, it will be of little conservation value at all. Contrary to popular belief, the re-introduction of a few quasi-feral apes has no significance for the conservation of the wild orang-utan when this primary effect is lost.

Institutional impediments

In view of the overwhelming wave of socio-economic forces demolishing the wildland forests and wetlands, and the scatter of inconsistent international influence, the government institutions for conservation in Southeast Asia are faced with problems in fulfilling their legislative obligation. What are essentially bureaucratic administrations must stand up for, and guarantee, the survival of protected species and areas[87]. The major issues with reference to past and present institutional constraints on effective nature conservation are complex, but may be summarised as follows:
- a legal framework for conservation which is unclear in relation to land-use rights, and in some respect ambiguous with reference to the protection of the living conditions and lives of protected species outside conservation areas;

---

[86] In 1995-96 orang-utan youngsters in Central Kalimantan were usually bartered by ship crews from major Asian ports for TV sets, video recorders or other electronic equipment.

[87] In practice, management consists mainly of administration, supplemented with attempts by individual staff to somehow 'participate' in development projects, while prophylactic protective measures, e.g. patrolling, are usually not implemented.

- a generally low priority for conservation, due to the unpopularity of protection in a social system accustomed to free resource use and expansive frontier development. Protection and conservation are commonly seen as constraining – if not opposing – developments. This false, short-sighted perception is supported by a general lack of awareness of ecological values, and a fundamental ignorance of the inherent value of endangered wildlife in general, and the orang-utan in particular;
- government organisation structures which are characterised by a rigid top-to-bottom command structure with insufficient delegation of bureaucratic responsibilities at the middle level[88], and by poor support for lower-ranking individuals in conflicts with other agencies;
- the current bureaucracies are not designed for active management; the administrative culture is unable to provide the appropriate terms of reference, and the bureaucracies are plagued by a deficient flow of information up the command chain, poor monitoring and control of performance and deficient evaluation of the effect of activities, so that feedback is virtually impossible.

These issues are aggravated because the authorities for protection and law enforcement must operate in a tangle of conflicting interests between the State and the regencies, as well as among regencies and communities. In addition, a hidden bias towards private interests commonly overshadows communal interests and a need to attain a stated objective (Hurst, 1990). Moreover, in the background looms the imperative of an economy held hostage by the voracious world market (e.g. for timber, palm-, patchouli-, and lumbang oils, coffee, rubber, maize, medicinal plants, rattans, birds and cassava), as well as political fear of the powers of a frustrated local populace. Together these constraints lead to a serious shortage of bureaucratic and financial support – if not outright sabotage – for protection. Furthermore, although the forestry organisation and culture in Indonesia have been subject to change under a new policy of the Minister of Forestry since the early 1990s, the acceptance of conservation as an issue of prime importance has been dangerously slow. The same applies to the eastern Malaysian states.

It is therefore hardly surprising that the history of forest management with reference to the distribution range of the orang-utan is one of inconsistent, and predominantly exploitation-oriented policies in Borneo and Sumatra. Concession rights were usually granted to politically powerful entrepreneurs, and indiscriminate exploitation was usually allowed even within Protection Forests and conservation areas. Under the influence of excessive financial profits from timber extraction[89] and the prospects of sharing in plantation dividends, the control and management mechanisms tend to become easily corrupted (Hurst, 1990). Indeed, in many instances the existence of a poorly functioning or corrupted authority in forest management and protection has been a serious liability to the conservation of forests and wildlife.

---

[88] The bureaucratic system exclusively delegates negating power (i.e. saying no) rather than positive responsibility.

[89] The official figures for foreign exchange earnings derived from the forestry industry in 1992 amounted to US$ 5000 million, from 34.7 million m³ of log 'production' (*Jakarta Post* 05.02.1993); the government revenue (IHH) from the processing and sale of this wood amounted to some US$ 118 million, while US$ 107 million was added to the reforestation fund (DR) (*Jakarta Post* 22.04.1993).

In the 1980s the Government of Indonesia paid closer attention to nature conservation, and raised the budgets and salaries of conservation workers to a reasonable level. However, the support of international development agencies soon negated this well-meant effort. First the agencies enforced an attitudinal change, emphasising the importance of a resourcist approach (replacing conservation with sustainable utilisation), and then they consistently downplayed contributions to the conservation sector in their financial support of 'development' in the forestry sector. The result was that nature conservation was relegated to the low-priority issues on the development agendas, both in the national and international context.

> During the Indonesian five-year development programme *Repelita IV* (1984-1989) the total budget allocation for nature conservation was $ 12 million, i.e. some $ 2.4 million/year. In 1988, the World Bank recommended that for *Repelita V* the allocation should be US$ 10 million, in order to cover the basic costs of a more effective conservation of Indonesia's unique biological diversity. It suggested that half of this allocation should be mobilised from donor grants throughout the world (Davis and Ackermann, 1988). In 1990/1991 the annual government (APBN) allocation was about $ 6 million (*Statistik DepHut* 1990/91); the organisation increased its manpower, but it was difficult to discern any improvement in its effectiveness in the field. At the same time, the World Bank and other donors arranged a sequence of major loans in the forestry sector but very little was invested to seriously improve the conservation organisation. Despite a purported emphasis on conservation, 90-95% of the international support for the sector was in fact a subsidy for improved 'production', i.e exploitation, of the wild-growing timber resource.
>
> Of all the multilateral support for Indonesia's forestry sector in 1990-95, that is $ 178,178,824 in loans, and $ 126,018,153 in grants, only $ 36.6 million (12%) was focused on conservation, including sustainable utilisation, and less than 9% was administrated through the Directorate-General PHPA. Of the grants, only some $ 6 million (i.e. 5%) was for so-called conservation, 50% of which is believed to have concerned the WWF Indonesia Programme (MoF Country Paper, Midterm Review, 1995) seeking 'cooperation with the stakeholders' (Jakarta Post Febr. 22nd. 1996). In this, the national conservation agency barely played a role.

The possibly well-intentioned attempts by the Indonesian Government to enforce sustainable use of the timber resource and to invest in forest regeneration (e.g. reforestation) were commonly frustrated by poor and biased control, inconsistent international advice and a complex bureaucracy facilitating the continuation of the *laissez-faire* policy and open access to State forest land for allegedly 'poor local people' and other stakeholders. The fate of legally protected wildlife and plants was, and is, consistently neglected at the crucial decision-making levels. The possible threats to orang-utan populations – or any other endangered wildlife – resulting from forest exploitation within a concession were never of serious concern in a political context. A subordinate conservation agency of deficient capacity and commitment (Worldbank, 1997), faced with ambiguous regulations and with powerful competitors in the forest- and land-use sector, is unlikely to alter that predicament.

With respect to the protection of all conservation areas within the distribution range of the endangered orang-utan in Borneo and Sumatra, it is therefore understandable, albeit unacceptable, that the current government organisation structures for

> The official Indonesian Forestry Policy 'on log production' states: 'The development of forest utilisation is based on the principle of benefit, sustainability and business.' It is claimed that the utilisation of the wildland forest estate is 'in accordance with ITTO guidelines, emphasising sustainability as well as productivity increase and the ecosystem's balance. The policy is directed to the continuation of forest production through increasing forest resource utilisation ... while considering forest functions as a source of revenue for national as well as regional development, as a source of income or jobs' (Sarijanto, 1996).
>
> Notwithstanding some 25 years of UNESCO education in the Man and Biosphere programme, and other international campaigns for conservation, it is important to realise that what is commonly understood as sustainable forestry in professional forestry circles has no bearing whatsoever on the conservation of a wildland forest structure of biological diversity. It is focused on the (economic) sustainability of timber production. It is concerned predominantly with a few tree species that were established wild, without any kind of investment, and which grow without management input, to be harvested, irrespective of other essential elements in the forest community. Management involves removal of lianas and destruction of all wild non-commercial trees, often by poisoning and liberation thinning, allegedly to give selected specimens an optimal chance of wood productivity. In the harvesting, measures may be taken to conserve exclusively a recruitment stock of seedlings and saplings of the few commercially valuable species (i.e. minimum of 25 trees/ha of minimum 20 cm diameter at breast height, DBH), but the rest of the forest is treated as unwanted 'vermin'. It must be evident that under such a regime for sustainable forestry there is no room for the conservation of biological diversity in general, or the orangutan in particular, unless this concept is formally redefined towards a more conservation-oriented or ecological view (see e.g. Lammerts van Bueren and Duivenvoorde, 1996; Lammerts van Bueren and Blom, 1997). Fortunately, the European Commission adopted a Council Regulation (No 3062/95) in 1996 defining sustainable forest management as: the management and use of forests and wooded lands in a way, and at a rate, that maintains their biodiversity, regeneration capacity, vitality and their potential to fulfil, now and in the future, relevant ecological, economic and social functions, at local, national and global levels, without causing any damage to other ecosystems. It is hoped that the EC can influence the timber and wood market in order to have this regulation enforced in the wood-producing countries.

conservation have been unable to deter or inhibit both encroachment and poaching (see e.g. *Six NGOs Question Status of National Parks* in *Jakarta Post* March 26th., 1997).

Despite public claims by the staff of the Indonesian Directorate-General PHPA that 'by regulation, human habitation is not allowed within any of the conservation areas' (Sumardja, 1992), the organisation evidently has great difficulty contesting cases of encroachment effectively through immediate assertive action, and ejecting illegal settlers. The organisation is not only sensitive to the situation's local political ramifications, but is also made acutely aware of the international humanitarian pressure against such actions, by way of the 'support' of the international conservation agencies. Yet, perhaps most important, the operational framework provides insufficient power to thwart or undo encroachment expediently. This problem has been aggravated by the evolving empowerment of regional authority which has gained momentum since the late 1980s in Indonesia. Unfortunately, not all regional authority in Kalimantan and Sumatra is altruistic and blessed with enough ecological insight to prevent hit-and-run exploitation of the natural resources in the region. The same applies to the East Malaysian states (Hurst, 1990).

In general it is difficult to find an example of good custodianship with respect to the wildlands of Borneo and Sumatra. It was almost non-existent in colonial times and after independence the few illuminating examples were too ephemeral to change the trend. When timber exploitation of wildlands was given into the hands of large corporations, often as joint ventures with foreign companies, some people saw in this an example of government-condoned piracy (Hurst, *op. cit.*). In the meantime, local communities have derived scant lasting economic benefits commensurate with the profits made from the timber exploitation. Hence, if people had been law-abiding previously with respect to State forest land, they readily sought new ways to share in the extraordinary profits.

It is not difficult to find examples where the ambitions of local communities have been manipulated, and where local authorities have consolidated their position and expanded their productive territory at the expense of State forest land. One of the means was to support claims for timber extraction, oil exploration and forest conversion, even inside conservation areas. In this way the Kutai Wildlife Reserve was reduced by more than 30%, and may, in the near future, be erased from the protected area system altogether. It is also happening with the Gunung Palung area. In addition, in some areas it has become common practice to commission local subsistence farmers to encroach upon land and stake out supposedly traditional claims. Large numbers of squatters (*perambah hutan*) are present within many designated National Parks (e.g. Tanjung Puting, Gunung Leuser) and other conservation areas (e.g. Gn. Niut, Danau Sentarum, Kayan Mentarang). Several of them admit to having property elsewhere (according to NGO reports). They carry on the typical expansion technique of slash-and-burn, moving from an exhausted base-plot on to new areas, and do not object to being manipulated as long as it is sufficiently profitable.

Behind the proximate impediments to conservation

A summary of the major impediments as outlined in the preceding pages may be condensed to four major issues, namely misconceptions, institutional deficiencies, ecological impediments and low funding priority.

1. Misconceptions:
    - Forest is an open-access resource, and free or empty land to be claimed for production[90];
    - forest is a collection of timber trees, mixed with useless organisms and additional biological diversity, some of which may be useful;
    - improved standards of living (development) lead to reduced pressure on conservation areas and protected species;
    - conservation is equivalent to sustainable utilisation;
    - local communities live in harmony with nature and should therefore participate in the management of conservation areas and sustainable harvesting of wildland forests;
    - protection and law enforcement are ineffective strategies for conservation;
    - protection and conservation thwart development;
    - orang-utans are a protected nuisance species competing for land and fruits.

---

[90] State forest is hardly recognised unless a concession 'right' has been issued for its 'utilisation'.

2. Institutional deficiencies:
- the current legal framework pays insufficient attention to safeguarding the living conditions of protected species, the necessary additional regulations and decrees to cover this gap have not yet been issued;
- the organisations and institutions for protection and law enforcement perform ineffectively (Worldbank 1997); integration in local government structures is problematic due to misconceptions and a poor compatibility of the socio-economic 'growth agenda' with the mission of conservation; government bureaucracies are apparently unable to implement effective nature conservation.

3. Ecological impediments:
- the current conservation areas are ecologically inadequate for orang-utan survival; they barely cover major centres of the distribution range, are too small, often too mountainous, and their lowland parts in particular are often degraded through logging and encroachment.

4. Low funding priority:
- in relation to projects with an expected economic return, the management of conservation areas is a venture of low status and virtually no financial benefits, even if tourism is involved.

If these major problem groups are to be tackled to ensure the survival of the orang-utan it is useful to recognise the underlying factors:

- Misconceptions are due to:
  - deficient education and, in the remoter regions, a mentality currently in transition from traditional tribal to modern civilised;
  - a resourcist view of life[91] and an anthropocentric bias; the ape is not valued as a unique relative, requiring considerable areas of untouched alluvial forest habitat to survive, wherever any (secondary) human interests are concerned;
  - ignorance, or consistent undervaluation, of the ecological services of a natural environment (wildland forests) in general and conservation areas in particular;
  - deficient, biased knowledge of subsistence populations in forest environments;
  - an unrealistic view of human mentality, which holds that an improved standard of living on the part of indigenous peoples will reduce, rather than increase, pressure on wildlands and natural resources;
  - poor extension and feedback pertaining to legal issues and conservation;
  - inconsistent and deficient law enforcement.

- Institutional impediments are due to:
  - misconceptions as outlined above at all levels of society;
  - a lopsided conflict of interests over land and resources between conservation and development due to opportunistically biased decisions;

---

[91] The fundamentalist idea that nature was created solely for human use, and is at the disposal of humans (Ehrenfeld, 1978; Hardin, 1982).

- a bureaucracy dependent on a ministerial culture strongly biased towards resource exploitation;
- a legal framework which has shortcomings for protection due to its resourcist orientation;
- a conservative bureaucratic culture unaccustomed to feedback and control, and unreceptive to scientific support;
- insufficient executive power within the government structure for protection and control;
- deficient education, training and discipline;
- acceptance and exploitation of inconsistent foreign technical assistance.

- Ecological impediments are due to:
  - misconceptions and ignorance concerning ecology in planning for land allocation and land use;
  - the conflict of interest over land and resources between development and conservation, so that conservation areas are leftovers in the land-allocation procedure;
  - deficient knowledge of the distribution of biologically diverse 'umbrella' species.

- Low funding priority for conservation is due to:
  - misconceptions concerning the value of biological diversity and of ecosystem functions in general, and the value of a close Pongid relative in particular;
  - the low expectation of returns of investment in conservation and ecotourism relative to the known revenues of forest exploitation and conversion;
  - the lack of visible proof of effect as a consequence of investments; protection usually means preventing an economic activity, allowing a natural system to regenerate or to develop 'on its own'.

It is perhaps superfluous to note that behind these factors lurks a host of socio-economic and political problems. These have an immediate impact upon the efficaciousness of conservation (see e.g. Dove, 1993), set in the international context of a voracious world market. These problems range from a skewed distribution of income and wealth due mainly to the abuse of power, and a growing cultural gap between urban and rural areas, to the widespread availability of consumerist advertisements in the media and the emergence of industrial forces and landlordship with powerful economic agendas. Added to the increasing demands for land and resources by an ever growing rural population, these factors comprise a formidable impediment to effective conservation.

However, the solutions to these socio-economic and political problems are the general responsibility of the government as a whole. The problems can be, and to some extent are, solved by means of an integrated approach but must not burden a specific nature conservation authority, and in particular one responsible for the protection of areas that are falsely considered as 'empty land' full of resources. To have custody over and to protect ecologically valuable areas is to create and maintain conditions for sustainable development, but should not be seen as 'depriving people of the consumption of natural resources.'

Section IV

The Future of the Wild Orang-utan

The future of the orang-utan is in the hands of humans who speak in euphemisms and adhere to ambiguous agendas. The photo shows the clear-cut and burned remains of what was prime orang-utan habitat up until 1996.

Only if law enforcement can be made to stop poaching effectively, and if land allocation allows the permanent forest estate to remain under forest cover, has the ape a chance of survival. Even up to the present, keeping an orang-utan in illegal custody seems only a minor offence. Confiscation of the animal is almost an administrative affair, having no tangible effect. Destruction of the creature's habitat is usually not even considered enough to warrant policing action.

Prospects of survival

The possibilities

In both Borneo and Sumatra, the illegal hunting of the orang-utan populations, as well as arson, deforestation and the accompanying fragmentation and habitat degradation due to selective logging, are the immediate, or proximate, causes of a dramatic, exponential decline in their numbers. These issues arise from the complex of impediments or ultimate causes (namely misconceptions, institutional deficiencies and ecological impediments) that were discussed earlier. Despite overwhelming odds, ranging from the world market down to the level of an individual person seeking to eke out a living, **the problems can be tackled if the will to help the orang-utan survive can be summoned, and a strategic plan is followed.**

Before addressing the tangle of diagnosed problems and impediments, it is useful to explore first the possibilities and conditions for improvement, in particular the potentially remaining habitat for the currently surviving orang-utan meta-populations, and the financial support for protection. One has to evaluate first the current protected area network in order to assess the survival prospects of the species, since (1) most of the current population fragments are outside the network, and (2) many established conservation areas are of dubious habitat quality.

Hence, in addition to the immediate need for an all-out campaign to arrest the illegal persecution (poaching) of orang-utans, two essential collateral issues with reference to the survival prospects of orang-utans should be addressed, along with one dealing with the protection of the existing conservation area network. The first issue concerns the safeguarding of the permanent wildland forest estate against the mounting demands for conversion and timber plantations, which are usually disguised under the euphemism of sustainable forest management and a growing need for agricultural production area. The second issue involves enhancing the protected area network by expanding existing conservation areas or by establishing new ones. These two basic issues are not mutually exclusive. Indeed, they must be pursued simultaneously to be of optimal benefit to the orang-utan.

The protected area system

A current conservation area system which covers less than an average of 10% of the range of the Bornean orang-utan is insufficient in extent and quality for the ape's survival. In Sumatra, over half the remaining range has recently been designated protected area (Rijksen and Griffiths, 1995). Several improvements to the State forest estate's conservation area network are possible, namely, the expansion and inter-linking of existing protected areas, and the establishment of new ones.

It is important to consider that land-allocation procedures are implemented at the provincial level, and the staggering deficiencies in protected area coverage with reference to the distribution range of the orang-utan must be redressed from that level upward to the central government. This implies that, in principle, full regional consent must first be acquired at the provincial planning level, before a proposal for a change in status of the State forest land can be considered by the Ministry of Forestry and the central Planning Agency.

Regarding both the conservation of viable populations and the efficiency (and efficaciousness) of protective management, it is recommended to establish large (i.e. > 10,000 km²) conservation areas for such sensitive 'mega-fauna' elements as the orang-utan (see e.g. East, 1981, 1983; Soule et al. 1979). On the basis of extinction probabilities of rainforest birds in the Amazon, Terborgh (1974) calculated a minimum reserve size of 2,590 km², which was accepted as an official guideline for the countries in the Amazon-Orinoco region. So far, only the mountainous Leuser Ecosystem in Aceh (northern Sumatra) complies with this concept.

A different approach to the problem of inadequate reserves is to reintroduce orang-utans into large existing protected areas that are currently outside the ape's range. Thus, a translocation or re-introduction project for the Kayan-Mentarang conservation area[92] could be considered in order to establish a new significant subpopulation, if poaching by the local human communities could be halted and the people could be resettled outside the realm of the reintroduced ape. In this way apes who are displaced by conversion of their habitat can be given a fair chance of survival.

Although the orang-utan's survival chances are greatest when forest reserves are well over 5,000 km², it is important to realise that the present situation demands extra care

---

[92] World Wide Fund for Nature consultants have recommended 'reserving' the area as a sanctuary for the Punan tribal community (Puri, pers. comm.); however, it is unlikely that the Indonesian government will allow tribal segregation and grant special land rights which are not specified in the legal framework for conservation areas. Note that during the 1930s a colonial discussion on the protection of tribal groups and their habitat was ambiguous (De Vries et al. 1939).

where smaller protected areas are being concerned. Any further loss of habitat, even in the smallest conservation areas, would cause an intolerable further reduction in the total Bornean orang-utan population (see Table IX). Hence, the governments should seek to expand small reserves to a realistic size and to establish new, ecologically viable conservation areas. One way to achieve meaningful mega-fauna reserves is to link existing conservation areas, including protection forests, through corridors of production forest in which a well-planned regime of selective harvesting is enforced. Several feasible options are still available. This means that the regional planning processes must begin to take serious account of ecology, and to integrate conservation in their aspirations for development..

The orang-utan in West Kalimantan has a protected home in five conservation areas, several of which may appear to be of significant size (i.e. > 1,000 km^2), but in fact are far below a 'safe' standard. However, an evaluation which takes into account the altitude, the actual forest cover, the habitat distribution, and the state of protection, reveals that considerable improvement or a linking up with surrounding production forest areas, or both, will be required in order to make such reserves significant for orang-utan conservation. For instance, the Gunung Niut and Gunung Becapa-Penrisen areas have been, and still are, subject to such encroachment and poaching pressure that the 900 km^2 area at present has hardly any remaining value for orang-utan conservation. Another supposedly major reserve, namely, the mountain range of Bentuang *(dan)* Karimun, may be large (i.e. some 6,000 km^2, or as some sources state, even 8,000 km^2), but is so mountainous that it contains less than 25% of suitable forest. Nevertheless, the area must be considered as the core region for a possibly greatly expanded protected area which should include the (proposed) Danau Sentarum Wildlife Reserve and the (mountainous) Protection Forests to the east, while extending across the national border into the Lanjak Entimau-Batang Aie Conservation Areas in Sarawak.

Central Kalimantan has the largest and most valuable expanse of continuous habitat coverage, yet only two poorly designed reserves are intended to offer protection to the ape. In reality the Tanjung Puting reserve forest has been degraded considerably by timber extraction and by encroachment, while the original orang-utan population has suffered the competitive impact of an invasion of at least 200 rehabilitant apes. Moreover, almost half the area was seriously damaged by fire during the drought of 1997.

East Kalimantan has currently only one reserve of supposed significance for orang-utans: the Kutai Conservation Area. However, it has not only been reduced in size where it required considerable extension for the eastern orang-utan population to survive, but it was also badly degraded through logging, oil exploitation, encroachment, drought and arson. Since the early 1990s, the reserve has been under mounting political and economic pressure to turn the entire area over for coal mining and gas exploration, and it is unlikely that the current drought- and fire-stricken forest area will remain of any significance for ape conservation.

In Sabah, orang-utans have found refuge in six conservation areas, of which only one (Crocker range) is larger than 1,000 km^2, yet due to its overall altitude it is ill-suited for orang-utan conservation. Even this area is no longer of sufficient quality to support a representative sub-population of apes, i.e. fewer than 30 (and even in Payne's optimistic estimates fewer than some 400). The other five conservation areas, except for the Danum Valley reserve within the extensive Southern Forest reserve complex, are too small and degraded to offer hope for the future of the ape. It is to be feared that much of the current Sabah population of orang-utans in protected areas is too small to survive without considerable additional conservation effort. Recently it has been proposed to add the Maliau basin (390 km^2) within the large Southern Forest reserve (NYSC) concession, and the badly degraded Mid Kinabatangan area (approximately 1,500 km^2 of 'forest' in at least two separate fragments) to the protected area network (Payne, pers. comm.). The former is still too small if the surrounding forest will be destroyed. For the latter, it will require effective protection and at least fifty years of uninterrupted regeneration before this degraded forest can regain its former significance as an orang-utan sanctuary.

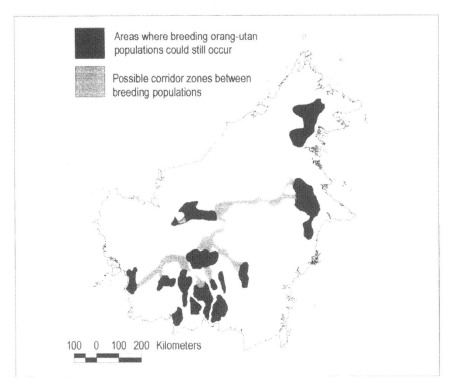

Ideal scenario for the situation in Borneo if State forest lands are managed properly and the habitat conserved; surviving orang-utan populations of a viable size in the year 2020.

The Sarawak situation is critical, although here it must be realised that the only conservation area larger than 1,000 km^2, the Lanjak-Entimau reserve, is joined across the border with the Bentuang and Karimun Wildlife Reserves in West Kalimantan. From an orang-utan standpoint it is therefore realistic to see the Lanjak Entimau Conservation Area as a minor extension of a much larger Indonesian range.

Finally, for a realistic evaluation of the current situation, one may consider the following pessimistic scenario: Assuming that all forest segments smaller than 1,000 km^2 as well as those currently considered 'critical' were eventually to lose their ape populations, and that expansion of, or addition to, the protected area network were impossible; altogether, the current reserves in Borneo would then provide no more than 3,860 km^2 of habitat in at least 12 fragments. This would guarantee the survival of no more than anything up to 3,500 apes, or 22% of the currently remaining meta-population of Bornean orang-utans; that is, if poaching were stopped and if the areas were protected such that the formally protected habitat soon regenerated back to its optimum carrying capacity for orang-utans.

TABLE XXIV

Overview of the relative conservation situation with respect to realistic propects in 1996: 'significant forest' stands for the total area of forest cover within the forest sectors (polygons) larger than 1,000 km^2 in the region, excluding the areas currently already considered critical; 'Cc units' reflect habitat of such quality as can realistically support the survival of one ape/unit (km^2); 'significant cons. area.' represents the accumulated habitat capacity of the conservation areas in the region which are currently larger than 1000 km^2; to be followed by a 'worst-case scenario' (Ws) percentage of apes surviving if the current situation with reference to conservation areas is to be consolidated; the Bukit Baka Reserve was extracted from West Kalimantan and has been added to Central Kalimantan (*).

Region	Significant forest in km^2	Cc units	Significant cons. area	Ws %
W. Kalimantan	27,338	4,360	2,360	30%
C. Kalimantan	58,067	7,197	1,186*	7%
E. Kalimantan	11,791	1,428	496	15%
Sabah	1,481	155	97	22%
Sarawak	1,877	278	all	100%
Sumatra	17,335	2,654	2,340	44%

In this pessimistic projection (Table XXIII), the ape populations in Sabah and in Central and East Kalimantan are likely to suffer heavily, while the Sumatran and the continuous West Kalimantan/Sarawak populations seem to have only a meagre chance of survival as well. It is relevant to reiterate here that the East Kalimantan and the Sabah populations are genetically different from those of West and Central Kalimantan, so the worst-case scenario would imply that at least one type of orang-utan would soon become extinct.

Obviously, in view of the current extensive possibilities for improvement of the

Section IV: the future of the wild orang-utan

conservation situation, it is unacceptable that more than 10,000 apes would be doomed to displacement in the near future, due to human neglect, ill planning and poor organisation. The present situation is ominous enough. Considering that forest fragments of less than 1,000 km^2 represent the lowest threshold of areas capable of maintaining a surviving population, if one detracts the fragments currently destined for conversion and reclamation, then soon altogether no more than at most 100,000 km^2 of forest would remain in Borneo, which is – or could regenerate to become – suitable for anything up to 25,000 km^2 of accumulated orang-utan habitat. Under the prevailing conditions this would support approximately 13,000 apes, but if allowed to regenerate would once again have the potential to support a population of anything up to 40,000 apes.

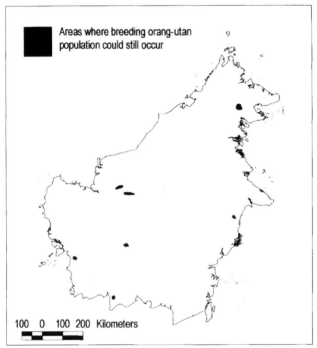

The predicted distribution range of the Bornean orang-utan in 2020 if the current laissez-faire policy regarding State forest land does not change.

The permanent wildland forest: integrating conservation of biological diversity and timber exploitation

The conservation of Indonesia's unique biological diversity in general, and the orang-utan in particular, will be impossible unless the entire State forest estate is considered. All lowland forest in Borneo and Sumatra has been given out for timber exploitation, and either has been, or is being affected by timber extraction. The current protected area network is inadequate, and its protection sadly deficient.

Conservation of the unique ape population is possible only if concession holders harvesting the timber from Production Forests can share responsibility and extend their management to include protective measures. Such steps will in the first place assure sustainability of productive capacity of the wildland forests, while at the same time preserving biological diversity. Accepting such a responsibility, however, will *require some major changes in the formal mandate accorded to concession holders* with reference to encroachment, poaching and trespassing for unregulated harvesting of non-timber forest produce. Without such changes, forest management sustainability is no more than an empty term, and Indonesia's unique biological diversity, including the orang-utan, is doomed. That the Indonesian government has acknowledged many of the constraints and pitfalls of sustainable wildland forest management (Suryohadikusumo, 1992) is perhaps an important condition for improvement.

In Kalimantan the TGHK procedure has allocated 212,000 km^2 of treecover (1994) as Permanent Forest estate (i.e. 101,700 km^2 of limited Production Forest, and 110,300 km^2 of Production Forest) and in Sumatra 96,000 km^2 (47,000 km^2 limited Production orest and 49,000 km^2 Production Forest). Because most orang-utans occur outside protected areas, any consideration regarding a reasonable chance of their survival must include the possibilities of a spatial integration of conservation into sustainable forestry. This implies that forest exploitation must be regulated and controlled so that much of the forest's original structure of biological diversity will be conserved (Lammerts van Bueren and Blom, 1997). It is significant that both the Indonesian and the Malaysian government seem to have signalled their commitment in this respect even in an international context, and, at least Indonesia has many of the appropriate regulating instruments in place.

Integrated forest management comprises the commercial extraction of wild-grown timber from a wildland forest, and seeks to combine sustainable exploitation and the conservation of the original biological diversity, especially including (formally) protected plants and animals in the concession area, as well as the ecological functions of the original forest. Empirical evidence indicates that for the sustainable development of human societies, wildland forests function better ecologically than any other type of land use (including man-made forests) in terms of water catchment, climate regulation, carbon sequestration and the maintenance of an original structure of biological diversity.

Ecological insight leads to the conviction that, along with the provision of the crucial functions mentioned above, such wildland forests also provide the cheapest possible 'production' of a wide variety of timber, since hardly any investment is required. If the production of only one type of timber over the shortest possible time were the objective, however, irrespective of the numerous other functions, it is evident that tree plantations are the economically desirable option. It should be realised, however, that economically managed timber plantations are 'ecological deserts' in comparison to wildland forests, and only slightly more ecologically valuable than modern agricultural fields or chicken 'batteries'. Their structure of biodiversity is deliberately impoverished, and exotic species usually comprise the dominant component.

It is good government to maintain and manage State forest because of its multifunctional values of benefit to the entire nation, and this should be under strict government control or that of semi-government organisations. The simple production of a timber crop in plantations, however, may be left to the private sector. Land must be allocated in such a way that every forest function is taken into account in order to safeguard the sustainability of development of adjacent productive and urban areas. Ignoring the multifunctional value of wildland forest and looking only at the possibilities for short-term cash profits will lead to disaster.

From the viewpoint of the conservation of biological diversity, the integration of conservation in a wildland forest area designated for commercial timber extraction implies a serious compromise. In the population dynamics of component organisms, this concerns fluctuations that are potentially dangerous and much greater than under natural circumstances. The regulated extraction of timber trees, however, may not be unlike natural gap-occurrence, albeit at a much greater frequency and on a more extensive scale (e.g. Whitmore, 1984). Natural tree mortality in Borneo is up to some 3 trees/ha/year (Rathert, 1997); tree-fall due to storms and landslips is almost insignificant. In comparison, in a concession area the extraction of timber results in the loss of anything between 4 and 25 full-grown commercial trees/ha/year in full reproductive vigour, over an area of up to one thousand hectares. In addition, much more 'collateral damage' due to timber felling and extraction occurs than when a tree falls due to natural causes, and remains where it fell, to decompose and be recycled into the system.

Despite the commercially oriented primary objective in permanent Production Forest, the baseline concept involving conservation of biological diversity is the paradigm that *protection against any **unscheduled and unauthorized** form of human impact* is the most effective strategy. In wildland forest areas designated for timber extraction (HPH), a compromise must be established in order to manage the obligatory tree extraction *as well as protection*. No doubt this requires a serious mental re-adjustment and a change in attitude on the part of concession area managers. This implies, for instance, that a logged-over forest is not 'left alone' to regenerate, but should be continuously protected against any unauthorized infringements. In other words, it means ongoing responsibility for the forest area, allowing only planned and authorized activities to take place anywhere within the forest estate. It is irrelevant to consider that such protection through 'policing' is difficult within the context of the prevailing free-for-all culture in Indonesia and Eastern Malaysia. If the Ministries of Forestry and Plantations were serious about integrated forest management in their permanent State forest estates, then protection and well-regulated exploitation is the only option, as, one hopes, the imminent strategy of timber certification will demand.

That forest regeneration after timber extraction is also most effective – and efficient – without further human interference and under a regime of strict protection is without question. Only in places where timber extraction operations have left large gaps (i.e. > 0.5 ha), should 'enrichment planting' be considered, preferably as soon as possible after extraction. In other words, in a Production Forest under exploitation, conservation of biological diversity in general, and orang-utan populations in particular, denotes continuous protection, before, during and after the operations. Such over-all safeguarding of the whole concession area, in conjunction with carefully planned and executed harvesting, are imperatives in the effective conservation of an area's structure of biological diversity in general, and the ape population in particular, thus making sustainable forest management a reality.

Indications for a successful integration of conservation and of commercial timber extraction may be derived from the following observations:

- *Impact on total area:*

  All areas of slope over average 40% inclination are to be avoided and not to be logged because of erosion risk. Areas with steep slopes irrespective of erosion risk, but of low yield and/or difficult access are also usually avoided (see Environmental Capability Guidelines – NMFP–1994; and Kepres 32, 1990); and finally some areas within a concession that have been formally classified as *Hutan Lindung*, are to be avoided. In concession areas of hilly terrain anything up to at least some 15% should in any event remain inviolate according to these regulations and be declared 'protected area' (*kawasan lindung*) according to Kepres 32 (1990). Identifying the area of 40% slope is dependent on accurate contour data, to be derived from realistic topographic maps, the obligatory aerial photographic coverage of the concession area and radar satellite imagery. Despite the current bureaucratic difficulties in acquiring such accurate material, it must be remembered that it does exist, and requires no more than goodwill and enforced regulations to be applied in appropriate planning.

- *Impact on forest structure:*

  The impact of timber extraction is localised and, according to the (TPTI) regulations for permanent forest (HPT), concerns no more than 3% of the concession area every year. This applies to all concession areas, which must consist of at least 50,000 ha. The timber operations take place in pre-selected blocks of some 1000-1500 ha/year (RKT), which form part of an officially accorded five-year plan (RKL) covering five adjacent blocks. A particular block may suffer a second round of timber extraction after no fewer than 35 years and preferably after 70 (see e.g. Kartawinata, 1978).

  The commercial stock of trees in a block is inventoried, and extractable trees (i.e. those of a girth over 50 cm DBH) are labelled (red) in the field, and indicated on a planning map. The rule is that at least 25 commercial stock 'mother-trees' of 25-50 cm DBH are not to be felled and must be protected from injury. Such 'protected trees' also include species on the protected tree list (e.g. *Eusideroxylon zwageri*) and are duly indicated in the planning maps as well as labelled in the field. In practice this means that in a concession with an average of 500 trees of DBH larger than 10 cm (comprising up to 150 species) in one hectare, anything between

28 and 50% of the average potential crop trees (of a DBH > 50 cm) is felled, although no more than some 90% of these trees may actually be extracted (Matikainen et al., 1998 reporting on the Inhutani Labanan-Berau concession in East Kalimantan). As a matter of fact, the economic harvesting baseline may be approximately 7 trees/ha. This is well below a general average of between 12 and 15 trees under the usual sustainable selective felling regimes, and considerably below what can be extracted in an average concession. Be that as it may, even with a low level of extraction, anything between 10 and 32% of the tree stock is more or less heavily affected, if not destroyed, due to insufficient thought being given to the surrounding vegetation.

- *Extent of area demolition:*
The total effect of timber extraction, mainly due to skidding, is tremendous, especially in undulating terrain. For instance, the extraction of tree trunks from a flatland plot resulted in the destruction of 14 % of the area, but the extraction of trees from a steep (43%) slope resulted in the skid-destruction of 32% of the area (Matikainen et al., 1998).

The impact of skid-track and road construction on the hydrology of a forest area has not yet been studied in great detail, but may, in some places, be devastating to enormous sectors of the forest, causing local dessication as well as changes in the hydrological conditions. Certainly this impact factor has rarely been a serious consideration in planning, though added to the felling impact it results in a total of some 60% of the forest being damaged. This level corroborates findings from studies in other areas (see e.g. Burgess, 1961; Johns, 1982).

Thus, in general, on the basis of observations it can be seen that the impact of timber extraction affects at least 60 % of the vegetation, but it is also undeniable that this severe impact can – and should – be reduced. Collateral damage is as bad for regeneration in a commercial sense as it is for conservation of biological diversity. Nonetheless, the limited data does suggest that some 40% of an area may remain inviolate as far as immediate demolition of the forest and compacting of the soil is concerned.

The possible after-effects of skidding and road construction on the forest structure are immensely important in terms of sustainability and conservation. A surveillance flight over several former concession areas, between Samarinda and the Sangkulirang peninsula in East Kalimantan, and across drought-stricken, logged-over forested areas in different stages of regeneration (and including the Kutai National Park), showed that the network of old logging roads and skid-tracks cause extraordinary desiccation of the remaining forest blocks during a dry season, as occurred in 1997-98. The tracks had apparently also been major inroads for the forest fires. All larger trees, in a belt up to some 400 metres wide along both sides of the tracks, had died. Areas without such tracks were still under a living (green) forest cover. In many areas where the track network is dense, this means that the entire forest of many thousands of hectares has died off, leaving a dangerous potential for a conflagration during future dry seasons. Thus, the skid-road network may, in combination with drought conditions, be a major cause for a massive, if not total, reduction of the habitat quality of an entire region.

- *Regeneration:*
  Empirical evidence indicates that gaps in the forest not exceeding a particular size (i.e. < 0.5 ha) are likely to regenerate to more or less original structural quality within one or two orang-utan generations. Nevertheless, the data from long-term research plots (e.g. the STREK plots at Berau, East Kalimantan: Rathert, 1997) indicate that where undisturbed (control) plots in the forest have retained a balanced growth and a dominance of certain commercial species for the last two decades, the plots under different regimes of timber extraction showed not only a sharp decline in standing volume (38%) of commercial timber after logging (due to removal of trees and to collateral damage), but also a dominance shift in the commercial crop of the near future. The dominance (in basal area) of the group of commercial trees was replaced by that of non-commercial trees. From an original proportion in basal area of 60%, the commercial species were down to 40% even after 18 years of regeneration. It is not known to what extent this shift in basal area species composition reflects a real change in tree composition and has an impact on the ecological balance (e.g. of animals), but it certainly has imminent economic significance. In any event, continued research is required to evaluate this reduction and to make restoration of the biological diversity structure effective.

- *Potential for protection:*
  A concession is a form of privatised right of land use. As such it could have a much stronger legislative force than that of a formal protected status, because violation of a private rather than common right is at stake. Law enforcement with reference to a protected status (*Cagar Alam, Suaka Margasatwa, Taman Nasional*) has been deficient, and has meant that the public often considers wildland forest in general, and conservation areas in particular, to be '*tanah kosong*' (empty land), which can readily be trespassed, looted or encroached upon with impunity. In general, a private right seems to be better acknowledged by the public, and the legal framework may offer stronger support. Hence, when advertising its objectives and goals appropriately, and when accorded the proper mandate commensurate with the objective of sustainability, a concession may proffer legislative opportunities for more effective enforcement against encroachment and trespassing.

- *Regulations concerning protection of genetic reserves and trees:*
  Some basic awareness regarding conservation of a potential renewable stock in its concession system has been instilled in forestry managers in Indonesia, although the regulations remain unclear. It seems that both the Ministry of Forestry and the Indonesian Timber Association (APHI) have issued a set of regulations; one proposing to allocate 100 ha per RKL (i.e. 5000 – 7000 ha) to be left inviolate for the conservation of reproductive stock, the other to allocate some 300 ha for the entire concession. In addition, some concessions (e.g. Inhutani) adhere to a regulation to 'protect' 25 '*pohon induk* or mother trees' per RKT. Although such measures may, to some extent, also favour some smaller components of the original biological diversity of a wildland forest, the measures are recommended in the first

place to help conserve the commercial timber production capacity. In many concession areas, plots have been established as 'genetic reserves' (*kawasan plasma nutfah*), in addition to a seed-tree unit (*kebon bibit*), but there seems to be little consistency in the pattern of establishment. The impression is that, especially in common, commercially oriented concessions, the major criteria are the relative difficulty of extracting any commercial crop and the low density of potential crop trees; in other words, sites of the lowest possible commercial interest are being set aside as 'genetic reserves.' Nonetheless, however insignificant the present establishment of *kawasan plasma nutfah* may be for the conservation of a wildland forest's original structure of biological diversity, or for an orang-utan habitat, the concept of reserving permanently inviolate areas has been initiated. A clearer understanding of the importance of orang-utan habitat characteristics in conjunction with obligations for future certification may therefore expand on this concept and make it significant.

- *Spatial planning and active management:*
Holders of a forest concession are legally obliged to create a plan stating their intended activities in a wildland forest, and to actively manage the area. In addition, employees carrying out a predetermined function are to be present in some sectors of the area. This is essentially the usual situation in a conservation area. And although the laws and regulations in Indonesia and Malaysia with reference to protection of forest land are as yet sadly ambiguous, concession management should certainly include implementing its responsibility to safeguard the forest area against trespassing, infringements, poaching and encroachment. Although some concessions may take this function more seriously than others do, the management's formal task in this respect is often frustrated by the collusion of authorities who should instead be expected to play a supportive role. In any case, accurate information is the key to planning, to operational efficiency and to effective protection of forest areas, and in every aspect of its dealings the concession management must strive to acquire, analyse and use up-to-date information. The active presence of authorised teams of people in a concession area (i.e. *in the field*) gives tremendous opportunities for effective information collection and protective management. This presence in the field, in combination with the right of land use, is a potentially more effective authority than that which the authorised agency for forest protection and nature conservation (PHPA/BKSDA) can muster for the official conservation areas.

- *Sustainable income generation*
The management of a concession is largely, if not entirely, supported by income from the harvesting operations. In addition, as well as yielding profit these also add substantially to the nation's and region's coffers in the form of taxes. This means that there is a steady and obligatory presence of active staff in the field, supported by sustainable income and able to tap resources for investments, and that the management is taken seriously at the regional (and national) level. If in the near

future a concession holder is obliged to pay serious attention to the conservation of biological diversity in order to be granted the required certification, measures for conservation, and for protection in particular, must then also be involved. Since, in principle, the conservation of a nation's biological diversity is the obligation of the government as a whole, it should be determined whether proven good performance in this field on the part of concession holders and their staff can be rewarded with a tax-reduction incentive. In this instance, of course, a government agency for conservation would need to undertake a consistent controlling function.

Regeneration

With adequate protection, many wildland forest areas which have been subjected to logging will regenerate in due course, i.e. anything between some 20 and 500 years, if they are protected against further looting and encroachment. Whether the same diversity and composition of species will be attained after succession depends on the impact of extraction and in particular the (possible) conservation of founder populations of vegetation communities, especially trees in the immediate surroundings. After selective logging, regeneration time is dependent on the degree of devastation and the ecological imbalance (e.g. an increase in herbivores) as well as on subsequent disturbance due to encroachment.

Riswan and Kartawinata (1987) calculated that after clear-felling it would take anything between 150 and 500 years for vegetation succession to reach the structural stage similar to that of the original primary forest. For selectively logged forest, regeneration is more rapid. This is not to imply that a wildland forest can be selectively logged with no serious or long-lasting impact upon the original structure of biological diversity: in this respect, 'no sound ecological basis for any selection system has yet been established' for forest use (Kartawinata and Vayda, 1984). However, with reference to the typical orang-utan habitat rich in fruit-tree species, empirical evidence suggests that a selectively logged-over forest can be largely restored as ape habitat in a period varying between 30 and 60 years. After demolition for small-scale *ladang* cultivation, some abandoned alluvial areas in the Alas valley (Aceh, northern Sumatra) regenerated to a diverse orang-utan habitat in approximately 40-50 years. It is significant that such rapid regeneration is facilitated by the close proximity of original, undisturbed forest, and by the limited size of the plots (< 10 ha).

Many of the staple resources of orang-utans are relatively fast-growing trees and lianas. Under forest conditions, trees of the families *Mallotus*, *Aglaia*, and *Tetramerista* begin to produce fruit after some eight years and, under optimal conditions, grow to mature size (>25 m) on alluvial soils in about 15 years. Even the strangling-fig trees, which provide extraordinary amounts of food, often during two seasons every year, can attain full maturity within some 15 years, and probably do not live much longer than some 60 years in any case.

Studies of the dynamics and regeneration of forest under the impact of timber extraction (Johns, 1985; Skorupa, 1987) suggest that through a strict regime of appropriate planning and the effective control of exploitation, the selective harvesting of trees in a large concession area (i.e. > 500 km^2) could provide opportunities for an, albeit reduced, orang-utan population to survive. It is believed that such a population could recover to its original size, in conjunction with the regeneration of the habitat, if it did not revert to below a threshold of genetic variation, and if some opportunity for genetic exchange with other demes were occasionally available. The quantitative value of such a threshold is unknown, but should not be estimated at below 50 adult individuals.

Thus, in summary, the survival prospects for orang-utans in forest designated for wild-timber production can be dramatically improved by integrating conservation with sustainable forest management. This implies better planning, applying a reduced-impact logging protocol (Smits *et al.*, 1992), careful timing and spatial organisation of the harvesting schedule, appropriate restoration or enrichment planting and a legal framework allowing for effective estate protection. These measures should be complemented by the establishment of reserved zones inside and between logging concessions, which will allow for the migration of displaced orang-utans and other wildlife. It is imperative, however, that (1) all operations are monitored and effectively controlled, (2) the opening up of forest land is not followed by an invasion of squatters determined to cultivate the land and scour the forest for minor forest produce[93], and (3) the forest is allowed to regenerate without forestry practices such as liberation thinning and poison girdling. In other words, conservation practices can be integrated within the commercial, sustainable extraction of timber from a wildland forest, if the area can be, and is, protected.

---

[93] The present survey, combined with the current knowledge relating to orang-utan ecology, indicates that the temporary disturbance caused by well-controlled selective logging is less harmful to prospects of the ape's survival than are activities for gathering minor forest produce (and associated poaching), encroachment by farmers, and subsequent occupation of the former forest land for plantations.

Conditions for survival

Recognition of global responsibility

It was noted earlier that behind the complex of proximate causes threatening the survival of the orang-utan, a voracious international market is a major driving force. Thus, the ape's survival should not be seen as a local problem, to be dealt with solely at the national level. A precondition for a successful orang-utan conservation strategy is global support and the deployment of independent, incorruptible expertise. Evidently the best chances for success are to be found in Indonesia, if only because it harbours the largest fragments of the remaining ape population.

The orang-utan is an 'umbrella species', representative of an entire ecological community in Sumatra and Borneo. It is also a powerful mascot for gaining the attraction of the public and hence for fund-raising, as the World Wide Fund for Nature has learned since the early 1970s, when it regularly used the image of the red ape for its campaigns. Where a general public fails to understand intellectual concepts such as 'ecosystem' or even 'rainforest' conservation, it readily accepts the orang-utan as the symbol requiring protection. In other words, orang-utan conservation or the protection of its significant subpopulations is readily understood as necessary, and in practice implies the conservation of the ecosystem to which it belongs. As a result, the notion of rainforest conservation could become so much more effective if international organisations, rather than rejecting species conservation as 'outdated', were to join in a concerted effort to support the protection of the ape, and to boost the Indonesian and East Malaysian conservation structures.

Since the early 1970s, the Indonesian government has been inviting international partnership in the conservation of nature, and it has recently specified its invitation with respect to the rainforests and the ape. Yet so far the Directorate-General PHPA has failed to assume an effective directing role in order to convene, coordinate and integrate the required support and to implement its Orang-utan Survival Programme. Indeed, the bureaucratic nature conservation authorities have shown no real interest at all in solving the plight of the orang-utan. It can certainly be surmised that if the international conservation community has not yet responded seriously to the ongoing invitation or provided significant support for the Programme, it is due in part to insufficient confidence in any collaboration with such a lax agency. After all, there is as yet no indication that under its current structure the national organisation could implement such a wide-ranging programme, or effectively coordinate international support.

There is no reason to doubt that Indonesia could as readily finance and organise effective conservation, as it did to establish, organise and maintain socio-political stability, and to develop a new aeroplane industry or a nuclear power plant, if only it could summon a political will[94] commensurate with its numerous public statements. Apparently this has not happened so far because the Indonesian government has expected the world to demonstrate its serious commitment in the form of financial and technical support. International contributions to the conservation of Indonesia's unique nature have been almost insignificant, and have been characterised by scattered, unprofessional aid-like actions, addressing the wrong problems. In any case, these actions have failed to generate any long-lasting serious commitment on the Indonesian side.

Until the dramatic collapse of Indonesia's political and economic structure, in 1998, it was in fact hardly appropriate for any call for partnership in forest conservation to have been because of a shortage of funds. Judging by the extraordinary wealth of the upper class, Indonesia appeared to be a rich country indeed, although conservation and sustainability of resource exploitation never featured prominently in its policies or in its priorities for financial allocation at the regional level. Perhaps the appeal was meant primarily to underscore the internationally agreed principle of global joint responsibility for environmental affairs. For example, in the address by the Vice-President of the Republic at the official opening of the second meeting of the parties to the United Nations Convention on Biological Diversity (November 1995), it was explicitly stated: '*Being the inhabitants of our planet earth, we should **share equally and fairly** the duty and responsibility for the utilisation and conservation of the biological diversity in a rational, efficient and continuous way, in a spirit of genuine partnership with a deep sense of belonging.*'

In 1991, the Indonesian Government completed its National Biodiversity Action Plan. It is resource oriented, and addresses the species level in terms of inventory and *ex situ* propagation only. Its three objectives are:
– slowing the loss of primary habitats of high biodiversity value;
– collecting and making available information on biodiversity, and;
– promoting sustainable use of biological resources.

The realistic objectives of the Action Plan thus indicate that the Government has been committed to 'slowing the loss', while at the same time promoting the use of biological resources. After all, the national priority, as explicitly stated in the Indonesian constitution, concerns sustainable use of the natural resources. Yet, in view of the alarming losses of biological diversity, Indonesia did indicate its willingness to accept additional international partnership and effective global support for conservation and improved protection. What this would have meant in practice, other than a desire for unconditional financial support, may never become clear. However, clearly revealed between the Action Plan lines was a lack of vision and leadership with respect to the real issues.

---

[94] The Government allocated its reforestation fund for pre-investments of some two hundred million dollars for drainage and demolition on of over 10,000 km² of peat-forest orang-utan habitat in Central Kalimantan (Presidential Decree *Kepres RI* no 82, 26 Dec. 1995) (see p. 216-218).

> Since the end of the 1980s, the speeches and addresses of Indonesian authorities to international forums have seemed to indicate that Indonesia is seeking serious partnership and assistance in conservation of its forests. In August 1989, at an international symposium on forestry in Seattle, the Minister of Forestry responded to the mounting threat to impose a boycott on the import of tropical timber by the industrialised nations of Europe and North America. He emphasised that 'all countries should be concerned about the preservation of forests in tropical countries', and urged the nations considering a boycott to 'improve cooperation in preventing the destruction of forests' and invited them 'to help preserve the natural forests in Indonesia' (*Jakarta Post* August 5th.1989).
>
> In December 1991, the President of the Republic of Indonesia, in his public opening address at the second Great Apes Conference in Jakarta, invited the international community to provide direct assistance related to the conservation of the orang-utan and its habitat (*Jakarta Post* December 19th.1991). In 1993, during the Global Forest Conference in Jakarta, and to an international audience representing 35 countries, the President repeated Indonesia's invitation for a 'global partnership to help preserve Indonesia's forest' (*Jakarta Post* February 18th.1993).

Since the Global Environmental Conference in Rio de Janeiro in 1992, many former developing countries have been willing to negotiate for partnership in the conservation of ecosystems and organisms which appear to be of global value. In Indonesia as well, the Government has seemed loathe to spend its resources unilaterally, where it may have felt that others should, and might be willing to, share in the burden. After all, in the late 1980s, the World Bank had already recommended to the Indonesian government that 50% of the approximately $20 million/year required for improved conservation should be borne by international donors (Davis and Ackermann, 1988: 20). The suggestion was never followed up, and failed to become even part of an international agenda.

Under the current (late 1990s) conditions of grave economic recession, the conservation of nature in Indonesia (and Malaysia) has become seriously dependent on external support. Yet it must be realised that if a repeated, and this time more justified, appeal for shared responsibility and partnership is only meant to perpetuate an outdated 'aid' situation with traditional 'counterparts', any international involvement will be wasted. If national agencies are not more receptive to a true spirit of partnership in which full commitment is assured and sustained, and responsibility is shared — as the Vice-President stressed — the passive hope for increasing gifts of money from international donors is a bad strategy to avert the exponential loss of wildlands and biological diversity. After all, the support in technical assistance, aid or loans that has so far been provided has been unable to generate sufficient national commitment (Davis and Ackerman, 1988; *Jakarta Post* March 16th. 1990) to improve the sustainable protection of any conservation area.

Until now, Indonesia has usually been provided with essentially 'free' technical assistance and equipment through a loan or 'aid' programme. Indeed, the government bureaucracy dealing with nature conservation has become conditioned to such easily acquired, yet ineffective support, while national commitment has eroded to the lowest possible level. In any human society, the simple acceptance of gifts without obligation leads to inertia and dependency, and never to commitment and responsibility; the field of nature conservation is apparently no exception to this. The Directorate-General for nature conservation and forest protection has not been able to deploy international support and technical assistance efficiently and effectively (World Bank 1997), mainly because that authority's primary interest did not seem to be the conservation of nature, but rather to wield its bureaucratic powers to the very limit of tolerability.

A newspaper clipping suggesting that the Indonesian government is concerned with the conservation of the orang-utan (Jakarta Post December 19, 1991).

If a renewed appeal for support came from a fully committed national authority seeking shared international responsibility, then the future for the ape, and other wildlife, would be quite hopeful. This would constitute a serious invitation for international investments to fulfil long-lease 'conservation concession' contracts with respect to internationally valued conservation areas. It would also imply a government's genuine desire to integrate conservation into serious attempts to establish sustainable harvesting of its forest resources. It could be an invitation to invest in the contractual service of international professional advisors and managers for wildlife conservation and monitoring, in accordance with current commercial and industrial models. However, there have been no indications so far that joint responsibility and effective partnership are on the agenda of the government bureaucracies dealing with nature conservation. Whereas the nature conservation agencies in Borneo and Sumatra can barely play any meaningful role in hindering the common scramble for land and resources, they have so far failed to note that if they were to create the appropriate conditions, international support could effectively alleviate the predicament.

The role of research

Since the 1960s, when O. Milton, R. Schenkel and F. Kurt were commissioned to gather information on the status of the orang-utan (and the rhino) in Sumatra, no serious initiatives for further updates of information have been undertaken by either the national or the international conservation organisations. Indeed, no serious range-spanning data-collecting surveys on the status of the ape have taken place since the 1930s in Kalimantan, although in 1989, Sugarjito and his colleagues in 1990 made a laudable attempt with the support of the Fauna and Flora International Society. All surveys in Sumatra were incidental, including those in the 1960s and 1970s, and were usually additional to a research project.

In Malaysian Borneo, the situation was different, mainly because an active national branch of the World Wildlife Fund had already been established during the 1970s. Backed by an influential local constituency, this national NGO commissioned a number of exploratory surveys in the service of nature conservation in general, and the distribution of the orang-utan in particular (e.g. Davies and Payne, 1982; Kavanagh et al.,1982; Davies, 1986; Payne, 1988). Unfortunately for the orang-utan, the organisations's influence has been less effective in Borneo's federal states than on the mainland.

Between the early 1970s and the late 1990s, conservation organisations in Indonesia and East Malaysia rarely, if ever, initiated, facilitated or supported applied research on orang-utan conservation on their own account, or even encouraged field researchers active in reserves to adopt a clear conservation-oriented objective in their studies. Although scores of expatriate field researchers have studied virtually all major aspects of rainforest ecology in Borneo and Sumatra, in most cases the initiative was extraneous and the management of wildlife and conservation areas has benefited inadequately, largely due to insufficient recognition of the opportunities and a remarkably weak national interest. As a consequence, much field research on the ape has been abysmally fundamental.

Nevertheless, empirical evidence (Brussard, 1982; MacKinnon et al., 1986) demonstrates that even long-term, pure ecological field research can have a tremendous conservation effect. Apparently the key is the establishment of a permanently manned facility. This gives the impression of a personal interest or territorial 'right' being involved, and effectively deters encroachment. The fact that a number of people, with an obvious interest in the area and its wild organisms, patrol and demonstrate concern about any kind of infringement upon that area, has an immense preventive power. Since this is so evident, why then have most conservation agencies not resorted to this practice and actively deployed at strategic sites scientists with a clear functional instruction?

> The Ketambe area, in Aceh, where a sequence of researchers has been studying the forest and its wildlife since 1971 (see Appendix 3), is the only area of alluvial rainforest in the Alas valley (inside the long standing Gunung Leuser wildlife reserve) which has remained free of encroachment and serious poaching. The ongoing presence and attentiveness of research students has been the major deterrent force preventing local people and authorities from seriously considering encroachment or conversion of the highly coveted alluvial forest area. The fact that international attention is focused and can be called upon to support claims against illegal exploitation, conversion and encroachment undoubtedly plays a major role in this protective function. It must be noted that the US -based Wildlife Conservation Society (formerly the New York Zoological Society) has apparently understood the value of research for active conservation, and has been establishing a network of research facilities 'for conservation' in Indonesia since 1994.

The effectiveness of conservation rests on up-to-date information. In order to be better informed, and to increase efficiency in the field, one would expect Indonesia and East Malaysia to seek the support of conservation organisations in order to make more efficient use of, and to actively deploy, field researchers in biology, ecology, and the human sciences, for monitoring and advisory services.

Deployment of applied research should then focus primarily on issues which contribute directly to a more effective conservation of the ape, or to any other major aspect of biodiversity. To this end, research must expand and diversify into, as well as integrate with, fields of interest other than the academic specialisations of biology. If international support were actively implemented, indigenous scientists would also have a better opportunity for training to become independent experts.

> Since the early 1970s, virtually all orang-utan-related field research has been focused on socio-ecology. Such research has had little direct bearing on the conservation of the ape, if only because it is carried out in habitat which is least disturbed. Thus, for instance, the long-term impact on orang-utans of logging and other major disturbances has been investigated insufficiently, despite a few pioneer studies (e.g. Johns and Marshall, 1992; van Schaik, Azwar and Priatna, 1995; Rao and van Schaik, 1997), while virtually nothing is known of the resource use and ontogenetic development of rehabilitated apes. The study of these apes is commonly held to be inferior in the established circles of field primatology. Indeed, it certainly lacks the romantic glamour of the pioneering studies of wild apes, but it is prescribed now that the traditional studies have reached their descriptive limits, because more detailed insight is required into the ontogeny of the ape and its prospects of survival under sustained human influence.

Thus, integrated applied research and monitoring must head any species survival programme. The studies should not be purely biological, with a secondary or obligatory reference to the protected area and its organisms, as is so often automatically planned, but strategic and able to address the real problems facing effective conservation.

It will be demonstrated shortly that a species survival programme for orang-utans involves establishment of new protected areas, the improved protection of existing conservation areas, the integration of conservation in sustainable forest utilisation, and the improvement of law enforcement and rehabilitation techniques. It also involves investments in educating and seeking the cooperation of local communities, where feasible. In each of these issues, integrated applied (strategic) research must play a pivotal role.

Section IV: the future of the wild orang-utan

Solving problems

As a result of the diagnosis, two major sets of problems were identified: first, the proximate causes of the decline of the orang-utan population, such as habitat destruction and illegal persecution, and second, the ultimate causes, such as (1) misconceptions, (2) institutional deficiencies, (3) ecological impediments and (4) financial impediments.

TABLE XXV
A summary of the major impediments to the survival of the orang-utan and of corresponding strategic actions to improve the situation.

Proximate impediments	Required solutions
• Illegal persecution	• Education in ethics and laws
	• Law enforcement
	Reintroduction
	Awareness campaign
• Habitat destruction	• Expand protected area network
	• Improve protective management
	• Improve sustainable forestry
	control logging operations
Ultimate impediments	**Possible solutions**
• Misconceptions	• Education and awareness campaigns
	• Guide appropriate integration of development and conservation
	• Deploy applied research and monitoring for feedback
• Institutional deficiencies	• Reorganisation and technical training
	Reallocation of organisation
	NGO involvement/control
	International support
	• Improve legal framework
• Ecological impediments	• Integrated planning, expand protected area network and protected areas
• Financial impediments	• Special tax to support conservation
	• International support
	• National NGO support

It will be evident that the driving forces underlying these causes, such as the exponential growth of the human population and the international market, are beyond the scope of measures to be proposed and discussed in this context.

It was elaborated earlier that options for improvement still exist in both Sumatra and Borneo. Hence, the orang-utan could be saved from imminent extinction if at least some of the major problems were tackled immediately. Of the ultimate causes, the misconceptions can, and should, be redressed by means of education, and by information dissemination through the media and other proven means of awareness building. The institutional deficiencies can be addressed by means of reorganisation, training and the involvement of NGO control. The ecological impediments could be ameliorated by improved integrated planning and a revision of land-allocation procedures. It is clear that by addressing the proximate causes of the decline and imminent extinction of the orang-utan, the road would be paved to tackle the ultimate causes of poor conservation of biological diversity in general.

In other words, three major strategic priority actions are required to launch an integrated programme that aims at improving the conservation of biological diversity, with the orang-utan as its attention-attracting representative or 'umbrella species'. First, illegal persecution (poaching) must be stopped by means of law enforcement and its accompanying media coverage; second, the Permanent Forest estate must be safeguarded against further deterioration and unsustainable exploitation; and third, the protected area network of relevance for orang-utans must be improved and expanded. Subsequently, the ultimate causes could be tackled.

Redressing misconceptions

Nature has amazing powers of resilience. Even animal populations can return from the brink of extinction when appropriate living conditions are restored. The reinstating of such conditions as will allow the orang-utan to survive involves protection against further human impact, both on individual apes and on its typical habitat requirements.

Orang-utan conservation must address a clash of inter-family interests, in which both the greed and the poverty of a number of highly mobile and culturally adaptable humans are pitted against the survival of an essentially stationary, culturally unadaptable ape relative. Whereas the human issue could, and should, be tackled by improved government functioning and an appropriate redistribution of economic means, the orang-utan's survival can be secured only by allowing it to keep its habitat, i.e. land and its natural ecosystem. Conservation of that habitat should not be seen as an economic loss, but rather as a safeguarding of the natural environmental conditions of a land-based ecosystem, which allow the human population to develop in a sustainable way.

The complex network of problems that plague effective protection of the rainforest areas in which orang-utans occur cannot be solved without first establishing a conceptual basis of rational arguments. Essential aspects of this basis are:

- Universal scientific opinion is that conservation of wildland forests and wild organisms is of primary interest for human survival (see also the World Charter for Nature adopted by the UN General Assembly) and a reflection of a civilised state having a universal ethical basis. The impetus of recreation and tourism corroborates this opinion. People in a civilised society apparently need to experience areas of wilderness, not for consumptive use but rather for a 're-creative' ethical reflection of their place in nature. Moreover, in more general terms, the forest ecosystem supports people's development much in the manner of 'capital in a bank', by way of the continuous provision of ecological interest. Sustainable welfare thus implies that people must protect this natural capital against individuals who seek to degrade it through consumptive use.

> Ecological interest accruing from primeval forest is accountable as (economic) 'services', namely:
> - an equitable 'regulated' climate conducive to optimum plant productivity
> - an evaporation 'pump' to recycle precipitation-forming rain clouds drifting to adjacent regions
> - a contained, regular flow of fresh water
> - the regulation of the oxygen/carbon dioxide balance
> - a reliable source of fish protein, especially for the poor
> - a source of pest-combatting organisms
> - a source of intellectual enlightenment and delight
> - a unique potential for a sustainable tourist industry
> - a controlling agent for erosion and flash floods
> - a store of genetic resources
> - a 'pump' and sink for carbon dioxide, polluting agents and dust
> - a habitat for fellow organisms, like the orang-utan

- An ecosystem functions as an integrated whole. Natural, or wildland, ecosystems have evolved to function optimally; when more aspects or sectors of a natural ecosystem are subject to cultivation (and domestication) its functioning becomes less efficient and some sensitive organisms disappear. A natural ecosystem must be larger than a specific minimum size so as to accommodate the requirements of its wild biological diversity structure. Moreover, elimination of any wild component of an ecosystem may seriously affect its ability to provide the essential ecological services for sustainable development.

- The tropical rainforest is an ancient, natural ecosystem, which originated and has evolved in the absence of human influence. Due now to their excessive numbers and technical might, people invariably have a degrading impact on a rainforest's delicate equilibrium. If the rainforest area is of sufficient size to accommodate viable populations of major representatives of its biological diversity structure, then, in technical terms, conservation of a rainforest ecosystem means protection of its periphery only, while, as a matter of principle, no management of its interior structure is needed, or should be permitted.

- History demonstrates that humans are antagonistic towards orang-utans. Orang-utans live(d) in forest areas uninhabited by people, and where people live in forest areas for some time, the ape is absent. While some wildlife is able to survive in close proximity to people, the orang-utan apparently can not. This can be easily understood when one considers the overlap in both species' ecological interests. Both favour the same habitat and many of the same resources, and so the ape is forced to expose itself to fatal conflicts. From the ape's point of view, it is immaterial into which cultural class people can be differentiated. Indigenous tribals, transmigrants, traders, farmers, army personnel, tourists, industrial tycoons, or bureaucrats are all absolutely identical in their commonly assertive approach regarding the ape's living conditions. In order to survive, orang-utans must therefore be protected; there is no way sustainable land use, other than **strictly controlled** timber harvesting and tourism, can be integrated, or made compatible, with orang-utan conservation.

- People desiring an area of land in order to subsist and increase their welfare readily convert a natural **habitat** of an orang-utan population, which the apes **need to survive**. Where people may have a number of socio-economic alternatives to fulfil their needs and desires, the ape has none. This principle applies even more strongly in the instance of timber exploitation and the conversion of forests into industrial plantations. No one will starve if timber extraction suddenly ceases in Borneo and Sumatra. It is, however, likely that if uncontrolled timber extraction and forest conversion continues, numerous people will soon have to face extended drought, frequent fires, flash floods and famine, and it is absolutely certain that the orang-utan will soon die out. A forest area for the conservation of a viable population of orang-utans should be larger than 10,000 km^2, and contain river valleys and/or peat-swamp forest. Such a size would also imply maximum efficiency and efficaciousness of protective management; smaller areas would require more active management.

- Sustainable development is achieved when the limits to the availability of land and resources are recognised and taken into account, ecological functions of wildlands (forests) are conserved and wild organisms are allowed their required space and survival conditions. As long as a growing body of people assumes that they have 'open access' to land and natural resources, there is no hope of achieving either ecological sustainability or socio-economic development. A frontier mentality inhibits development and entraps people at a stationary subsistence level, which is detrimental to the environment as well. The exponential human population growth will soon exceed the availability of land, and mounting frustration and poverty will overwhelm any human desire for or interest in natural replenishment of the region. Therefore, the establishment of strictly protected conservation areas is a major (psychological) step on the road to sustainable development.

- The human societies within the distribution range of the orang-utan are social and economic systems in transition. Rudimentary ethics, poor ecological understanding and deficient integration of social regulating functions in such societies are giving rise to ecologically damaging, and economically inefficient, developments. Allowing such people to partake of the forest resources as so-called stakeholders will not in the least support conservation of the structure of biological diversity in general (Dove, 1993) and the survival of the orang-utan in particular; on the contrary, degradation of the forest and extinction of the ape will be accelerated.

> Solving the underlying complex of socio-economic and institutional problems may boost improvement of the conditions for better conservation. This is, however, the primary function of government as a whole and far beyond the scope of a conservation agency's management responsibility. Nonetheless, **the concept** of conservation must be integrated within all aspects of the general developmental process. This does not imply that conservation areas and organisms are to be sacrificed for development, as is common practice under recent so-called Integrated Conservation and Development Programmes (ICDP). In such misguided programmes, local 'stakeholders' are invited to 'participate in the management' (Golder, 1996), and illicit land- and resource-use of a protected area is often legitimised by such programmes under the euphemism of 'sustainability'. Integration is not achieved by spending resources for conservation in development processes, or by allowing exploitation within conservation areas; ecological integrity and biological diversity is doomed if conservation needs to be corrupted in the interests of development. It is a fallacy to expect that socio-economic development for the mitigation of poverty will make conservation self-enforcing; a higher material standard of living will not result in reduced pressures on natural resources (Hannah, 1992; Brandon, 1997).
>
> The integration of conservation and development means that ecological principles and an existential right of wild species are accepted as major guidelines for government, in all aspects of development planning and implementation. It also means that law enforcement for effective protection of wildland areas and wild species is a matter of fact. Integration implies that development must also serve the interests of wildlife conservation in order to keep the world liveable.

When one scrutinises the diagnostic breakdown of problems affecting the possible survival of the orang-utan, solutions readily present themselves (Table XXV). Taking these arguments as the lead, it is possible to address specific solutions to the complex of problems identified in the chapter Diagnosis. The list of essential issues makes it easier to propose possible solutions, and to devise a blueprint depicting the aims of more or less interdependent project activities.

Evidently, misconceptions play a major role in all basic impediments to effective conservation. These can be corrected by means of an integrated approach involving education, the dissemination of information, extension, training and law enforcement. Audio-visual media have the most effective impact in this respect, and it is essential that media coverage of law enforcement is prominently included.

Education with respect to ethics, ecology and the natural heritage of Southeast Asia can be much improved in Borneo and Sumatra, both in the schools and in a more general sense. It is therefore recommended that the conservation agencies pay due attention to this and quickly develop an efficient service in this regard.

However, solutions may be more easily proposed than implemented in a developing society preoccupied with myriad problems other than the survival of 'forest people' – *orang hutan* – in wildlands still full of supposedly free resources. Education with an emphasis on wildlife conservation and the development of ecological insight is a weak issue.

> In Southeast Asia, the education of people in ethics and ecology may be most efficiently achieved when deploying a teacher-training approach for secular and religious leaders as well as for the regular educational staff of schools and local universities. In addition, in some cases, the participatory approach which was developed in Africa by Takayoshi Kano and Suehisa Kuroda, to involve local assistance in tracking bonobos, may be deployed. Jane Goodall also used the same approach for simple research issues concerning chimpanzees. It appears to raise widespread public awareness of the need for protection of the apes and a sense of revulsion relating to their persecution. An alternative approach may be to advocate the potential dangers of poaching and eating apes, because contact with the blood and tissues could facilitate the transmission of disease agents which can be fatal to humans, for instance, prions causing degenerative diseases (e.g. Kuru, Creutzfeldt-Jakob, Scrapie) and viruses causing immuno-deficiencies (AIDS), hemorragUe (Ebola) and hepatitis, as has been demonstrated with reference to chimpanzees in Africa.

Under the given circumstances, it may well be most feasible to give priority to spreading simple fear of the hazardous health consequences of ape handling, consumption of its meat and the destruction of its habitat, and to instilling respect for law enforcement. After all, the crux of sustainable State development is obligatory allegiance to the legal framework. The issues to be drilled in are primarily (1) the appropriate perception regarding State forest land, namely, that it is essentially off limits to people without a licence and (2) the acknowledgement that protection of wildland forest integrity is essential for maintaining ecological services in the interest of all. The misconception concerning the free-access resource or *tanah kosong* image pertaining to State Forest land can be tackled the hard way if it is done indiscriminately, consistently and accompanied by the appropriate dissemination of information.

For rural people having difficulty accepting the notion of the State acting in the common interest, a collateral solution to this problem may be provided by the observation that Southeast Asian society has a well-developed sense of, and respect for, ownership. Although land cannot be owned in Indonesia, according to the Constitution, it can be handed over under (long-)lease which is, in the view of the public, equivalent to a private, inviolable right. If a forest area could be accorded this status of private right of use, the conditions are created in which fear of punishment would hinder infringement of that area. In other words, a programme for conservation can increase its chance of success, if it is to establish from the beginning an image for the area which proclaims a clearly acknowledged territorial right. The right of 'land use' (in Indonesia *hak pakai*) must be held by a publicly recognisable, respected body, possibly involving local 'participation' of persons without a consumptive interest in the area or its resources, which subsequently enforces territoriality[95].

---

[95] Establishment of a territorial 'right' can also be achieved within the framework of the Indonesian Constitution by delegating custody of the area to (1) a special unit of the Military Service (ABRI) which already has the official function of safeguarding the integrity of the country, or (2) a special Conservation Foundation, preferably with strong international connections. There is no doubt that in the Indonesian context the armed forces are best prepared to fulfill such a multi-faceted task.

Thus, it is to be expected that the erroneous image of free, common land can be replaced by awareness that State forest land best serves the long-term interests and welfare of the community, by a complex of measures which include:
- the establishment and enforcement of a territorial right, to be accorded to a publicly acknowledged (professional) body with an unambiguous conservation objective
- consistent and effective protection of the area, accompanied by
- widespread dissemination of information on the state of affairs.

It is possible that similar 'respect' for an acknowledged 'right' can be applied regarding more effective protection of the orang-utan. Since State authorities are constrained by bureaucracy to apply such a concept effectively, this is where a crucial function is waiting to be implemented by a Foundation prepared to defend the ape's legal and ethical rights. If sufficient awareness can be raised with respect to the protective claim of such an assertive agency or a foundation in relation to orang-utans, it is expected that people would come to hold the ape in respect. Unfortunately, the existing 'green NGO's' in Indonesia and Malaysia have not yet seized this opportunity.

Since most orang-utans occur in forested areas outside the conservation area network, and all such forest areas are allocated for timber exploitation or conversion, it is vital that special attention be given to the management of these areas. It is realistic to acknowledge that if this large sector (i.e. 82%) of the current ape population is doomed and to be written off, even the orang-utans within the conservation areas may not be able to survive.

It is, however, unrealistic to believe that all the forest areas in which orang-utans are found can soon be accorded a protected status. Therefore, where feasible, the survival chances of the apes in such forested areas should be maximised, despite persistent sustainable forest utilisation. It is obvious that State forest under a Permanent Forest status, as Protection or Production Forest, can still offer appropriate living conditions, albeit to reduced numbers of apes (Payne and Andau, 1988; see also Johns, 1985). Improved regulations for sustainable (e.g. selective) forestry and reduced impact logging rules, in combination with careful planning, and a willingness to preserve the structure of biological diversity of an area, can be a sound basis to conserve a population of orang-utans in areas where forest preservation is not feasible.

In a timber concession, the harvesting schedule (Five-Year Workplan, or RKL) can be organised in such a way as to inflict minimal damage (i.e. some 30-40% degradation) upon the most important habitat patches, while particular forest sectors may even be set aside to be kept inviolate.

The permanent State forest estate: The integration of ape conservation and sustainable forestry

The integration of wildlife conservation and of sustainable extraction of timber from wildland forest in the permanent forest estate (HPT, HP, HL) requires in the first place the resolution to allow some compromise in the predominantly profit-oriented objective of forest exploitation in a timber concession. In theory this should not entail such a major mental adjustment, because the concession regulations for permanent production forest stipulate that exploitation must be (economically) sustainable. This means that regeneration must be guaranteed, and that the area must remain under a forest cover. That such regulations have not always been taken seriously is mainly due to deficient or corrupted control, and to poor enforcement of the regulations. Fuelled by the spirit of 'reformasi' and the impending timber certification procedure, however, changes in this situation are likely to occur in the near future. Principal goals in integrated forest management with special reference to sustainability and conservation of biological diversity are:

- to reduce logging damage and avoid fragmentation of inviolate forest sections *as much as possible*;
- to reduce skid-track networks to minimise hydrological impact and desiccation *as much as possible;*
- to facilitate regeneration of the original structure of biological diversity characterised by a dominance of commercial trees;
- to ban commercial stock-'improvement' practices (liberation thinning, *menjarang*, poisoning, liana cutting, etc.) altogether;
- to establish, at all levels of government, that sustainable forest management (and conservation of biological diversity) is possible only under a regime of clear and effectively enforced regulations which apply to everyone, including local people ('*masyarakat*');
- to seek better and unambiguous regulations providing a mandate to enforce responsibility for sustainable forest management and for protection of the forest in the designated area;
- to protect the entire estate from all forms of unauthorised infringements (e.g. collection of forest produce) impairing the structure of biological diversity;
- to plan for and establish the conservation of significant areas including '*kawasan lindung*' and slopes, and any known sites of special value for genetic stock, as well as protected species in contiguous landscape structures.

It will be evident that only minor investments would be required for such integration, as most of the goals are already obligatory for good traditional forest management. The required innovations and additions to the regular management concern adjusted spatial planning, taking conservation into serious consideration, effective protection of the whole concession, consistent control of operations and abrogation of 'improvement'.

Section IV: the future of the wild orang-utan

Many forestry officials and concession holders may believe that the establishment of anything up to 1000 ha of 'genetic reserves' and *'kawasan lindung'* per concession is a sufficient compromise to accommodate the conservation of biological diversity in general, and the habitat of orang-utan in particular. This is nonsense. Such locations are of no significance whatsoever, except perhaps to safeguard some genetic stock of a few commercial trees. Most important in the integration of conservation and commercial timber exploitation is a change in attitude with respect to the entire concession area. As a matter of fact, *the whole area should be considered protected, and be managed appropriately*. Thus, some sectors could be left unaffected permanently, the largest sector would be unaffected for most of the time and eventually only small sectors of up to a thousand hectares would suffer (the then well-controlled and minimised) impact of timber extraction for no more than one year. In areas designated Permanent Forest estate, the crucial question should not be, 'Where does one find and delineate some relatively small and barely significant areas for the conservation of biological diversity?', or, 'Where does one exclude certain plots from exploitation, to be maintained as obligatory genetic reserves?', but rather, 'Where does efficient management provide the best opportunities for sustainable harvesting *in conjunction with* the effective protection of wildland forest conditions?' Thus, the whole area must be considered for integrating conservation measures in management for harvesting, while within that area certain refuge sections may be set aside for sustained preservation and extra protection.

> The designation of some small, isolated sectors to be kept inviolate does not guarantee the conservation of biological diversity, except perhaps for a few small organisms. Experimental research has shown that in such 'islands' many species inevitably become extinct due to genetic erosion within their too small, isolated populations, while for some larger organisms like the orang-utan, such islands are too small in any case to house any significant population (Soule, 1986; Leigh, et al. 1993; Turner et al. 1995; Turner, 1996). For instance, a 100 ha forest can house no more than at most two family units of strictly territorial gibbons, while such a small forest patch would simply be too inadequate to sustain even a single orang-utan.

In addition to the professional economic baseline of knowledge necessary for forest exploitation, the following guidelines are also imperative:
- understanding the ecological functions of a wildland forest for the surrounding, adjacent productive areas
- acknowledging the value of the wildland forest in terms of its structure of biological diversity, i.e. respect for the living system and each of its components, rather than consideration of the market value of a few trees
- knowing major aspects of the life history and ecology of a careful selection of 'indicator' or 'umbrella' organisms, and maintaining an image of representative sectors of the forest structure in the most important areas (land systems).

The rest is a matter of appropriate planning and of subsequent management implementation with special attention to effective protection. Even if the original populations of many species is to some extent reduced temporarily or permanently, the likelihood of survival of the entire original 'stock of species' (i.e. the structure of biological diversity) in the standard concessions of a minimal size (500 km^2) is high, as long as protection against non-authorised encroachment and forest-produce exploitation is effective.

In practical terms, the planning should make optimal use of the geographic information system (GIS) approach. One may imagine the procedure as a sequence of information layers superimposed on an accurate geographical basis, from which particular relationships can be highlighted and others suppressed at will. The following layers are important with respect to the conservation of biological diversity in general, and of orang-utan populations in particular, with reference to production forest:

- accurate topography, with appropriate contours
- administrative borders
- land systems (land structure / ecological units) based on geomorphological data
- pattern of forest vegetation classes image (satellite or aerial photo interpretation)
- soil and hydrology patterns
- > 40% slope (patches) occurrence (*petak* level)
- established *kawasan lindung* locations and formal Protection Forest (*Hutan lindung*) areas
- existing tracks, skidding tracks and roads/buildings, posts, etc.
- cruising data on abundance (dominance) and rarity of tree species
- realistic forest suitability classes
- planned harvesting and extraction pattern
- formalised RKT planning
- real timber extraction pattern (real skid-roads, locations of removed trees, remaining future stock)
- enrichment planting plan
- crude distribution patterns according to habitat requirements of orang-utan and other umbrella species,
- specific sites of importance (research plots of extraordinary information value, established genetic reserve plots, caves, wetlands of special importance, etc.)
- realistic land-suitability classes
- regional RTRWP/TGHK planning data
- transmigration areas, precise extent (boundaries according to SK)
- villages in near surroundings (and/or inside) the concession area
- demographics and indication of possible impact (economic parameters)
- traffic/trade flows and markets.

It will be evident from the list that many of the basic information layers also feature prominently in the traditional planning for efficient timber extraction. Indeed, the planning for conservation management of an orang-utan population (and the structure of biological diversity in general) is, in this case, an integrated extension of this planning process, with an emphasis according to the enumerated goals (notably reduced- impact extraction), and including the development of a strategy for effective protection.

In planning for special '*kawasan lindung*' areas to be kept permanently inviolate and under extra protection, particular attention should be given to the well-known fact that ecotones, i.e. transition zones between two or more extensive areas of different environmental characteristics, are crucial not only as orang-utan habitat, but because they also usually harbour the highest levels of biological diversity, e.g. valleys, edges of limestone or other geological outcrops, and so on.

Taking these factors into consideration will result in thematic maps of the forest area, which can, for instance, exclude the immediate land requirements for timber extraction and other possible uses, and focus on areas for which protective management must be developed. Indeed, the GIS thematic layers of information of greatest significance for conservation are those which facilitate modelling and visualise potential and real threats (i.e. communities, infrastructure, etc.), so that protective strategies can be developed and implemented expediently.

The major problem in the integration of conservation and exploitation is the weakness of an authorised (and legal) basis for effective protection of the concession estate. One might expect that for the harvesting of natural resources on State Forest land a concession contract (*Peraturan Pemerintah* PP 21/ 1970; Forest Agreement; SK Menhut 236/Kpts-II/95) between the Minister of Forestry and the concession holder would delegate full responsibility for the sustainability of the tree cover for a period up to 35 years, and would be a sound, unambiguous basis for management. This, however, is not the case. The obligations stipulated concern mainly bureaucratic affairs, and though appearing to address responsibility for the area and its permanent forest cover, in reality they provide no mandate for fulfilling such a responsibility. Because the sustainability of forest management and the conservation of biological diversity is *impossible* without a formal mandate to prevent poaching, encroachment and trespassing, adjustment of the laws and regulations in this respect is a major issue if anyone is serious about the integration of both sustainable forestry and conservation and harvesting. It is believed that semi-government organisations, like Inhutani in Indonesia, can and should play an essential role in this respect, and should have unreserved international support.

In the reality of concession management, several authorities are scrabbling about in the no-man's-land of fundamental responsibility, while unauthorised people usually flock into the area for the uncontrollable extraction of (protected) forest produce as well as to poach. Thus, large sections within the concession boundaries may be converted for transmigration, without any prior notification, discussion or negotiations with the concession holder, and concession roads are readily used as a public thoroughfare. Large contingencies of any nearby population will try to subsist primarily by looting the surrounding forest estate, rather than by turning to agriculture. This applies not only to transmigration communities (see e.g. Kartawinata and Vayda, 1984), but also to the rapidly developing local population, with the result that an indeterminate number of unspecifiable people (*masyarakat*) will soon have made a wasteland of the forest. Some good examples are in evidence all over Kalimantan and Sumatra. Under such conditions the protection of the area cannot possibly be guaranteed, so that conservation of the biological diversity is impossible.

It is essential that the legal basis for protective management is made sound and clear to all stakeholders in the concession's wide surroundings. The primary problem is ambiguity over land allocation and land use, as well as resource extraction with reference to forest produce, and in particular protected (tree) species. Identifying the appropriate questions and clarifying the issues must be a major project effort, which will certainly require serious legal advice. It is important in the Indonesian situation that the questions to be answered by the legal advisors are goal oriented. Finally, the presentation of the clarification will require the services of public relations professionals to facilitate among local populations an increased awareness of the changing situation.

The regulations pertaining to concession management stipulate the obligation to manage the estate such that sustainability of production and the conservation of the original forest functions is assured. Protection of the forest against encroachment and (timber and animal) poaching should be done by deploying one forest guard (*jaga wana*) for every 3,000–6,000 ha, but the mandate of such a functionary remains extremely vague. One wonders how a responsibility for protection and sustainable management can be fulfilled when State representatives in any potential conflict pertaining to forest land use have often dealt with the matter ambiguously, and have sought the paths of least resistance and of maximum economic gain, rather than considered the long-term common-interest benefits. In cases concerning wildland forests, the authority often yields to claims by individuals, with reference to the constitutional issue of 'greatest welfare of the people' or unspecifiable traditional rights (i.e. *adat*), as long as the claims do not conflict with an immediate economic interest of the State. It is not uncommon that the prospect of some sharing in the land-use profits has biased the outcome of a conflict.

In Indonesia, virtually all the forested areas of outstanding biodiversity value are State forest land (*Hutan Negara*), in the custody of the Ministry of Forestry and Plantations. Earlier it was emphasised that ecotones, and in particular the interfaces of watercourses and land, are the zones with the highest levels of original (and often endemic) biological diversity. In colonial times, one of the very first protective regulations in forestry forbade the damaging of the forested fringes of watercourses. Yet, Indonesia's current legal documentation pertaining to forest lands states: 'All land within two kilometres of a river is available for use by local inhabitants'. As a consequence, a local government wields considerable authority in this area, overruling the concession-holder's right, and in effect all the alluvial valleys and well-watered lowlands, indeed the prime habitat of the orang-utan, are most readily falling victim to conversion and settlement. Even a conservation area status can hardly prevent this destructive development process. The result of such ambiguity, however, is that much State Forest land and its resources are up for grabs to whoever can wield any authority or claim some vague 'traditional right'.

Nowadays, all but the conservation areas on State forest land are formally given in concession by the Minister to a third party, either the private sector or a semi-government agency (*Perum*), for exploitation of the timber resource and, in some instances, the land. The concept behind this policy may have been sound, but under the current regime of ambiguous regulations, its implementation is too often corrupted and abused for hit-and-run profit making. For one thing, the tax revenues from the profits have not been sufficiently rechannelled to develop and support the regional and local economies. Hence local people have learned to compare the growing wealth of those entrusted with authority over the forest areas with the meagre allotments for their own subsistence, and have come to resent the system. If the spirit of 'reformasi' of 1998 is seeking to redress this imbalance, it is hoped that it will **not** try to do so along the lines of greater 'participation' and the involvement of local consumers in the supposed sustained harvesting of forests (see e.g. Grumbine, 1990). Instead, the real problems of misuse of power, collusion and corruption should be grappled with in order to restore the essential confidence in an authority governing the sustained use and conservation of the wildland forest estate.

> One example of the ambiguity which erodes commitment to the protection of a wildland forest area is the regulation for protected tree species. Some are protected because they are of special value in terms of traditional use; for instance, *Koompassia excelsa (Mangris)* because it often serves as a support for wild forest bees, *Eusideroxylon zwageri (Ulin)* because it used to be basic, rot- and water-resistant building wood for traditional houses, and *Shorea pinanga (Tengkawang)* because its mast fruiting yielded a commercial crop of illipe nuts and attracts herds of bearded pigs to be hunted. While the concession management is obliged to be careful to avoid felling or damaging the formally protected tree species, local people, when organised into a cooperating group, are entitled to trespass into the concession and use the trees, which, in case of the rare ulin tree (iron-wood) means consumptive (which will prove fatal to the tree) use, for a market and without taxation. Perhaps this ambiguity stems from political opportunism coupled with the fear of a backlash to abolish supposed 'traditional rights', and from the false myth that autochtonous rural people live 'in harmony with nature.' From the empirical evidence, it is obvious, however, that such a protective status, exempting an undefined category of consumers, has no significance whatsoever for the conservation of the tree species, while it corrupts the possibilities for protection of an area. Such corruptive ambiguity must be banished by means of official legal action, and be redressed in a new regulation, because effective protection and the conservation of biological diversity is impossible within an ambiguous framework.

The (typically western) romantic ideology of 'power to the people' in terms of resource use is unrealistic in the face of a free market, and leads to looting as well as to the utter destruction of the most valuable original components of the structure of biological diversity (see e.g. Kramer *et al.*, 1997). In this regard, one must give sombre thought to the fact that many unique and extremely valuable animals and plants have been exterminated through the uncontrolled harvesting by local people, driven by international market forces. For instance, the Sumatran rhino, a large number of birds (Straw headed bulbul, Hill mynah, White-rumped shama, Orange-breasted thrush, White-crested laughing thrush and Green magpie), certain butterflies, freshwater fishes, palms (notably the smaller *Cicas convoluta*) and all large-flowered orchids are locally extinct due to looting by local professionals.

Upgrading the organisation for conservation

The current government structures for the conservation of nature in Borneo and Sumatra are fully subordinate to the exploitation-oriented systems in which they are embedded (Hurst, 1990). This means that effective conservation of wildland areas and protection of species is terribly constrained. It can become effective only by integrating conservation in all aspects of government and by assigning a powerful independent government agency the task of asserting conservation as a long-term national or common-interest goal.

A government agency for nature conservation is formally charged to defend protected areas and wildlife. In Borneo and Sumatra such an obligation would prescribe that no wild orang-utan could or should be sacrificed without the government agency having done its utmost to prevent this. One would expect that under this formal obligation there is no room for allegiance to, or even sympathy with, the desires of other government departments, industrial interests, or the unsubstantiated claims of local people when these clash with the primary, long-term national interest for which conservation stands.

This would also denote that as an essential part of the modern State structure, a government agency for nature conservation should be consulted in all instances where the country's natural resources are subjected to impact. This involves all human activities which degrade the ecology, such as the exploitation of timber, coal, oil, gas and water, as well as transmigration, translocation, swamp reclamation, irrigation and infrastuctural works. In cases where the impact on the ape's living conditions could be minimised, the agency should ensure that the impact is absolutely minimal. Where large areas of habitat are planned for conversion, the agency should have the legal obligation to veto such actions, and to subsequently present all related information to the higher state authorities, so that a weighted decision can be made concerning the possible sacrifice of any given number of apes. Considering the special protected status of the orang-utan in Indonesia, such a decision can ultimately be made only by the President of the Republic.

Under the current conditions, this appears to be an illusionary scenario. Although an evolution in this respect may be underway, a simple line of formal duty, as may be derived from the legal framework (the basis for the authority of a conservation agency) in Indonesia and Malaysia, is as yet insufficiently reflected in the activities of the organisations.

> As in any human situation with insecure functional support, Southeast-Asian bureaucracies are also characterised by a culturally ingrained sense of respect and subservience, demanding allegiance to any potential power structure. This inhibits a solid commitment to the formal duty especially in the field of nature conservation. It applies on all levels throughout the structures dealing with wildlands, if anyone with both influence and economic interest in the area or its wildlife stakes a claim. As a consequence, conservation interests, if considered at all, are usually subordinate to commercial or local political interests and to extended family allegiance, in all areas where protected wildlife is present.

This predicament of course shatters any hope of sustainability of harvesting and the conservation of biological diversity in general, or the protection of endangered species in particular. Indeed, despite international technical assistance to managing authorities, the Sumatran rhino even vanished from many protected areas before their very eyes (Rabinowitz, 1994). It is by no means unlikely that the orang-utan is next to follow.

In view of the problems and deficiencies identified earlier in the discussion of the survey results, the question of strengthening conservation organisations needs to be raised. One may contemplate three major integrated ways:

- First, the Government structure for nature conservation must be made independent of exploitative interests (see also Barber *et al.* 1995).

  In Indonesia, the stronghold of the orang-utan, this can be done by either reversing the roles in the Ministry, turning the latter into a Ministry of Conservation, with a subordinate role for sustainable forestry, or by moving the responsibilities for wildlands and wildlife on State land to the State Ministry of the Environment, adding Conservation as its main function. However, the way to make protection most effective in the Indonesian context is to add the responsibility for the ecological stability of the State to the primary function of the Armed Forces. The fact that military personnel is often implicated in forest demolition, poaching and illegal trade adds a powerful incentive to the arguments for assigning the highest conservation role to this powerful State organisation. Indeed, the army's regular function to guarantee and safeguard harmonious living conditions for the people should in any case primarily include ecological conditions. This would mean the establishment, integration and training of a special superior and powerful army unit for a protective role in important conservation areas[96].

- Second, the legal framework and guidelines for protective management of conservation areas must be redefined.

---

[96] In Senegal (West Africa) a special army unit is in charge of conservation, and with great success; in Nepal a Ghurka army unit has been instrumental in the effective protection of the last rhino- and tiger reserves (Martin and Vigne, 1995).

Earlier it was mentioned that the legal framework and guidelines for management of conservation areas are ambiguous. In order to increase efficiency and effectiveness, a functioning conservation department should seek to revise the legal framework in order to make feasible the protection and conservation of the natural forest environment's ecological integrity.

- Third, the public and private sector must be involved.

  For all permanent State forest land in Indonesia, except for the conservation areas, the management is delegated by concession (HPH) contract to a private industrial organisation or a 'government enterprise'[97] (BUMN) in order to extract the timber. Conservation areas, however, are currently largely in the custody of the State, as represented by the Directorate-General of Forest Protection and Nature Conservation (PHPA). Hence it would appear to make sense to delegate conservation to professional third parties to be supervised *(mengawas)* by the government. The establishment of the Leuser International Foundation as the professional third party for conservation of the Leuser Ecosystem (Rijksen and Griffiths, 1995; Kepres 33/1998) is seen as an opportunity to develop successful conservation models (World Bank, 1997). A government agency for conservation could actively pursue the establishment of more such 'conservation concessions' by respectable third parties.

A government structure can thus play a stronger role in enforcing the existing regulations and in assisting in the development of better legislation. It may greatly increase its role in education and extension for conservation and regain public respect by consistently implementing protection laws and by inspecting management according to the conservation regulations for concession holders, or for any other agency to which the custodianship of a wildland area has been entrusted.

Conservation of the ecologically sensitive wildland forest and its wildlife means finding the most effective and efficient ways to prevent people from consumptively using the area and anything that might be considered a resource or commodity in that area. Successful conservation therefore starts from a basis of socio-economic knowledge of a local situation, taking realistic account of the positive and negative forces acting upon the wildland area and its structure of biological diversity, notably the developing human communities in the immediate environment. In its implementation, conservation requires a strategic approach of integrated coordination on many levels of action in order to undo or to reduce the impact, involving the whole scale of relevant social regulating structures. In Indonesia, it is obvious that the current government agency is incapable of executing effective conservation of the parks and reserves (World Bank, 1997). It is shackled to the

---

[97] Large areas of Production Forest are in the custody of semi-privatised Indonesian State Forestry Corporations, e.g. Perum Perhutani for Java, and PT. Inhutani for Sumatra and Kalimantan; however, unlike conservation area management, that for Production Forest is nowhere in the direct custody of the government.

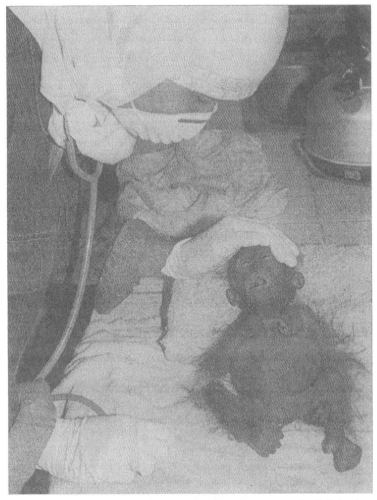

As long as the media fail to expose the grim reality of apes being persecuted and displaced, the ape will remain the object of sentimentalism, but will not be conserved.

ministerial structure and constrained by its bureaucratic culture to seek effective integration with the gamut of regulating and consumptive structures on the local level.

As a matter of fact, the objective or overall purpose of conservation is simple, namely, to protect a wildland and its organisms against human interference. It is important, however, that the objective retains its simplicity and is not adulterated by irrelevant collateral issues, such as references to participation or sustainable development. The strategic approach to effective conservation is complex enough; an ambiguous or adulterated objective, seeking to accommodate development ideologies, renders conservation impossible.

The government's main task is to ensure that the objective is reached, i.e. a control function, but its bureaucratic structure may not be appropriate to design and implement the strategic approach, especially if it is subordinate within a predominantly resourcist or consumerist culture.

In Indonesia it would probably better befit the role of the current government agency for conservation if it were to supervise and facilitate implementation of conservation activities by a third party. Thus, conservation areas could be allocated as a 'conservation concession' for a given period of at least 50 years. Such a third party could be a special unit of the Armed Forces, a semi-government enterprise, a foundation or even a private company. Sponsorship could be shared by international conservation agencies, and under such supervision there should be no objection that management also includes international partnership if and where special skills are required.

Funding

Solving the constraints and reorganising effective conservation of the orang-utan is a long-term programme requiring considerable financial resources. Five options for a long-term supply of financial means may be considered simultaneously:
- regular government allocation
- a trust fund
- international support
- national NGO/Foundation support
- special restoration and compensation tax for exploiters of natural resources (oil, gas, coal, timber, industries on converted forest land) in rainforest on State forest land.

The Indonesian government has repeatedly acknowledged its primary responsibility to conserve the orang-utan. This responsibility, however, has so far been expressed insufficiently in effective action. The formal concern is mainly reflected in (limited) financial support, for conservation in general and policy matters in particular. Hence such support tends only to maintain an inflated bureaucracy.

Considering that the orang-utan is a most attractive umbrella species, representing a large number of wild organisms in Sumatra and Borneo, it would be sensible if the government allocated to the protection of the ape a considerable percentage of its annual financial allocation for conservation. After all, adequate protection implies adequate protection of the ape's habitat, including endangered plants and animals within the region. With reference to the orang-utan, species conservation is in reality ecosystem and biodiversity conservation.

In the mid 1970s, a kind of trust fund was established in Indonesia (Indonesian Wildlife Fund) at the instigation of H.R.H. Prince Bernhard, and with the support of the *Wereld Natuur Fonds* (WNF). The Fund is still in existence, possibly with very deep coffers, and is under the exclusive custody of (ex) high forestry officials. So far, however, it has not contributed visibly to the improvement of Indonesia's conservation organisation. It has supported occasional surveys and small-scale projects concerning wildlife management, but its effectiveness and disbursement is virtually impossible to scrutinise. If the IWF is a typical example of Indonesia's interpretation of a trust fund, this concept is not the way to boost conservation. It is expected that a trust fund based on both international and national contributions can only function when governed by an international staff.

At an international level, the plight of the orang-utan could arouse considerable compassion as well as financial support. International conservation organisations have frequently used the image and dilemma of the red ape to raise substantial funds. This means that the ape is an excellent symbol for internationally shared responsibility concerning the wildland forests in western Southeast Asia. If such campaigns could be coordinated, and were closely examined by the public supporters, the resulting funds could really be deployed for the conservation of the ape. An international foundation with a solid basis in Indonesia should be commissioned by the Indonesian Government to fulfil such a coordinating role.

Within Indonesia the potential for fund raising should have been excellent, since the wealth of the upper social classes was tremendous until the country's economical and political collapse in 1998. Still, it is commonly believed that Indonesian citizens have spirited a capital of at least US$ 80 billion (House of Representatives, in *Jakarta Post* December 16th. 1997), and possibly as much as US$ 200 billion, into banks outside the country. Several NGOs with a focus on wildlife conservation do exist, but their fund-raising capability is either suboptimal or entirely opaque.

The basic environmental principle of 'the polluter pays' has become increasingly accepted in global politics. Indonesia is in the forefront with its levy to have the forest exploiter pay for the replanting of deforested land. In 1993, the Reforestation Fund (*Dana Reboisasi*), which was established in the early 1980s, had accumulated US$ 1.3 billion, which is far more than the total Global Environmental Facility (GEF) funds available for biodiversity conservation worldwide. The target for Pelita VI is US$ 1.7 billion; the revenue for 1994/95 amounted to approximately US$ 353 million (Sarijanto, 1996); the annual interest earnings are at least US$ 400 million (*Jakarta Post* December 23rd. 1994). However, according to the Indonesian Environmental Forum (WALHI), and international agencies, the efficiency and effectiveness of the disbursement of the Reforestation Fund is unfortunately substandard, if not controversial.

That the current disbursement structure of the Reforestation Fund is perhaps suboptimal and skewed is no reason to doubt the principle that conservation should be funded by taxes levied on the exploitation of natural resources. Apart from a few political decisions to re-allocate funds for developments unrelated to forest restoration, a major issue of contention concerning the Reforestation Fund should be the emphasis on reforestation. In practice, reforestation in Indonesia often refers to the establishment of industrial tree plantations for purely commercial purposes[98], rather than to the restoration of a secondary, more or less natural, multifunctional wildland forest fit to conserve its original biological diversity structure and major ecological functions. Because one would expect that a modern industrial venture, such as a timber or pulp-wood plantation, would be obliged to generate its own investment costs, such subsidies can readily be challenged as an inappropriate allocation of funds.

> According to the Asian Development Bank, the disbursements of the Reforestation Fund (DR) during 1989-93 were less than 16% of the funds received over that period. Only 2% of the disbursements were allocated for the Directorate-General PHPA; 45% were deployed for the Secretariat General (which is responsible for training, for example) and 33% were allocated for the DG Forest Production to be spent in the development of timber estates, which displace natural forests (i.e. conversion forest) and are frequently established with exotic species. No funds were allocated for restoration (i.e. protection of regeneration) of wildland forest; in the period 1990-1994 a total of 480,000 ha of tree plantation was established on critical – badly eroded or grass savanna – land (*Jakarta Post* December 23rd.1994).
>
> In 1994, the then President decreed that approximately US$ 190 million from the Reforestation Fund must be provided as a soft loan to support the state aircraft corporation. Indonesian NGOs filed a lawsuit against the decree, arguing that such support was illegal within the legal framework in which the fund had been established (Pura, 1994 and *Jakarta Post* December 23rd., 1994). The lawsuit was rejected, but the case exposed the fact that the Reforestation Fund was controlled mainly by the then President and the Ministry of Finance, rather than by the Minister of Forestry (Barber et al. 1995), and that it came to be deployed for anything but its original objective, forest restoration.
>
> In 1995 this was corroborated when the then President again decreed that some US$ 227 million had to be used for the PLG drainage and reclamation project in Central Kalimantan, allegedly for boosting wet-rice production (*Far Eastern Economic Review*, 1996), but in reality creating an ecological disaster on a national scale. In 1996, the reforestation fund was drained once again, this time of some US$ 100 million for the establishment of a pulp and paper plant in East Kalimantan (*Earth Island Journal*, 1997).

It would, however, be more fitting to have the exploiter of State forest land pay a special, annual fee for the **regeneration, or ecological compensation**, of the ecosystem functions which his activities have degraded. It is therefore strongly recommended to restructure the reforestation levy (DR) so that it becomes an 'eco-tax' on the exploitation and conversion of all State forest land to the amount of up to 20% of all revenues for the full period that the forest system is exploited or that the land yields a marketable product. This would apply to tourism and to the extraction of timber and forest produce, as well as to conversion; the percentage of taxation would be dependent on the measured degree of damage done to the ecosystem, with conversion being taxed at the highest rate. Such an eco-tax would easily cover the financial support required for the permanent protection of Indonesia's few conservation areas and the country's unique wildlife.

---

[98] It is conceivable that this interpretation leads to subsidies for the utter destruction of natural or wildland forests through the establishment of, for instance, oil-palm and rubber plantations.

Well-regulated, high-quality tourism could generate income and collateral awareness for real orang-utan conservation; feral ex-rehabilitant apes in restricted areas could play a role in this respect.

THE ORANG-UTAN SURVIVAL PROGRAMME

Introduction

In response to the crisis relating to orang-utan protection during the late 1980s, the Indonesian Ministry of Forestry designed an Orang-utan Survival Programme. It was prepared through an extended collaborative process that involved in some way virtually all of the world's orang-utan experts. Major sections of the programme (conceived during the International Conference on Forest Biology and Conservation in Borneo (Kotakinabalu, 1990) were presented and discussed at several international gatherings of experts; for example, during the Great Apes Conference (Jakarta/Pangkalanbun, 1991), then at the IUCN-CBSG Orang-utan population and habitat viability analysis workshop (Medan, 1993). The whole programme was then finally presented by the subdirector of Fauna and Conservation, Widodo S. Ramono, and discussed at the International Conference on Orang-utans, The Neglected Ape (March 1994), in California (USA). The programme was officially endorsed during a meeting of specialists at the Ministry of Forestry (Indonesia) in 1995.

Objective

The overall objective of the Orang-utan Survival programme is:
To conserve the wild orang-utan in its natural habitat.

This objective implies: improving the protection of the orang-utan and its habitat requirements by boosting law enforcement and the conservation of lowland and hill rainforest, in particular alluvial and peat-swamp forests.

Major goals of the Programme

Effective conservation of the endangered orang-utan concerns the integration of three major approaches:
(1) protection or law enforcement;
(2) establishment and improved conservation of reserves or sanctuaries for the protection of subpopulations and their habitat;
(3) better planning and enforcement of regulations concerning sustainable forest management.

Hence the programme has the following aims:
- to improve the protection policy and conservation management with reference to the orang-utan.

To this end, a special centre and unit should be established for the coordination of conservation with reference to the ape. The scatter of ideas, initiatives and donor support ensuing from present and future media coverage of the misery to which the ape has been and still is subjected must be scrutinised, integrated and coordinated in order to become effective for ape conservation. Applied ecological research must be deployed to assist in the conservation of habitat and to increase public awareness of the multifunctional and intrinsic values of rainforest and orang-utans. A combined research/training strategy must be designed to serve policy making and to support the effectiveness of protection of all State forest lands. This also includes seeking a revision of the legal framework so that the living conditions of protected species can be safeguarded.

- to establish a wildlife database; to continuously collect and process data on the distribution and status of the ape and its habitat;
  To this end, a network of informants will be developed, including the government (forestry and nature conservation) structures, NGOs, and academic environmental study centres. The data will be collected and processed in a central unit, and subsequently be disseminated widely. Training will be provided for the collection of both direct and indirect information, and for the appropriate processing and storage of the data.

- to stop further illegal hunting of and trade in orang-utans;
  To this end, law enforcement will be boosted, by supporting detection and prosecution or by deploying legal advisors, or both. In addition, the instances of transgression will be used for propaganda and public education purposes, with reference to conservation issues.

- to improve the conservation of the ape in its wildland habitat;
  To this end, ways and means will be explored, and, when feasible, implemented, for improving law enforcement with reference to (1) the protection of the ape, i.e. to stop illegal hunting, and (2) the conservation of its habitat, i.e. to prevent unplanned encroachment. It will also establish the new rehabilitation procedures, and design as well as control ape-viewing for tourism and educational/awareness purposes.

- to improve forestry planning and timber extraction operations so as to include orang-utan conservation, involving the ways and means to minimise environmental damage due to timber extraction;
  To this end, experiments will be conducted and monitored and timber concessions and other industries dependent on rainforest resources will be informed and advised about improved ways and means of sustainable forest utilisation according to EC Council guidelines. This also involves support for a better control of existing regulations in timber exploitation.

- to extend the protected rainforest area network by recommending the establishment of new formally protected wildlife sanctuaries;
  To this end, accurate information and proposals will be submitted to the authorities.

- to explore and advocate the acknowledgement of special habitat (gene-pool) reserves within the Permanent Forestry estate, through local participation of agencies outside the formal conservation structure;
  To this end, advice and assistance in the planning and designation of carefully selected, representative sectors of rainforest for 'orang-utan sanctuaries' or 'refuges' will be provided.

- to facilitate large-scale habitat restoration projects in logged-over areas of Production Forest so as to (1) link fragmented habitat patches (or stepping stones), and (2) restore the diversity of fruit trees (keystone resources) in degraded habitat;
  To this end, working relationships with national and international donors and agencies for reforestation will be established, and advice as well as assistance in the restoration of natural forest and orang-utan habitat will be provided.

- to pay due attention to the translocation of orang-utans and other protected species that are being trapped in small remnant pockets of badly degraded forest due to surrounding habitat destruction;
  To this end, a specially trained unit will be deployed, and a special fund established.

- to develop material for ecological education and extension concerning the survival options of the ape;

- to secure sufficient additional resources for effective protection and the conservation of sanctuaries;

- to establish international relations with Malaysia for orang-utan conservation and effective protection of trans-frontier reserves.

Section IV: the future of the wild orang-utan

TABLE XXVI

Overview framework of the Orang-utan Survival Programme (OUSP):
(the column 'Fin.' (finished) refers to the achievement of the aim in assessed percentage in 1997).

Aims or expected results	Projects	Activities	Fin.
1. Causes of decline of orang-utans diagnosed	1.1 Survey and problem analyses	1.1.a Survey Kalimantan	85%
		1.1.b Survey Sumatra	90%
		1.1.c Problem analysis	99%
	1.2 Workshop for integration of conservation in sust. forest use	1.2. Bringing together policy, law, research, concession holders etc.	-
2. Programme organised for coordination and implementation	2.1 Establishment of Foundation	2.1. Fundraising/lobbying	2%
	2.2. Establishment of OUSP staff bureau	2.2. Coordination/organisation of protection	10%
	2.3 Establishment of orang-utan protection team	2.3 Training special units	-
3. Feedback established	3.1 Monitoring	3.1 a Assessment effect programme	-
		3.1 b Monitoring quality reintro /quarantine	10%
	3.2 Applied research	3.2 a socio-ecology of reintroduction	-
		3.2 b Effects of logging	10%
		3.2 c Pathology and cure in captive apes	5%
4. Law enforcement functioning and awareness raised	4.1 Facilitation of detection	4.1.a Boosting operation	20%
	4.2 Awareness campaigns	4.2.a Awareness campaign Kalimantan	1%
		4.2.b Idem Sumatra	-
		4.2.c Preparation education/extension materials	-
5. Orang-utan reintroduction projects operating	5.1 Quarantine and re-introduction	5.1.a Quarantine and reintroduction stations Kalimantan	70%
		5.1.b Quarantine and reintroduction Sumatra	5%
		5.1.c Emergency translocation	-
	5.2 Applied research and conservation centres for protection	5.2.a Establishment of Meratus reintroduction centre	10%
		5.2.b Establishment of Sumatran conservation centre	-

Aims or expected results	Projects		Activities	Fin.
6. Protected areas network for orang-utans established and management functional	6.1	Establishment of new orang-utan sanctuaries and development of area protection plans	6.1.a Sumatra (Angkola)	5%
			6.1.b Kalimantan (peat swamps)	1%
	6.2.	Upgrading of protective management and extension of existing conservation areas	6.2.a Kalimantan	10%*
			6.2.b Sumatra (Leuser Ecosystem)	90%
7. Habitat outside protected areas better conserved	7.1	Improved timber harvesting	7.1 a Ecol. assistance in planning	-
			7.1 b Control of operations	-
	7.2	Habitat restoration	7.2 a Enrichment planting	-
			7.2 b Protection of regeneration	-
8. Tourism developed	8.1	Planning	Where needed – after survey/evaluation	5%
	8.2	Management support	Where feasible	-
	8.3	Control	Where needed – after evaluation	-

* *The percentage for upgrading of protective management and extension of existing reserves in Kalimantan refers mainly to the ODA technical assistance project for Danau Sentarum, and to some lesser extent to the WWF/ITTO Bentuang dan Karimun project, the WWF/EPIQ/USAID sponsored establishment of a new protected area in the Sebuku-Sembakung region, as well as to the UNESCO Kutai project.*

Structure of programme implementation

General

Following the call by the Indonesian government (Bandung Conference 1993) for international support for the conservation of wildland forest and its resources, the programme is to be implemented as an international partnership. Partnership implies shared responsibilities, serious remuneration and a differentiation of tasks and functions according to specific talents, skill and expertise. It is essentially different from traditional 'technical assistance' and 'development aid'. It is to combine the regular national allocation for conservation and additional financial support, national expertise and international technical expert support for the issues enumerated above. The programme comprises all aspects of institutional development in the service of species protection and may be linked to other such programmes for increased efficiency, e.g. the Indonesian Rhino Conservation Programme. The orang-utan in fact serves as a major 'flagship species' for conservation of the high biological diversity of the lowland rainforest in western Indonesia and Malaysia.

> The Orang-utan Survival Programme (OUSP) may be seen as addressing two major issues, namely, protection of the ape and protection of its living conditions (i.e. its habitat). These issues require diverse structural activities in a large number of locations in Sumatra and Kalimantan, to be supported by different national and international sponsors. Such a large, dynamic and expanding programme requires coordination and an organisation which can handle acquisition, information dissemination, extension, training, guidance, advisory services, management and the maintenance of a network of relations, both nationally and internationally.
>
> The bureau is to coordinate and organise the Orang-utan Survival Programme, to acquire and manage information, to conduct applied research and monitoring, to seek and administer financial support, and to direct, coordinate and assist every serious attempt to improve the survival chances of the orang-utan.
>
> In terms of organisation, the OUSP bureau consists of a basic staff of director, deputy director and secretary. The bureau staff is supported by at least two senior expatriate advisors, one on matters of practical implementation, the other on strategic issues. The bureau is supported by a foundation, its staff appointed by the Minister of Forestry and supervised by the Directorate-General PHPA.
>
> The four major programme components of the bureau are:
> (1) applied research information management and guidance for
> (2) boosting wildlife protection (by the government),
> (3) conservation area protection, and
> (4) guidance for integration of conservation in sustainable forestry.
>
> For the ongoing, active protection of the orang-utan, and for the active conservation management of the forest areas in which the ape occurs, the bureau will provide the necessary information, guidance and physical input. This includes the redesigning of conservation areas and recommendations for the reorganisation of the management structure of forest area habitat. In addition, the bureau will actively coordinate and control quarantine and reintroduction facility management, and organise a monitoring system for feedback. Thus, the bureau will provide relevant information, on-the-job guidance and training, assistance in the organisation of the management and, where appropriate, the restoration of habitat.

The programme is expected to have the highest chances of success under the *aegis* of a Foundation (*Yayasan*) with an influential board. The Foundation is to act as the receptacle for national and international sponsorship and to support an executive bureau with an international staff for coordination and implementation of the programme. In accordance with the guidelines of the government concerning the protection of species, the programme is to be carried out under the supervision of the Directorate-General PHPA.

Concerning the integrated multisectoral nature of the programme, the supervision of programme execution by the Directorate-General PHPA will be under the regular scrutiny of a multidepartmental Steering Body, including representatives of, for instance, the Armed Forces, the State Ministry of the Environment and the Directorate-Generals of the Ministry of Forestry.

Reorganisation of rehabilitation: reintroduction

For the orang-utan to survive, it must be protected according to the legal framework, and all poaching must be stopped[99]. For orang-utan protection, law enforcement, which involves confiscation of live specimens, must be supplemented with rehabilitation. The new approach to rehabilitation, as laid down in a formal ministerial guideline (*SK Menhut* 280/Kpts-II/95), is based on IUCN/SSC recommendations and has been designed to comply fully with the Indonesian legal framework relating to confiscated protected animals, while avoiding the major pitfalls and risks of the traditional rehabilitation process.

Law enforcement for the protection of orang-utans concerns confiscation of illegally kept specimens and ape remains, as well as subsequent legal punishment of the indicted person in the form of fines and custody. When carried out consistently, law enforcement becomes pro-active, and has a powerful educational value. Fear of the consequences of an illegal act does prevent people from transgressing. This educational value can be boosted by means of information dissemination, media coverage and formal education for awareness. If, however, law enforcement and confiscation are not consistent, the whole concept of rehabilitation is of no conservation value whatsoever, and may turn out to be a liability.

In order to break with the outdated tradition, the new approach to rehabilitation is called reintroduction, and goes well beyond the simple return to the wild of some ape individuals. In addition to preparing confiscated orang-utans for a feral existence, it repopulates forest areas of suitable habitat in which wild conspecifics no longer occur. Consequently, reintroduction makes effective use of an attractive key species

---

[99] Even in the cases where large-scale conversion of forest causes orang-utan populations to be doomed, culling or catching of 'doomed' apes will never be allowed without the appropriate formal permit, to be issued by the President of the Republic.

to improve the conservation of newly selected lowland rainforest areas. Its principles are the following:
- apes must be professionally tested for contagious diseases, medicated and thoroughly quarantined before being considered for reintroduction;
- quarantine is rigidly separated from reintroduction (and socialisation);
- reintroduction of ex-captive apes is carried out in areas where wild conspecifics no longer occur[100]; it includes a period of socialisation under feral conditions;
- reintroduction is undertaken in a forest area which has been scrutinised for its suitability as a habitat;
- the specimens are raised together as a group of up to 20 individuals, and after their return to the wild;
- the whole group is left at the site where it was reintroduced, that is, the re-introduction site itself is left and when more apes are available a new one will be established in another location;
- visitor attendance is not allowed at any stage of the process before the apes are entirely independent of provisioning and can live successfully in the wild state;
- the attendant staff for provisioning and monitoring must be proven to be free of contagious diseases and must operate under strict terms of reference concerning close contact and attendant behaviour with respect to different age classes;
- the reintroduction process is regularly evaluated by an independent body.

In 1991, the Minister commissioned the Wanariset Samboja (forestry research) station to set up a reintroduction programme as a model experiment to test the new approach. After careful appraisal, the basic principles of reintroduction were formally adopted by the Indonesian Government in 1995 (Keputusan Menteri Kehutanan No. 280/Kpts-II/95), to supplant the outdated traditional rehabilitation procedure. It was strongly recommended that the station be subjected to a thorough evaluation after 5 years, i.e. after 1996.

In light of the current situation in which considerable populations of orang-utans have been fragmented under pressure of poaching and habitat destruction, it is unwise to even consider the rehabilitation and reintroduction of zoo-bred apes, or the return of illegally exported orang-utans. This is unnecessary as well as uneconomical, in view of the numbers of surviving wild orang-utans in dire need of better protection. Effective alternatives for survival of the species are available.

Economic constraints also dictate that acceptable solutions must first be found for the problem concerning confiscated orang-utans that cannot be reintroduced, due to chronic illness, a mental fixation on dependency[101], or a dangerous, unpredictable attitude towards humans.

---

[100] It is conceivable that reintroduction is also indicated in regions where it can save an area of natural forest including an **isolated** and depleted population of wild apes.

[101] According to the legal framework, such specimens may be 'destroyed' or be assigned to permanent custody; an ethically acceptable option would be to reintroduce such apes by placing them on an isolated island under a regime of continuous provisioning; for healthy, wholly domesticated specimens another option would be to distribute them to zoos.

The first priority, however, must be to save as many as possible of the many *wild* subpopulations in Borneo and Sumatra which currently live in deplorable conditions, due to forest degradation, fragmentation and what may be called refugee-overcrowding of remaining habitat. This implies an effective stop to poaching and a total ban on trade, followed by the appropriate reintroduction of confiscated survivors, and, simultaneously, a concerted effort to safeguard and restore large areas of suitable habitat.

> Economic use of the available funds also dictates that elaborate testing for subspecific identification has a lower priority than testing for contagious chronic diseases. Great care has been – and is – taken to reintroduce ex-captive orang-utans in groups of their own subspecific identity, in both Borneo and Sumatra, on the basis of crude identification techniques (e.g. the visual and tactile hair test). Fortunately, from the confiscation procedure it is also often possible to reliably verify where the specimen was caught, and a simple visual check is usually sufficient to confirm its origin. This applies even for specimens confiscated outside Borneo and Sumatra. A consistent differentiation within the regions of Borneo is almost impossible in practice, and should not be attempted.

The widespread demolition of forest patches due to arson during the 1997 drought, and the subsequent threats to the surviving ape refugees, constitute a major predicament warranting emergency action. However, it could divert the attention and meagre resources away from the actions necessary to secure the conservation of intact wild populations that could have a viable future. The Survival Programme urgently needs a strong organisational direction.

> The captive populations of orang-utans in zoos and institutions have been differentiated since the 1970s, and were segregated at the instigation of the Jersey Wildlife Conservation Trust (i.e. the Zoo of Gerald Durrell) (Mallinson, 1978) in order to avoid further hybridisation of the subspecies. The existing hybrid population of some 180 apes (and 31 of unknown origin) will 'be phased out' and in some cases ape individuals have been sterilised to further prevent reproduction.
>
> Irrespective of why zoos pursue such a segregation policy, it should not be related to the false idea that captive-bred orang-utans are – or may one day be – of significance for the conservation of a wild (or even rehabilitant) population. In terms of the main (if not exclusive) function of captive orang-utans in zoos, namely, awareness building for a civilised (i.e. ethical) view of wild species, it is even questionable whether segregation is justifiable.
>
> It is recommended to bring these 'unwanted' hybrids together, and to try to rehabilitate them in an appropriate tropical forest area, for instance in Peninsular Malaysia, or on Bali, or, indeed, on a suitable forested island anywhere in the tropics or subtropics. It is probable that these apes can be managed so as to make a second generation fully feral, independent of provisioning.
>
> It is relevant to note here the initiative of the Centre for Orangutan and Chimpanzee Conservation (established by Patty Ragan), which owns an area of subtropical forest in Wauchula (South Central Florida, USA). There, ex-captive, unwanted orang-utans and chimpanzees are being offered 'enriching habitats with large spaces for climbing and running' (Prime Apes, newsletter 3 [1]; Summer 1997). Whether the respective areas of available habitat are large enough to allow feral populations to thrive, and to what extent false King Kong sentiments play a guiding role in this project, should be matters of grave concern, however.

Establishment of new conservation areas

A programme designed to facilitate orang-utan survival has to consider many thousands of square kilometres of contiguous forest for protection and restoration, rather than be content with the current composite fragments of a few thousand hectares of 'leftover' land that constitutes much of the present protected area network.

As a result of continuous updates of information pertaining to the distribution and status of orang-utans, the programme will be able to focus on better and more effective conservation of subpopulations in their habitat. This involves the establishment of new protected areas (reserves and sanctuaries) of relevance for orang-utan conservation, as well as the integration of better planning and control of protective measures in selective logging techniques for sustainable forest exploitation in the permanent forest estate.

Obviously the protection of established conservation areas must also be upgraded considerably; the majority of conservation areas harbouring orang-utans have no protection whatsoever, sometimes notwithstanding a considerable staff of PHPA officials. In some instances established reserves must be redesigned and extended in order to give the local orang-utan population any chance of survival. This will involve the reorganisation or establishment of an effective protective structure, the preparation of master plans as well as annual management plans for every major area, execution of the plans, and monitoring of the effects of management for organisatory feedback. It particularly requires the instillation of discipline to carry out effective patrols. In view of the poor original design and considerable degradation of many existing conservation areas, special attention must be given to habitat restoration, the development of corridors between fragmented habitat patches, expansion into surrounding forest estates and the organisation of functional buffer zones.

On the basis of the present survey results, the following proposals pertaining to the extension of the conservation area network and habitat protection for the (at least) six separate and different types of orang-utans are considered of priority in the Orang-utan Survival Programme:

1. The West Kalimantan population

- **West Kalimantan coastal swamps**
  The focus of attention for establishment of new conservation areas are (10) the Kapuas swamps, and, despite the massive destruction by arson during the 1997 drought, (11) the Sukadana–Kendawangan coastal swamps.

- **Lower Schwaner range**
  The Berangin–Sebayan area (9) forms the lower end of the Schwaner mountain range, and, despite widespread destruction due to arson during the 1997 drought,

must have some potential left for the establishment of a very important orang-utan reserve. It contains the protection forests known as Bukit Perai (1,620 km^2) and Bukit Ronga (2,600 km^2) which form the watershed of the rivers Kapuas and Pawan (covering some 16,224 km^2).

- **The Bentuang Karimun and Danau Sentarum complex**
  The Danau Sentarum Wildlife Reserve (DSWR) has almost become 'habitat island' in an area of reclaimed and cultivated land. In order to prevent the isolation of the unique subpopulation of orang utans, and many other species, from other populations in the large and barely disturbed mountainous Bentuang dan Karimun Nature Reserve (BKNR), a wide corridor zone between the two reserves can and should be established. The fact that a road is planned to traverse this corridor should not be a major impediment. The swamp forest east of DSWR, between the Leboyan and Embaloh Rivers, still appears to be of good habitat quality, after the selective logging operations, and harbours many of the larger mammal species, including orang-utans. The zone between this swamp forest east of DSWR and the foothills of BKNR is several kilometres wide, and when this area was inspected in 1995 most of it was covered in shrub vegetation and secondary stands of forest. Protection and professional management of this connecting zone will rapidly improve the ecological function of this area as a corridor between the two reserves. It is proposed to establish a special regional planning unit in West Kalimantan for this purpose and to integrate conservation with the existing development plans.

2. The Central Kalimantan population

- **Peat-swamp orang-utan refuges**
  Peat and flood-plain forests are the most important types of high-quality habitat for orang-utans. The largest relatively undisturbed subpopulations of orang-utans in the mid 1990s occured in the extensive peat forests of Central Kalimantan (Meijaard, 1997). However, this province at present has hardly any conservation areas of significance for orang-utans. The Tanjung Puting reserve is badly degraded as well as fully isolated. Large areas of relatively undisturbed peat forest occur in the flood plains of the rivers Katingan, Kapuas, Kahayan and Sebangau in Central Kalimantan. Although most of the 15,000 km^2 of the swamp forest KAKAB-PLG area has been reclaimed, it is desirable to save the aapproximately 5,000 km^2 of original swamp-forest remaining in the area from further desiccation, looting and arson. This would require in the first place to block and refill the major east west drainage canal near Palangkaraya. A project proposal to this extent is under consideration at the WWF Indonesia Programme.

  For dealing with the conservation of remaining forests in this important province, it was proposed (in 1996) to establish an Ecological Planning Unit within the BAPPEDA structure of the regional government in order to add an ecological and forestry dimension to the regional planning processes, so that no assets and ecological functions in the service of development would be wasted. This Unit will prepare an Integrated Conservation and Development project (after the model of the Leuser Development project) in order to conserve some 1.5 million hectares of the Sebangau catchment, and to boost the development of its surroundings. Other priority areas requiring attention from the Ecological Planning Unit are (12) Jelai-Lamandau-Arui, (16) Katingan flood plain and (26-27) the central Schwaner range.

- **The Schwaner mountain range and its foothills.**
The foothills and valleys of the Schwaner mountain range (26–27) and adjacent Kahayan-Miri catchment are an extremely important ecotone habitat for the orang-utan. Much of the area is State forest land, currently in the custody of timber concessions. Due to its essential water catchment function the forests in this region should be conserved, and any future transmigration project or industrial cash-crop cultivation in this area is ill-designed. Several areas have the potential to be established as functional national parks, contiguous with the current designated Bukit Baka-Bukit Raya National Park, notably the Batikap I, II and III Protection Forests, including some of the lowlands (e.g. the Melawi valley in West Kalimantan, and the water catchments of the rivers Mendawai, Sampit and Kahayan).

3. The East Kalimantan population

- **Tinda-Hantung (Sambaliung) hills and Telen-Wahau area**
The western and northern foothills fringing the central flood plains, west of Muara Kaman, and including the water catchments of the rivers Belayan, Senyur, Kelinjau and Telen (37-38).

4. The northern Bornean population (including Sabah)

- **Sembakung-Sebuku**
The Sebuku-Sembakung water catchments along the boundary with Sabah (and extending as a trans-frontier reserve into the Southern Forest) (43-44-45-46). Fortunately, some major sector of this area was established as a wildlife reserve in 1998, on instigation of the WWF Indonesia Programme; the protective management of this area must however be a matter of serious concern.

- **Southern forest**
For the northeastern population of orang-utans, the best prospects of survival are to be found within the Southern forest reserves in the custody of the Sabah Foundation, in particular some considerable sectors of the very large timber Production Forests, adding up to a total of some 13,000 km^2. The small forest islands of the Danum Valley Conservation area (438 km^2) and the Malinau–Mount Lotong area (390 km^2), situated in a wide expanse of badly degraded secondary vegetation, are of little use to safeguard the ape's survival.

- **Kinabatangan river**

  The WWF-sponsored establishment of a patchwork of remnant forest and lowland swamp areas along the river Kinabatangan as a reserve may also become of some significance for orang-utan conservation, mainly because the Kinabatangan district is still sparsely populated by humans. The area proposed for sanctuary is some 306 km^2, fragmented into eight segments along approximately 80 km of river, linking forest reserves and timber concession territory belonging to the Sabah Forest Development Authority (Lackman-Acrenaz, 1997 unpubl. report).

5. The northern Sumatran population

- **Aceh**

  The Leuser Ecosystem requires sustained international attention and the effective support of the Leuser International Foundation. An uninterrupted implementation of the Masterplan (Rijksen and Griffiths, 1995) by the Indonesian-EU Leuser Management Unit partnership for conservation of the Leuser Ecosystem could guarantee that some 50% of the orang-utan range north of Lake Toba is safeguarded. In order to protect some 90% of the current Acehnese ape population, the Leuser Ecosystem should be extended northwestward, to include all the uninhabited mountain range, which is essential for water catchment and erosion control in Aceh province, as well as for the conservation of all other unique megafauna components of northern Sumatra (elephant, rhino, tiger, clouded leopard, bear). In particular the water catchment of the river Woyla region between the upper valley of the river Teunom in the northwest and the river Meulaboh in the south is a last stronghold of the orang-utan and should be given immediate protected status.

6. The southern Sumatran population

- **South Tapanuli–Batang Gadis**

  Several large forest areas in Tapanuli regency still contain orang-utan populations and should be conserved; the highest priority for attention should be given to the approximately 4,000 km^2 uninhabited Batang Gadis – Siondop valley and the sparsely inhabited, remote coastal flood plain, between Sibolga Bay and the Natal-Jembatan Merah road. Part of the unique Siondop valley has already been allocated for conversion into timber plantation. Nevertheless, conservation of this still forested area of significant size (i.e. over 2,000 km^2) could give a fair chance of survival to the unique southern type of Sumatran orang-utan, descendant kin of the very first type specimen of *ourang outang* described for western science. A proposal applying for a conservation ststua of the area has been submitted to the Ministry of Forestry in 1997. The extraordinary delay in processing the proposal suggests that alternative interests in the area are at play.

- **Batang Toru**
  Another important forest area of approximately 1000 km², in the low mountain region between Sibolga, Tarutung and Padangsidempuan, i.e. the western water catchment of the river Batang Toru, should also be conserved.

- **Coastal and northwest Tapanuli**
  North of Sibolga Bay is a coastal lowland forest area of prime importance, namely, the swamp catchment of the Lau Tapus, along the coast of Tapanuli Tengah. This swamp area of some 120 km² should have its status upgraded to wildlife sanctuary and, for better protection, could be added to the Leuser Ecosystem.

Conservation of habitat outside conservation areas: integration of conservation and sustainable forest management

Most orang-utans occur in wildland forests designated for production, i.e. outside the current protected area network. This implies that the 'umbrella species', which constitute the vast quantity of the unique original biological diversity of Southeast Asia, are at the mercy of commercially oriented concession holders, who have so far been free to operate like pirates with a treasure chest, looting the timber and taking no responsibility whatsoever for the biological diversity structure that makes up a wildland forest and could assure its regeneration. Moreover, the regulations so far have effectively prevented a concession holder from assuming any protective responsibility for such a forest, as an undefined and unruly 'public' (*masyarakat*) is formally allowed to take, unimpeded, whatever it likes from the forest. In the euphemistic newspeak of today, this is usually referred to as 'participation in management for sustainable use by traditional people'. Ultimately this has led to the total demolition of all but the more remote and inaccessible sectors of State forest land, and to the accelerating demise of the orang-utan. If the shrinking estate of wildland forest destined for production cannot be made to further function as habitat for orang-utans, the ape is doomed.

The question is whether, and how, conservation can be integrated into commercial forestry.[102] After all, conservation and exploitation are on opposite ends of the profit scale; exploitation results in cash, conservation yields intangibles. Integration implies that a compromise is accepted on both sides; but since conservation may be seen as an investment in long-term sustainable harvesting and only requires the cost of improved protection, the net loss for the exploitation (profit) side is minimal.

It should be evident that under the current regime of regulations for harvesting and with the imminent pressure for certification of sustainable forest management in

---

[102] Note that the concept of conservation can, and should, be integrated into the sustainable exploitation of a wildland area, but that sustainable utilisation **can not**, and should not, be integrated into the conservation of a protected area.

wildland forests, the opportunities for integration of conservation and commercial exploitation in Indonesia are real:

1 the impact of extraction on a total concession area (i.e. > 50.000 ha) leaves some forest inviolate, as some sectors, e.g. slopes above 40° inclination, must be exempted, or are inaccessible or do not have enough commercial trees to warrant entering (or a combination of all three factors); this is reflected in the established concept of *kawasan lindung*; the impact can be reduced through better planning and control of implementation;

2 the impact of harvesting on the forest structure is according to a planned strategy; it affects only 1000 ha/year at a time, and moves along in blocks to possibly return after 35 years; this means that much of the area, if protected, can remain untouched for most of the time;

3 the extent of area demolition by timber extraction (felling and skidding) is limited, and can be reduced considerably through better planning and control of implementation; this is also extremely important for regeneration of a commercial stock, or sustainable production;

4 the regeneration of a tree cover is virtually complete in wildland forest after some 50 years if the extraction damage is minimised, protection is assured, and no 'improvement' or thinning takes place;

5 the potential for effective protection is higher than for established conservation areas, due to the psychological effect of 'private right';

6 the regulations concerning (genetic and other non-utilisation) reserves within a concession are already established in the current forest management culture;

7 spatial planning and active management are well established in the forest management culture, only the control and feedback is still deficient;

8 a sustainable income generation from exploitation can assure the sustainability of effective protection.

All these factors mean that integration of conservation in sustainable exploitation of wildland forest is feasible. The imminent international pressure for certification of wild-grown timber may promote the required integration.

Be that as it may, it is believed that under a careful regime of sustainable forest management, in accordance with IUCN principles, orang-utans can survive in wildland forests which have been designated Permanent Production Forests (*hutan produksi terbatas*). The major conditions for such sustainable forest management with special regard to orang-utan conservation are:

> Certification, however, is not the ultimate solution to avert an ongoing destruction of the wildland forests. On the contrary, even the threat of certification seems actually to increase the conversion of wildlands into (timber) plantations. The key to avoid an impending certification is already widely used in Indonesia; it is in formal land allocation for plantation development. While the production of high-quality timber is probably most efficient in a plantation system, a certificate for the sustainable production of timber trees from such plantations is readily granted and does not entail the (hazardous) responsibilities for conserving the structure of biological diversity of a wildland forest. Indeed, for timber industry landlords, the current trend in Indonesia is to convert, as soon as possible, all but the most inaccessible wildland forests into plantations; the Ministry of Forestry adjusted its name in 1998 to include 'and Plantations' in order to cover the new conversion policy which anticipates impending certification problems. Thus, now that the wild-grown timber stock is virtually depleted, and it will take at least twenty years before any significant new supply may be expected, it seems that the threat of timber certification and its imposed responsibility for ecological sustainability will soon result in the utter demolition of all wildland forests except for the few ill-designed conservation areas (Sarwono Kusumaatmadja, pers. comm.), thus creating vast ecological deserts.

- the will to integrate conservation into the sustainable commercial exploitation of wildland forest, and acceptance of the responsibility for the conservation of the unique original biological diversity of a whole concession area;

- effective protection of the entire concession area against intrusions for the exploitation of non-timber forest produce and for encroachment;

- selective logging according to the regulations, extracting no more than 10 trees/ha (or up to some 30 m^3/ha) and avoiding the creation of gaps larger than 500 m^2 (see Brouwer, 1996);

- careful planning of annual harvesting schedule, in order to minimise the socio-ecological shock-wave effect;

- careful planning of mechanical log extraction (skid trails) so as to avoid hydrological disturbance and excessive desiccation;

- establishing undisturbed refuges (i.e. *kawasan lindung*) and prohibiting forest disturbance along watercourses wider than 2 m for a 300 m wide zone;

- prohibiting pre-and post-felling 'refining treatment', such as liberation thinning, liana destruction and poison girdling;

- prohibiting poaching (and strict regulation of the possible hunt for terrestrial herbivores);

- promoting awareness campaigns in surrounding communities and integrating them into the regional government structure.

Professional forestry relationships must be established with concession holders for effective integration of conservation in their planning of operations from the first (URKPH) proposal or HPH forest use plan, to the last (URKT) yearly work plan proposals. The programme can also facilitate a better control of the operations in forest areas where orang-utans occur. In addition, training courses for concession personnel in sustainable forest management may be initiated, and publications relating to awareness can be disseminated to back up law enforcement.

In State Forest land of permanent forest status under concession, the programme will seek the establishment of significant refuges (gene-pool sanctuaries, and *kawasan lindung*) and such management as conserves the crucial conditions for orang-utan survival. By integrating the forest exploitation planning of adjacent concession holders, it is possible to link up reserved forested refuges so that at least mega-fauna like the orang-utan have options for dispersal over larger areas.

Awareness: an appeal for general public support

Applied ecological research will collect and process relevant information for improved management and should play a role in the professional production of extension and educational materials. The programme will also process the information in a manner to generate greater ecological awareness. The current fund-raising campaigns launched by the major international conservation organisations may rake in money, but cannot generate the required awareness, because they cash in on sentiment and appear to be designed for an infantile public. Extension and educational programmes for different target groups have to be developed, notably for local schools and universities, the local government and the private sector dependent on forest resources (e.g. tourism, timber trade, rotan trade), but such programmes should be factual and of sufficient intellectual appeal.

The private sector and NGOs with reference to forest exploitation can participate in the extension of appropriate conservation activities. It should also be encouraged to facilitate training in, as well as provide much of the material required for, such an extension.

Development of ecotourism with respect to orang-utans

'Most countries in the world now sell their culture, society and natural environment for consumption by visitors' (Pleumarom, 1994). Tourism can bring in considerable amounts of foreign currency, if well regulated, and is often seen as being supportive of local development, as well as a major incentive for conservation, of the culture as well as of the wilderness. The validity of this assumption is questionable, however; the economic benefits from tourism do not necessarily generate sufficient incentives to support conservation sustainably (Brandon, 1996), while the unrestrained growth of tourism quickly destroys all major assets.

> Tourism's economic benefits and their supposed effect for conservation have been substantially overrated. It must be realised that some 70-80% of tourist money goes to nonindigenous or foreign-owned tour operators, airlines and hotels, and is expended on imported food and beverages (Pleumarom, ibid.). For instance, the ape-viewing attraction at the designated Tanjung Puting National Park (Central Kalimantan) is structured so that the bulk of the revenue bypasses Indonesian society and the conservation of the area and its apes, because it is funnelled into the (essentially US-based) Orangutan Foundation International.
>
> It may not be surprising that when local people as well as guards see conservation areas being protected for 'use' by foreign visitors and entrepreneurs, yet fail to benefit from the income that tourism obviously yields, they develop a strong antipathy towards the concept of conservation. But it must be realized that even when tourist revenue is a significant contribution to the income of local communities and conservation area staff, the effects for conservation can also be either negligible or outright negative, if proper regulation and controlled management is lacking. Thus it was found that in Tangkoko Duasaudara Nature Reserve (Sulawesi), for instance, the financial benefits which accrue to the park staff utterly fail to act as an incentive to provide better protection for the control of poaching (Kinnaird and O'Brien, 1996). This applies as well to the tourist attraction of ape-viewing in Sumatra (Bukit Lawang – Bohorok), where the destructive pressures by local communities upon the park have increased, as the local economics of tourism entirely superceded protection.
>
> As a consequence, the development of ecotourism in the service of nature conservation requires careful planning, consistent regulation and effective enforcement (Brandon, 1996).

Ecotourism contrasts with common or mass tourism in that it is aimed at attracting fewer people at levels that do not cause significant ecological or cultural disruption, or both (Brandon, 1996).

The case of the mountain gorilla project on the border of Rwanda, Zaïre and Uganda has demonstrated that well-planned protective management combined with controlled tourism can effectively protect these great apes (Vedder and Weber, 1990; Aveling, 1992; Sholley, 1992). The project was designed with a strong emphasis on factors outside the park boundary, and gives great attention to protection. It is implemented by the local authority, with occasional surveillance by expatriate advisors. The combined effect of regularly reenforced international attention, by both international conservation agencies and a growing stream of tourists, and the considerable revenue generated by this attention[103], has made the government and local communities aware of the 'value' of the gorillas. Indeed, this awareness was even sustained during the ethnic upheavals of the early 1990s, and the gorillas survived.

Tourists are taken in small groups of up to six persons to observe a habituated gorilla group for a period not exceeding one hour/day. Bookings must be made well in advance. However, the key to success is the initial control of touristic development. The Rwanda Mountain Gorilla Project acknowledges that tourism is difficult to control (Macfie, 1992) and should not be seen as the major means to

---

[103] In 1990, over 5,000 tourists visited the Virunga park to see gorillas. These visits yielded over $1 million in revenue.

generate financial support for the ape's conservation, because the economic incentive corrupts the primary conservation objective.

As soon as apes are able to be observed, enterprising visitors and film-makers swarm to the site, and the pressure for tourism development is mounting as entrepreneurs immediately recognise the opportunities. Without appropriate and adequate controls, however, the developments mushroom until the asset is destroyed. The Gorilla Project has learned that a most serious incentive for the strict control of tourism is safety: The intensified proximity of visitors and apes increases the risks of injury and the transmission of serious diseases. Visitors without the controlling agent of a disciplined guide will readily expose themselves to dangerous interactions with the apes, because they invariably fail to appreciate the wildlife mentality. Orang-utans beyond adolescence can be extremely dangerous, with a physical strength anything up to eight times that of a well-trained human athlete, and an orang-utan bite can result in a dangerous wound or, more commonly, severed fingers.

Free market forces and competition will readily destroy the assets for ecotourism (Brandon, *op. cit.*). If there are no limits to attracting visitors and no attendant constraints, the carrying capacity of an area is readily overshot and the mental development of the apes 'spoiled' in the lust for hit-and-run profits, as can be observed, for instance, in the former orang-utan rehabilitation stations at Bukit Lawang (Bohorok) in North Sumatra and at Tanjung Puting (Central Kalimantan).

It is doubtful whether the touristic value of the 'solitary' orang-utan can match the spectacle of a social gorilla group, but the habitat of orang-utans has certainly more to offer than the (usually secondary) habitat of the gorilla. It is possible to habituate wild demes of orang-utans to the presence of small groups of tourists, as was demonstrated at the Ketambe research station in the 1970s. However, the persistent risk of exposing a wild deme to possible disease agents and regular stress from continuously changing groups of humans means that this should be avoided, except in special circumstances. This consideration led to the timely decision to discourage tourism, stop rehabilitation and focus on better controlled research at Ketambe in 1978.

Consequently, for the touristic development of orang-utan viewing after the model of the mountain gorilla, it is recommended to deploy feral rehabilitant orang-utans from the new reintroduction process. In this way the risks to the wild population are kept to a minimum. Furthermore, it must be realised that Indonesian federal law appropriately prohibits tourism in nature reserves.

It is envisaged that after due monitoring and evaluation of the reintroduction process, a special sector of habitat can be given a formal status such that it can be used for guided tour-groups (ecotourism) wishing to see reintroduced apes. As in the mountain gorilla programme, the guided groups of tourists should not exceed a fixed number of persons per day in a particular area, and bookings must be made in advance.

Orang-utans have a powerful distinguishing ability, and can immediately evaluate the mental state of both ape and human individuals, as well as differentiate between the age and sex classes. Evidently the close evolutionary kinship between human and ape is readily recognised by orang-utans. Since their social structure is largely dependent on long-lasting inter-individual relationships, and because the apes possess an excellent memory, the poached orang-utan juveniles come to be imprinted primarily with the peculiarities of a human (or domestic), rather than a conspecific society. Due to such imprinting a rehabilitant ape may, in later life, confuse the roles required for either society; an ex-captive orang-utan will have acquired a mental map of its role in a captive (i.e. domestic) situation, and during rehabilitation will establish an accessory map of the rehabilitant situation, which may or may not extend into that of a feral condition, dependent on the ape's willingness and mental ability to become feral. It is interesting that many individuals rarely confuse the different roles and can readily switch according to the circumstances (Russon, unpubl.), but some orang-utans become fixed in a primary quasi-human role, with dangerous consequences for both ape and human.

In addition, orang-utans of both sexes during their late adolescence in general become status conscious and assertive, and they will attempt to establish a high status, especially in triadic interactions in which a member of the opposite sex is present, or in instances where food is (made) available. Gradually increasing intensities of physical dominance, including gnawing and, eventually, biting, as well as rape, are the ways that an orang-utan commonly asserts its intentions or position. Females are much more dangerous than males in their assertive actions, mainly because they do not usually reveal their aggressive intention and await the opportune moment to make a premeditated attack.

A consistent and respectful attendance which minimises ape-human contact can successfully rehabilitate most orang-utans to a feral life devoid of dangerous confrontations with humans. However, if an orang-utan individual has been raised under inconsistent or cruel circumstances it will turn readily into an incorrigible, unpredictable and extremely dangerous creature, thus posing a threat to any human attendant, as well as to its fellow apes. As a consequence, the human attendants in the rehabilitation process are exposed to the greatest risks, especially if their behaviour is inconsistent and non-respectful, since they inevitably develop a long-lasting relationship with the individual apes.

Because tourists are newly encountered and usually operate as a group, their attendance to a usual society of well-rehabilitated feral apes bears little risk of dangerous confrontations, especially if they keep a respectful distance, give way and are properly guided by a trained guide, respected by the apes. A human (tourist) coming within the visual sphere of a resident orang-utan must realise, however, that he/she is infringing upon the creature's private sphere, which may incite an aggressive reaction on the part of the ape if it is not somehow assured of the visitor's nonassertive intentions.

This kind of well-controlled kind ecotourism should be developed not earlier than three years after the group of apes in that sector has become fully independent of food provisioning, and no individual can be expected to seek contact with visitors.

The still extant rehabilitation stations of the 1970s at Bohorok and Camp Leakey/Tanjung Puting stations have been officially closed for further rehabilitation since 1996. It is hoped that both will soon be restructured and thoroughly reorganised in order to upgrade the tourism-related viewing facilities, in accordance with the mountain gorilla model. This includes continuous provisioning, in order to discourage the rehabilitants from dispersing into the wild population. After due reorganisation, special temporary permits, or annual contracts, may be provided to the private sector for implementing a management plan in which provisioning and care of the current stock of 'rehabilitant' orang-utans is regulated. The reorganisation, including strict terms of reference, can make these former rehabilitation stations a more adequate 'tool' in the service of proper rainforest and orang-utan conservation.

Ecotourism for orang-utan conservation should not facilitate close contact with fully dependent quasi-wild apes, neither on the part of the tourists nor of the attendants. It is extremely dangerous for human and ape, and devalues that very element of wildness which compels a tourist to seek an encounter in the first place. The point of integrating tourism with conservation involving apes is so that visitors representing civilised cultures may experience this somehow mystical spirit of our *wild* relative. That spirit can easily be destroyed by close contact, and turns either into dependency or aggression, or both. A display of feral apes being provisioned by a human attendant handing out food in natural rainforest surroundings has no educative value whatsoever. It merely exploits the King-Kong sensation at the expense of the ape. If visitors want to see a show which demonstrates misguided charity or asserts human superiority at the expense of wild and dangerous animals, they are better advised to visit an outdated circus. It need not be concealed that the provisioning of reintroduced apes is in fact a poor redress of the shameful inadequacy of protection of our closest relative.

In cases where continuous provisioning of rehabilitant orang-utans is indicated (as in the former rehabilitation centres at Bukit Lawang/Bohorok and Tanjung Puting/Camp Leakey), it is recommended to stop handing out the food, but instead to organise the provisioning through a mechanical box system, similar to that which was developed for wild chimpanzees at Gombe Stream (Tanzania) by Goodall in the early 1970s. In this way the apes can be prevented from further associating humans with the food provided, and hence a safe distance between humans and apes can be better maintained.

Section V

Appendices and References

A sub-adult female Sumatran orang-utan of the dark, Southern type.

A sub-adult male Sumatran orang-utan, apparently of mixed descent.

Appendix 1

Table XXVII

List of vernacular names of the orang-utan (after Yasuma, 1994 and pers. dos.).

Vernacular name	Ethnic or cultural identity; [region]
Hirang (utan)	Kayan
Helong lietiea	Kayan and Kenyah; [Modang, Long Bleh]
Kaheyu	Ngadju; Southern groups, [east Central Kalimantan]
Kahui	Murut; [Northern west Kal. groups, south Sarawak]
Keō, Ke'u, Keyu, or Kuyuh.	Ma'anyan and Bawo; Southern groups, [Kanowit, South-Kal. prov. and east of S. Barito]
Kihiyu	Ot Danum; [north Centr. Kal. prov.]
Kisau or Kog'iu	Orang Sungai; Northern groups, [east-Sabah]
Kogiu	Kadazan; Northern groups, [north Sabah]
Koju	Punan: Northern groups, [S. Blayan]
Kuyang, Kuye*	Kenyah, Kayan and Punan: [Apo Kayan, Badung, Bakung, Lepok Jalan, Lepok Yau, S. Tubuh, S. Lurah, Melap, lower S. Kayan]
Maias or Mayas	Bidayuh, Iban and Lun Dayeh; [Sarawak north of G. Niut, north West Kal. prov., north Sarawak and north East Kal prov.; also western Malayu; West-Kalimantan prov.]
Oyang Dok	Kenyah; [Badeng, S. Lurah]
Orang Hutan or Orang-utan	modern Indonesians, islamized dayaks; transmigrants [Indo-Malay archipelago] and Dusun [North Sabah]
Tjaului	townspeople [Balikpapan, Samarinda]
Ulan/Urang Utan	islamized Dayaks and coastal Kenyah; [Tidung regency, northeast-Kalimantan]
Uraagng Utatn	Benuaq, Bahau and Tunjung; [S. Kedangpahu, along S. Mahakam] also transmigrants [western East-Kalimantan]
Uyang Paya**	Kenyah; Apo Kayan, [north East Kal. prov.]
Mawas	Batak, Malays; [Sumatra] and the Batek [in W. Malaysia]
Maweh***	Acehnese, Gayo; [Aceh prov.]
(Le tjo)	Minang east of Barisan [Rokan, Riau, central Sumatra]
(Atu pandak/rimbu)	Rawas
(Sedapa)	Palembang [north Palembang area]
(Sebaba)	Pasemah [southeast Bengkulu]
(Gugu)	Serawai [south Bengkulu]
(Li koiyau)	Riau Malay [Batang Hari]
(Umang)	Karo Batak [Dairi]
(Orang pendek)	Rejang Lebong Minang [west Barisan, north Bengkulu]

The names between brackets refer to orang pendek, the southern type of orang-utan.
*Kuyah or kuyal in the Kayan language may indicate 'a monkey' in general

** *Paya is a vernacular term for trouble, misery, nuisance, implying 'nuisance people' as the local name for the red ape.*

*** *A large harbour town along Aceh's east coast is called Lhok Seumaweh, 'bay of the orang-utan'*

Three basic roots can be discerned in the list of Bornean names, one modern and two traditional; the names derived from *orang hutan* are supposedly of modern, Malay origin, e.g. *urang, uyang, uraagng, ulan, oyang* and, perhaps, *kuyang*. The name '*mawas*' and its derivates are evidently of Sumatran origin, surviving even in the myths of the aboriginal tribes of the Malay peninsula (Schebesta, 1928). Considering that the Iban or 'sea dyaks' of southwestern Borneo are of Sumatran origin, it is interesting to note that they apparently named the Bornean orang-utan after its Sumatran brethren. This suggests that until some four centuries ago the distribution range of the Sumatran subspecies still extended as far south as the Iban's original homeland, i.e. along the banks of the river Batang Hari in the northern section of what is currently South Sumatra province.

The presumably original Bornean names have a more or less common root structure, namely, '*Ka-u*', '*Ke-u*', '*Ki-u*', '*Ko-u*' and '*Ku-o*'. Is this root perhaps derived from a human interpretation of the most common vocalisations of orang-utans in the presence of humans? For instance, the loud 'kiss sound' of the ape, which is the usual vocalisation of a wild, anxious orang-utan confronting humans or other species disturbing to the orang-utan. Or, perhaps, even the 'lork call', or 'bark' (see e.g. Rijksen, 1978).

It is interesting that at least two Sumatran names for the *orang pendek,* from Riau province, i.e. *Li koiyau* and *Le tjo,* are based on the same root. The *tjo* sound is undoubtedly derived from this common 'kiss sound' vocalisation.

Appendix 2

What's in a name?

The scientific name of the orang-utan has given rise to considerable confusion; the name *Simia satyrus* which was originally given by Linnaeus to the ape in 1758 was changed erroneously in 1927.

> In 1925 the International Commission on Zoological Nomenclature published Opinion No 90 (Smithsonian Miscellaneous Collections 73 [3]) to announce that it had failed to reach a two-thirds majority vote for suspension of ten generic names, among which *Simia* Linnaeus 1758, so that 'the Law of Priority is to be applied in these cases.' Apparently it had been suggested that the name *Simia* had been coined earlier for the Magot (*Macaca sylvanus*), and hence could not be applied to the orang-utan (or chimpanzee). A report by Allen, referring to *Simia troglodytes* Gmel., Linn. S.N. v.i: 26, 1788 (i.e. the chimpanzee) and to *Simia satyrus* Linn., 1758 (i.e. the orang-utan), lists all the names which could not be suspended and should be treated according to the rule of Priority.

Reasons for a possible change are not specified, apparently because 'no good purpose can be served by publication of the arguments for and against suspension', according to the Commission on Zoological Nomenclature. However, Stiles and Orleman, in a publication entitled *The Nomenclature for Man, the Chimpanzee, the Orang-utan, and the Barbary Ape* (in Bull. US Public Health Service Hyg. Lab.; 145; pp. 1-66), reveal that doubt was raised concerning the two Dutch references which Linnaeus had used to describe the orang-utan. It was noted that de Bondt had supposedly described a human patient suffering from 'hypertrichy' (hairiness), and that Tulp apparently had described 'a chimpanzee from Angola.'

It appears that the International Commission on Zoological Nomenclature at the time possessed incomplete historical information regarding the Dutch 'discoverers'[104] and their publications.

The first European to report on the red ape was the 1st century historian Pliny the Elder. Then, in the 13th century the Venetian adventurer Marco Polo, in his travel diary concerning the description of the coastal region of Pasè, in Aceh, noted: 'there is a medium-sized ape here which has a face like that of a man.' The first detailed description, however, was made by the Dutch physician Jacob de Bondt (Bontius), inspector of surgery in Batavia (1592-1631). When his description of the Bornean ape was finally published in the *Historiae Naturalis et Medicae Indiae Orientalis*, twenty-seven years after his death, it had been preceded by the description of a preserved specimen, by the physician Nicolaas Tulp (Tulpius), the Mayor of Amsterdam.

---

[104] The ape has probably been known to the civilised world since antiquity; Tulp had already acknowledged that the Satyr described by Plinius was most probably the orang-utan, notwithstanding some minor inaccuracies in the early descriptions. It is also interesting to note that the Arab encyclopaedic cosmography *'Ajā 'ib* refers to the narrative of the ninth-century traveller Ibn al-Faqīh al-Hamadhānī describing the fully haired, tree-dwelling aboriginal 'humans' on the island of Rāmnī (i.e. Sumatra) (Kruk, 1995).

> The publication in the Bulletin of the Health Service drew special attention to the 'potential danger which might arise in medical and public health work because of continued confusion' over the taxonomic names. This is nonsense, because if any confusion ever occurred in the New World in the early nineteenth century (e.g. Owen, 1835; Knox, 1840), it had already been cleared up in Europe more than a century earlier, around 1790, through the publications of Camper, Vosmaer, de Buffon and Linnaeus (Dobson, 1953-54).
>
> In any event, eight of the nineteen Commission members voted against suspension, while a postscript to the Opinion noted that 'additional data have been obtained by the Secretary which persuade him that it is by no means clear, under the Rules, that *Simia* should be transferred to the Barbary Ape.'
>
> At the Budapest meeting in 1926, one Commissioner was appointed to prepare recommendations in this case, and in 1927 the Commission, having taken careful note of the recent publication of the US Public Health Service, unanimously adopted the resolution to (1) entirely suppress the name *Simia* and (2) to enforce the Law of Priority, which apparently meant that the orang-utan was henceforth to be called *Pongo pygmaeus* (Hoppius, 1786). The decision was published as Opinion No 114 (Smithsonian Misc. Coll. 73 [6], 1929).

Until the late nineteenth century, the modern western name, orang-utan, which is derived from the Malay *orang hutan,* was never used locally to denote the ape. The Malay word means 'person (or people) of the forest', and was (is) commonly used derogatorily to indicate a person of a savage, crude or 'primitive' nature. It was coined by de Bondt, albeit spelled wrongly, as *ourang outang*, which would translate as a person with a debt. The name *orang hutan* was probably the most convenient synonym found by his Malay interpreter for any one of the ape's vernacular names given to de Bondt, perhaps something like *maias, keo* or *kahui*, as the ape is known in southwestern Kalimantan.

When de Bondt's description of East Indian animals was published posthumously in 1658, the editor, the physician Willem Piso, or the publishers, the brothers Lodewijk and Daniel Elsevier, decided to enliven the text with what they probably believed to be a fair representation of the unknown 'ourang outang'. The illustration in fact was a poor copy of a fictitious tropical 'wildman' or 'Ape' of earlier publications, such as can be found in B. Von Breydenbach's *Journey to the Holy Land* (1483-84), and notably in K. Von Gessner's *Compendium Historiae Animalum* (1602) (Spencer, 1995).

Unfortunately, the added illustration seems to have incited confusion even in the great Linneaus, because in his *Systema Naturae* of 1758 he refers to de Bondt's *Homo sylvestris* under the Genus Homo (p. 24: *Homo troglodytes* and including the synonym *Homo nocturnus*) as occurring in Aethiopia, Java, Amboinea and Ternate. With reference to the type locality, he must have misread de Bondt's treatise, which clearly stated that the *Orang-Outang* came from the county of Sukadana in (west) Borneo, rather than from Ethiopia, Java or the Ambon archipelago. In any case, a similar mistake must have meant that some influential twentieth-century scientists falsely queried whether de Bondt had indeed referred to the red-haired ape of Borneo, and, incredibly, it was soon believed that the physician had referred to a hairy human patient (Napier and Napier, 1967; Röhrer-Ertl, 1983; Spencer, 1993). It is perhaps noteworthy in this respect that all the ethnic groups of western Indonesia are among

Appendix 2

the least hairy humans in the world, and such 'hypertrichy' as is attributed to de Bondt's alleged 'patient' has never been recorded in the region.

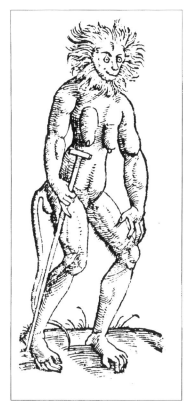 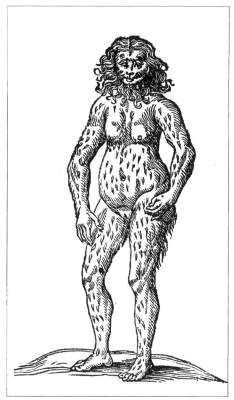

The posthumously added illustration of an 'ourang-outang' in de Bondt's publication, as compared to von Gessner's 'ape' (right); photocopies of the original illustrations.

It is inconceivable why anyone should doubt that de Bondt described the ape; the chapter *Caput xxxii* is the last of his descriptions of animals of the East Indies, and none of the chapters refer to different types of humans in the archipelago (or to their diseases). De Bondt gives the type locality from which the ape was taken, and notes that he had seen it walking upright, which he certainly would not have specifically remarked upon had it concerned a human. Moreover, he provides details of the ape's typical withdrawal behaviour, and in the closing sentence of the chapter he enumerates other primate species. It is also noteworthy that the other illustrations of the animals described all are of miserable quality, and were either copied from other sources or drawn from badly deformed preserved specimens.

The Dutch anatomist Petrus Camper, who was the first to describe accurately a dissected orang-utan (1782), never doubted the primacy of de Bondt in his description of the orang-utan. Camper had in fact been born in a household which had been used to pet orang-utans; Camper's father Florentinus was a Minister of the

Dutch Reformed Church, who had lived in Batavia from 1702-13. Valentijn (1726: p. 242) describes how he had seen only two captive orang-utans in his life, 'the last with Mr Camper, the seigneur of Ouwerkerk aan de IJsel', Petrus Camper's father.

Be that as it may, the second, although first published, description of the orang-utan is much more important, because it was used by Linnaeus as the type-specimen of *Simia satyrus*. It was given by another Dutch physician, the major of Amsterdam, Nicolaas Tulp. Incredibly, his description was also rejected by the learned Commission of twentieth-century taxonomists. In 1641, Tulp described, in Dutch and Latin, what he called the *Indiaansche Satyr* (Indian Satyr). He obtained the dead specimen for dissection from the menagerie of the *Stadhouder*, Prince Frederik Hendrik van Oranje. He mentions that it is the *Homo sylvestris* or *Orang-outang* of which (his brother-in-law), Samuel Bloemaert, who was a merchant and staff member of the United East Indies Company (VOC), was able to give details concerning its lewd behaviour. The sexual prowess of the creature seemed to identify it as the Satyr which, in antiquity, had been described by Pliny as living in the 'Indies'.

Despite the clumsy illustration depicting the (dead, and somehow propped up) specimen, it shows virtually all the characteristics of an orang-utan (Rijksen, 1982), e.g. the flat face lacking prominent brow-ridges, the 'hooked nose', the longish hair standing out from around the face, the sparsely haired pot-belly and breast and the long wavy hair on the upper arms and the calves. Tulp's remarks about the ape's behaviour also point to the orang-utan rather than to any other ape. For instance, he mentions that the ape often walked upright, and habitually covered itself with a blanket when going to sleep (p.737), which is typical behaviour for a captive orang-utan but unusual for a chimpanzee.

From a taxonomic point of view, it is significant that the Indian satyr in Tulp's illustration shows no trace of the 'knuckles' on the fingers of the hands, which are so characteristic for the chimpanzee[105]. It is highly likely that Tulp would have made mention of such a characteristic if his specimen had possessed such typical knuckles, where he elaborates in detail on the amazing likeness of the orang-utan's hands, fingers and thumb to those of man (p. 372). In addition, the extremely sparsely haired breast and belly, which are specifically mentioned twice (p. 372 and 375; in addition to the illustration), and the 'sharp hooked nose', would be atypical for the chimpanzee, but fit the description of orang-utan characteristics. In contrast to the wide prominent nostrils of the chimpanzee, it is the delicate bridge of the orang-utan's nose that catches one's attention. Perhaps most intriguing, however, is that in the Dutch edition Tulp casually notes, 'its foot has no nails' (*desselfs voet heeft geen nagels*; p. 375), a description which can only apply to the orang-utan, as it typically often lacks nails on the 'thumb' of the foot (see e.g. Camper, 1782).

---

[105] cf. Dobson (1953-54); however, Reynolds (1967) saw 'webbing' of the fingers in the illustration and felt it demonstrated that Tulp had described a bonobo (p. 44)

> A possible source of confusion may have been Tulp's remark that the back of his specimen was covered with "black hairs", which is virtually the only piece of description in the whole treatise which might point to another ape, notably the siamang, or, indeed, the chimpanzee, although nineteenth-century descriptions of Sumatran orang utans often refer to their 'black skin' and dark "blackish brown hair" (Hagen, 1890, van Balen, 1898). Tulp described a freshly deceased specimen which had survived an arduous sea voyage, then been kept in captivity in Holland's cold, wet climate, and consequently must have been in very poor physical condition. The hair of captive orang-utans in poor condition becomes brittle, short and dull, lacking any red or maroon lustre. As Tulp's specimen came from the Angkola region, where reportedly very dark individuals are common, it is probable when describing the hair colour of his specimen, the physician got the impression that it was so dark as to be called 'black', in comparison to the usually described 'red' of the *Orang Outang*.

Tulp noted that his specimen originated from Angola. Hence, some scientists may have believed that Tulp had confused Africa and the East Indies, and had described the chimpanzee (e.g. Dobson, 1953-54)[106] or the bonobo (Reynolds, 1967). No one from among the Commission scientists seems to have taken seriously that Tulp, who was mayor of the world's international trade centre at that time, and in any case unlikely to be confused by geographical locations, refers explicitly to his **Indian** specimen, and mentions that a similar ape, called the *Quoias Morrou*, is known from the West African country Sierra Leone (p.375).

> In the seventeenth and eighteenth century, the chimpanzee was usually designated by the name Pongo (*Simia troglodytes* Tyson 1699), while the orang-utan was often called the Jocko (e.g. Buffon, 1789; see also Dobson, 1953-54), although most of the European naturalists outside Holland commonly and persistently confused the two different apes. Unfortunately, the German colonial naturalist F. Baron von Wurmb compounded the confusion when he published in 1781[107] a '*Beschrijving van de groote Borneoosche Orang Outang of de Oost Indische Pongo*' (description of the **large** Bornean orang-utan or the **East Indian** Pongo), because he wanted to distinguish 'his' large adult orang-utan specimen from the usually small, adolescent apes thus far described. In his ignorance, he probably believed that it concerned a separate, second species for the East Indies. Consequently he misused the name of the giant African ape that had first been described by the 16th century traveller Andrew Battell during his travels along the west coast of Africa (Purchas, 1613), although he specifically mentions the East Indies' type locality of his specimen.

Actually, the type locality mentioned by Tulp is the present district of Angkola, which was usually written as Angola in the seventeenth-century records of the United East Indies Company (VOC). This region in North Sumatra lies south of

---

[106] On June 1st, 1698, Edward Tyson described what, according to his illustration, must have been the first chimpanzee, to the Royal Society in London, and one year later he published a treatise on the comparative anatomy of this African ape under the title *Orang-outang, sive Homo sylvestris: or, the Anatomy of a Pygmie Compared to that of a Monkey, an Ape, and a Man*. If any one of the taxonomic 'discoverers' erred in the identity of apes, it was this eminent English Fellow of the Royal Society.

[107] Because it was said that Von Wurmb did not 'create' a Latin name, but, inappropriately, used the common name of the African ape among eighteenth century travellers, taxonomists reject Von Wurmb's priority (Groves and Holthuis, 1985) and accord priority instead to B.G.E. de Lacepede who, in his ignorance, and after Von Wurb, mentioned the Orang-utan under the now apparently latinised name Pongo in his *Tableau des divisions* ...of 1799.

Lake Toba just inland from the major coastal township of Sibolga[108], a small harbour town which had been a mediaeval trade centre for forest produce (to China) and was one of the last landfall sites for taking in fresh water before the seventeenth-century VOC merchantmen crossed the Indian ocean to Mauritius. Kalff (1925) presented evidence that orang-utans were still to be found in the environs of Padang in the eighteenth century. The present surveys confirmed that in 1996 the ape still occurred in Angkola, and even further south along the western foothills of the Barisan mountain range around the equator.

Ignorance of seventeenth century geography may have meant that the early twentieth-century taxonomists believed that Tulp's Indian satyr specimen originated from Angola in western central Africa. However, the country which became known as Angola late in the nineteenth century was actually called Manicongo, Ndonga or Ndola in the sixteenth, seventeenth and eighteenth centuries. In the late sixteenth century, the Portugese had a fort at the mouth of the river Zaire (or Congo), which was called Angolia, but the name Angola is not to be found on maps before the nineteenth century, when the Portugese began to colonise the southwestern coast. Hence there is no reason to doubt Tulp's statements concerning a Sumatran type locality for his orang-utan specimen.

The great Swedish naturalist Carl von Linné (Linnaeus) accepted Tulp's description in his treatise *Systema Naturae* of 1758 (p. 25) and described the orang-utan of the Netherlands East Indies under the name *Simia satyrus*. He acknowledged, like Tulp, that a very similar ape occurred in Africa. When, the following year, Linnaeus acquired for the first time the small skull of a juvenile orang-utan from G. Edwards[109], he described it as *Simia pygmaeus* in the dissertation he wrote for his student, Christian. E. Hoppe (Westerlund, 1889). However, for the rules of systematic nomenclature, the 1785 edition of the *Systema Naturae,* which features the name *Simia satyrus,* is the basic reference, and in all the subsequent editions of his famous *Systema Naturae* von Linne retained the original *satyrus* as the specific name for the red ape.

In 1927 the name *Simia satyrus* was officially suppressed in Opinion 114 of the International Commission on Zoological Nomenclature 'according to the Law of Priority' (Groves and Holthuis, 1988). Apparently the Commission was unaware that Von Linné had correctly used the first accurate reference for his identification of the orang-utan, namely, Tulp's publication. If they had been aware, it may be supposed

---

[108] The mountainous region right above the old port town of Sibolga to the East is known as Anggoha, while the district of Angkola is the major valley of the Angkola river at a distance of 75 km southeast of Sibolga which was and still is the first inhabited valley on the old trade route to central Sumatra.

[109] The illustration of the *pygmaeus* ape which was published by Edwards in 1758 is also hardly recognisable as an orang-utan, although it shows the typical long hands and feet; like Tulp's, it may well be an original illustration, whereas the illustration in Beeckman (1718) is obviously a poor copy of Tulp's illustration, although no one ever doubted Beeckman's description.

> CAROLI LINNÆI
> Equitis De Stella Polari,
> Archiatri Regii, Med. & Botan. Profess. Upsal.;
> Acad. Upsal. Holmens. Petropol. Berol. Imper.
> Lond. Monspel. Tolos. Florent. Soc.
>
> # SYSTEMA NATURÆ
> Per
> *REGNA TRIA NATURÆ,*
> Secundum
> CLASSES, ORDINES,
> GENERA, SPECIES,
> Cum
> *CHARACTERIBUS, DIFFERENTIIS,*
> *SYNONYMIS, LOCIS.*
>
> Tomus I.
>
> Editio Decima, Reformata.
>
> *Cum Privilegio S:æ R:æ M:tis Sveciæ.*
>
> *HOLMIÆ,*
> Impensis Direct. LAURENTII SALVII,
> 1758.

Frontispiece of the 1758 edition of Systema Naturae in which von Linné described the Anthropomorpha, including de Bondt's and Tulp's orang-utans.

that there would have been no reason to reject the name *Simia satyrus* for the orang-utan, as first mentioned in his *Systema Naturae* of 1758.

After 1927, however, the red ape was given an old, corrupted, Congolese name for the gorilla, '*mpungu* (i.e. Pongo) as its generic name, although it is telling that the taxonomist Chasen explicitly violated the Opinion, 'contrary to my usual practice', and continued to use *Simia satyrus* (Chasen, 1940).

In order to apply the Law of Priority correctly, we propose restoring the name given to the orang-utan in von Linné's *Systema Naturae* of 1758, namely *Simia satyrus* Linnaeus 1758 (Linnaeus 1758, 1766). Since Tulp definitely described a Sumatran orang-utan, it would perhaps be apt to propose that the Sumatran subspecies have the addition *S. s. indicus* Tulpius 1641 (as Linnaeus in fact did), rather than *abelii*. After all, the name *satyrus* fits the 'man of the forest' best.

Addendum

Transcript of Chapter 32 of *Historiae Naturalis et Medicae Indiae Orientalis* (p. 84-85)

(Jacobus Bontius)

Ourang Outan *sive Homo silvestris.*
Hircipedes Satyros, Sphingas, Faunosque petulcos,
Nec pueri credunt: tamen hoc mirabile Monstrum
Humana spectra facie, tum moribus illi
Assimele in gemitu, tum fletibus ora rigando

Plinius, ille Naturae Genius, lib.7, cap.2, de Satyris dixit: Sunt & Satyri, subsolanis in Indiis locis & montibus pernicissimum animal; tum quadrupedes, tum & recte currentes humana specie & effigie, propter velocitatem nonnisi senes aut aegri capiuntur.

Ast quod majorem meretur admirationem, vidi ego aliquot utriusque fexus erecte incedentes, imprimis eam (cujus effigiem hic exhibeo) Satyram foemellam tanta, verecundia ab ignotis fibi hominibus occulentem, tum quoque faciem manibus (liceat ita dicere) tegentem, ubertimque lacrymantem, gemitus cientem, & caeteros humanos actus experimentem, ut nihil ei humani de esse diceres paeter loquelam. Loqui vero eos easque posse, Iavani aiunt, sed non velle, ne ad labores cogerentur: ridicule me hercules. Nomen ei indunt *Ourang Outang*, quod hominem silvae significat, oesque nasci affirmant e libidine milierum Indarum, quae se Simiis & Cercopithecis detestanda libidine miscent.

Nec pueri credunt, nisi qui nondum are lavantur.

Porro in Insula Borneo, in Regno Succodana dicto, a nostris Mercatoribus propter Oryzam & Adamantes frequentato, homines montani caudati in interioribus Regni inveniuntur, quos multi e nostris in aula Regis Succodanae viderunt. Cauda autemillis est prominentia quaedam ossis Coccygos, ad quatuor, aut paulo amplius, digitos excrescens, eodem modo, quo truncata cauda Canum, (quos nos *Spligiones* vocamus) sed depilis.

Simias denique, Cercopithecos, Bavianos plurimos Sylvae Iavanorum alunt.

Addendum 2

Transcript of *Geneeskundige Waarnemingen IIIe Boek, Zesenvijftigste Hoofdstuk* (Observationum Medicarum Lib. III, Caput LVI)

(Nicolaas Tulp)

Een Indiaansche Satyr (Satyrus Indicus)

(p. 370) Schoon het buiten het perk der Geneeskunde is, zal ik evenwel aan dit webbe nog weven den Indiaansche Satyr, by ons geheugen van Angola gebragt, en aan Frederik Hendrik, Prins van
(p. 372) Oranje vereert. Deeze Satyr nu was viervoetig; maar van de menschelyke gedaante, welke het vertoont, wordt het van de Indianen *Orang Outang* genoemt, of een bosch mensch, gelyk van de Afrikanen *Quoias Morrou*. Gelykende in langte naar een kind van drie jaren, en in dikte naar een van zes.

Zyn lighaam was nog vet, nog mager, maar vierkant; evenwel zeer wel in staat, en zeer snel. Maar zyne ledematen waren zoo vast, en zyne spieren zoo grof, dat hy alles en durfde en konde. Van voren overal glad, maar van agteren ruig en met zwarte hairen bezet. Het aangezicht vertoonde een mensch, maar de neus spits en hakig, eene gerimpelde en tandeloze oude Vrouw.

Maar de oren verschilden niets van de menschelyke gedaante. Gelyk ook niet de borst aan beide syden versiert met een opzwellende mam (want ze was van 't Vrouwelijk Geslagt,) de buik had een diepe navel, en zoo wel de onderste als de bovenste ledematen zulk een naaukeurige gelykenis met een mensch, dat men naauwelyks een Ey gelyker aan een Ey ziet. Ook ontbrak aan den elleboog het vereischte gewricht niet, nog aan de handen de order der
(p. 373) vingeren: zelfs niet de menschelyke figuur aan de duim: of de kuiten aan de beenen, of het steunsel der hiel aan de voet. Welke nette en geschikte gedaante der ledematen oorzaak was, dat zy veeltyds regt over eind ging, en niet minder deftig optilde, dan gemakkelyk vervoerde allerhande zeer zware lasten.

Zullende drinken nam zy het handvat der kan met de eene hand, maar de andere onder de grond van het vat leggende, vreef zy daar meede af de nattigheid aan, de lippen overgelaten, niet minder handig dan of men een zeer naaukeurigen Hoveling gezien had. Welke zelfde handigheid zy ook waarnam als ze naar bed ging. Want neigende het hoofd in het kussen, en het lighaam behoorlyk met dekens toedekkende, dekte ze zig niet anders, dan of een zeer verwyft mensch daar zou neder leggen.

Ja de Koning van Sambaca[110] verhaalde eertyds aan onzen Zwager Samuel Bloemaart, dat deze Satyrs, inzonderheid de mannen, in het Eiland Borneo, zoo groot een vertrouwen des harten hebben, en zulk een sterk samenstel van Spieren, dat ze meer dan eens geweld gedaan hebben op gewapende mannen; ik laat staan op de zwakke sexe van Vrouwen en Dochters.

Naar welke zy somtyds zulk een brandend verlangen hebben, dat zy haar geroofd hebbende,
(p. 374) meer dan eens geschaakt hebben. Want zy zyn zeer genegen tot het byslapen, ('t welk hun met de dartele Satyrs der ouden gemeen is,) ja somtyds zoo brooddronken en dartel, dat de

---

[110] i.e. Sambas on the coast of West Kalimantan.

Indiaansche Vrouwen zig daarom meer dan voor vergift, wagten voor de bossen en waranden, waar deze onbeschaamde dieren schuilen.

Al het wel gelyk waaragtig verhaalt word van deezen Sartyr, zo is 'er waarlyk niets waarschynelyker, dan dat de Satyrs der ouden naar hunne afbeelding gemaakt zyn. Welke *Plinius* voor zyne Lezers zullende beschryven, uitdrukkelyk schryft, "dat het een viervoetig dier is, op de schaduwagtige bergen der Indianen, zeer snel, met een menschelijke gedaante, maar met geiten voeten, ruig over het geheele lighaam; hebbende geene menschelyke gewoontes, zig vergenoegende in de schuilhoeken der bossen, en vliedende de samenwoning der menschen *lib. vii. cap. iii,* zie *Pausanias lib.* i."

(p. 375)

Van welke tekenen schoon enigsins verschilt de Satyr van den zaligen Hieronymus, evenwel komt hy enigsins overeen met de versierselen der Digters. "Het was, zegt hy, een mensch met een kromme neus, en een voorhooft ruw door de hoornen, eindigende de uiteinden des lighaams in geiten voeten." Welke zelfde gedaante de Digters klaarder zullende uitdrukken, hunne Satyrs noemen dartel, onbeschaamt, twee gedaantes hebbende, tweehoornig, en somtyds dartele Godheden der bosschen.

Welke bynamen der ouden aan het rigtsnoer der waarheid getoetst zynde, zal men zien dat ze niet geheel gedwaalt hebben. Want dit dartel dier word ook nog gevonden op de schaduwagtige bergen van Indien; gelijk ook in Africa tusschen Sierra Liona en den uitstek der berg, alwaar misschien die plaatsen zyn, welke *Plinius* zegt *lib. v. cap v.* by nagt te glinsteren door de gedurige vyeren der Aegipanen en de dartelheid der Satyrs. Het heeft behagen in spelonkige afscheidingen; het vlied de menschelyke gezelschappen, en word niet te onregt genoemt geil, ruig, viervoetig, en menschelyke gedaante vertonende en voorzien van een kromme neus

Maar desselfs voet heeft geen nagels, nog het voorhoofd hoornen der geiten, of het lighaam overal hairen; maar alleen het hoofd, de schouder, en de rug. Gelijk de andere delen glad zyn, zoo zyn de oren geensins scherp, gelyk *Horatius* verkeerdelyk versiert heeft, maar gebogen, en om het met een woord te zeggen, waarlyk menschelyk. In 't kort, daar is of geen Satyr in de natuur der dingen: of zoo 'er een is, zal het buiten twyfel dat dier zyn, dat hier in de plaat van ons is afgebeeld.

(p. 376)

Welke beschryving *Plinius* misschien ondernomen heeft, maar om dit te konnen doen, zyn den zeer naarstigen Man misschien in de weg geweest, aan de eene syde de vernuftige versierselen der Digteren, zoo betovert door de aanlokselen van Circe, dat ze ik weet niet welke zinnen zouden bedriegen: maar aan de anderen kant, de uitlandige volken; zoo ver afgelegen, dat het nodig was in een's anders voetstappen te treden, en liever gemakkelyk goedtekeuren, dan naauwkeurig te onderzoeken naar de natuur der gebeurde en van allen reeds aangenomene zaak.

Waar aan indien iemand geloof weigert, die vervalt ligtelyk in de angels van lastering, en verheft het gezag van zyne eige geschiedenis, verschillende van het gemeene gevoelen. Welk ongelyk eertyds aan zig toegebragt, *Herodotus*, anders geen onaardig Schryver, bynaar nog niet verduwt heeft. Zoo veel dan in zoo grote duisterheid heeft konnen geschieden, heeft Plinius wel de beste waarnemingen uitgekozen, en getragt zynen Satyr met waarheid aan de Nakomelingschap overtelaten. Maar hy heeft moeten zorg dragen voor zyn' goeden naam, en niet te vry afwyken van het verouderde gevoelen, veel min van de verdigte fabelen der Poëten. Welker dikke duisternis de aanbrekende opkomst van deezen Indiaansche Satyr misschien zal opklaren.

(p. 377)

Appendix 2

(p. 378)

Wagt u evenwel dat gy door dit woord verstaat de verwarde namen van Pan, Aegipan, Faunus en Boschgoden: waar onder, schoon Pan, Aegipan en Faunus, onder malkander overeenkomen, verschilt de Boschgod evenwel van die, en de Satyr van allen. Hoort *Vergilius lib. i georg.* Pan de bewaarder der Schapen; en een weinig daar na, en Sylvanus nemende de tedere Cypres van hare wortel. *lib.ii, georg.* die de Boschgoden kent, en Pan, en den ouden Boschgod; en de Gezusters der Nymphen, en *lib.i Metamorph.* de Fauni, en Nymphen, en satyrs en op de Bergen wonende Boschgoden, zijn Godheden der Boeren, en *lib. i. fastorum* de Jeugd genegen tot Maaltyden en de wellust der satyrs. Waaruit klaar blijkt, dat Pan, die dezelfde is met Aegipan, (beide dog zyn voor hun halve deel gelyk aan een geit) uitdrukkelyk onderscheiden word van den Boschgod, schoon het *Plutarchus* anders heeft toegeschenen; en van die en alle andere Satyrs, welke wy hier vertoont hebben.

Deezen ook (als men in de heilige Boeken gissen mag) geven ook op meer dan eene plaats te kennen *Moses* en *Esaias*, onder de namen *Sagnir* en *Segnirim*, waar mede niet zoo zeer betekent word de duivel, gelyk sommigen het verklaren, als een Boschdier, en ruig gelyk de Geiten, springende niet alleen in de Bosschen, maar zyne Makkers ook toeroepende: waar voor de Koning Jerobeam gezegt word Priesters en Altaren ingestelt te hebben. Welk zelfde dier *Mosis Deut. cap xxxii vers 17.* met de Chaldeen *Schedim* noemt, van de schrik, die het de Reizigers aanjaagt.

Al het welk de Engelsche Vertalers juistelyk ziende overeenkomen met den Satyr, hebben zy by Profeet *Esaias* het woord *Sagnir* misschien niet kwalyk verklaart door Satyr. Zie *Levit., cap. xvii, vers vii. 2 Chron. cap. xv. vers xv. Esaias cap. xiii. vers 21.* en *cap. xxxiv. vers xiv.*

(p. 379)

Waarby ik nog voeg, dat hetzelfde Dier misschien beschreven word van *Plutarchus,* in het leven van Sylla, alwaar Hy zegt "digt by Dyrrachium is Appolonia, en digt daar by Nymphaeum. Eene heilige Plaats, die uit een groenend Wout en Velden, gedurig verspreide aders van Vuur uitbraakt. Daar zegt men, dat een slapende Satyr, hoedanig een de Beeldwerkers en Schilders aftekenen, gevangen, en tot Sylla gebragt is; dat veele Tolken gevraagt hadden wie hy was, als hy eindelyk geluid had gegeven, wel niet een menschelyk of klaar, maar schor; en geheel verward uit het briessen van een Paard, en het bleten van een Bok; dat Sylla verbaast was, en een afkeer van 't Monster had."

APPENDIX 3

The criteria for critically endangered, endangered and vulnerable taxa
(after IUCN, 1994)

Critically Endangered (CR)
A taxon is Critically Endangered when it is facing an extremely high risk of extinction in the wild in the immediate future, as defined by any of the following criteria (A to E):

A. Population reduction in the form of either of the following:
   1. An observed, estimated, inferred or suspected reduction of at least 80% over the last ten years or three generations, whichever is the longer, based on (and specifying) any of the following:
      (a) direct observation
      (b) an index of abundance appropriate for the taxon
      (c) a decline in area of occupancy, extent of occurrence or quality of habitat, or both
      (d) actual or potential levels of exploitation
      (e) the effects of introduced taxa, hybridisation, pollutants, competitors or parasites
   2. A reduction of at least 80%, projected or suspected to be met within the next ten years or three generations, whichever is the longer, based on (and specifying) any of (b), (c), (d) or (e) above.

B. Extent of occurrence estimated to be less than 100 km^2 or area of occupancy estimated to be less than 10 km^2, and estimates indicating any two of the following:
   1. Severely fragmented or known to exist at only a single location.
   2. Continuing decline, observed, inferred or projected, in any of the following:
      (a) extent of occurrence
      (b) area of occupancy
      (c) area, extent or quality of habitat, or both
      (d) number of locations or sub populations
      (e) number of mature individuals
   3. Extreme fluctuations in any of the following:
      (a) extent of occurrence
      (b) area of occupancy
      (c) number of locations or subpopulations
      (d) number of mature individuals

C. Population estimated to number fewer than 250 mature individuals and either:
   1. An estimated continuing decline of at least 25% within three years or one generation, whichever is the longer or

2. A continuing decline, observed, projected, or inferred, in numbers of mature individuals and population structure in the form of either:
   (a) severely fragmented (i.e. no estimated to contain more than 50 mature individuals)
   (b) all individuals are in a single population.

D. Population estimated to number fewer than 50 mature individuals.

E. Quantitative analysis showing the probability of extinction in the wild is at least 50% within ten years or three generations, whichever is the longer.

Endangered (EN)
A taxon is endangered when it is not Critically Endangered but is facing a very high risk of extinction in the wild in the near future, as defined by any of the following criteria (A to E):

A. Population reduction in the form of either of the following:
   1. An observed, estimated, inferred or suspected reduction of at least 50% over the last ten years or three generations, whichever is the longer, based on (and specifying) any of the following:
      (a) direct observation
      (b) an index of abundance appropriate for the taxon
      (c) a decline in area of occupancy, extent of occurrence or quality of habitat, or both
      (d) actual or potential levels of exploitation
      (e) the effects of introduced taxa, hybridisation, pollutants, competitors or parasites
   2. A reduction of at least 50%, projected or suspected to be met within the next ten years or three generations, whichever is the longer, based on (and specifying) any of (b), (c), (d) or (e) above.

B. Extent of occurrence estimated to be less than 5000 km^2 or area of occupancy estimated to be less than 500 km^2, and estimates indicating any two of the following:
   1. Severely fragmented or known to exist at no more than five locations.
   2. Continuing decline, observed, inferred or projected, in any of the following:
      (a) extent of occurrence
      (b) area of occupancy
      (c) area, extent or quality of habitat, or both
      (d) number of locations or subpopulations
      (e) number of mature individuals
   3. Extreme fluctuations in any of the following:
      (a) extent of occurrence
      (b) area of occupancy

(c) number of locations or subpopulations
(d) number of mature individuals

C. Population estimated to number less than 2500 mature individuals and either:
   1. An estimated continuing decline of at least 20% within five years or two generations, whichever is the longer or
   2. A continuing decline, observed, projected, or inferred, in numbers of mature individuals and population structure in the form of either:
      (a) severely fragmented (i.e. no subpopulation estimated to contain mote than 250 mature individuals)
      (b) all individuals are in a single population.

D. Population estimated to number fewer than 250 mature individuals.

E. Quantitative analysis showing the probability of extinction in the wild is at least 20% within 20 years or five generations, whichever is the longer.

Vulnerable (VU)
A taxon is Vulnerable when it is not Critically Endangered or Endangered but is facing a high risk of extinction in the wild in the medium-term future, as defined by any of the following criteria (A to E):

A. Population reduction in the form of either of the following:
   1. An observed, estimated, inferred or suspected reduction of at least 20% over the last ten years or three generations, whichever is the longer, based on (and specifying) any of the following:
      (a) direct observation
      (b) an index of abundance appropriate for the taxon
      (c) a decline in area of occupancy, extent of occurrence or quality of habitat, or both
      (d) actual or potential levels of exploitation
      (e) the effects of introduced taxa, hybridisation, pollutants, competitors or parasites
   2. A reduction of at least 20%, projected or suspected to be met within the next ten or three generations, whichever is the longer, based on (and specifying) any of (b), (c), (d) or (e) above.

B. Extent of occurrence estimated to be less than 20,000 km^2 or area of occupancy estimated to be less than 2,000 km^2, and estimates indicating any two of the following:
   1. Severely fragmented or known to exist at no more than ten locations.
   2. Continuing decline, observed, inferred or projected, in any of the following:
      (a) extent of occurrence
      (b) area of occupancy

(c) area, extent and/or quality of habitat
(d) number of locations or subpopulations
(e) number of mature individuals

3. Extreme fluctuations in any of the following:
(a) extent of occurrence
(b) area of occupancy
(c) number of locations or sub populations
(d) number of mature individuals

C. Population estimated to number less than 10,000 mature individuals and either:
1. An estimated continuing decline of at least 10% within ten years or three generations, whichever is the longer, or

2. A continuing decline, observed, projected, or inferred, in numbers of mature individuals and population structure in the form of either:
(a) severely fragmented (i.e. no subpopulation estimated to contain more than 1,000 mature individuals)
(b) all individuals are in a single population

D. Population very small or restricted in the form of either of the following:
1. Population estimated to number fewer than 1,000 mature individuals.
2. Population is characterised by an acute restriction in its area of occupancy (typically less than 100 km^2) or in the number of locations (typically fewer than five). Such a taxon would thus be prone to the effects of human activities (or stochastic events whose impacts are increased by human activities) within a very short period of time in an unforeseeable future, and is thus capable of becoming Critically Endangered or even Extinct within a very short period.

E. Quantitative analysis showing that the probability of extinction in the wild is at least 10 % within 100 years.

Appendix 4

Table XXVIII (a)

Overview of the estimated numbers of orang-utans in the distribution range up until mid-1997, before the devastating impact of arson during that year's drought; based on the survey results as shown in Table XII (a-f).

Re	W. Kalimantan polygons	Forest – km^2	Cc unit	Estim. density	Total number
1	Sambas (27+27+446)	500	37	2.5	93
2	Mempawah (10+12+23+129)	174	10	2.5	25
3	Gunung Niut (36+61+98)195	195	15	1.8	27
4	Ketunga (19+26+268)	313	19	1.8	34
5	D.Sentarum-Bent. Karimun	5869	734	2.8	2055
6	Kapuas Hulu (1962+595+41)	2598	195	0.6	1170
7	Madi plateau-Melawi	2856	257	1.9	488
8	Bukit Baka	474	35	1.6	56
9	Berangin-Sebayan	4047	364	1.8	655
10	Kapuas swamps	6798	407	2.9	404
11	Sukadana-Kendawangan	7768	583	2.9	1688

Total W. Kalimantan – 6,695

Table XXVIII (b)

Re	C. Kalimantan polygons	Forest – km^2	Cc unit	Estim. density	Total number
12	Jelai-Lamandau-Arui	12516	782	2.2	1720
13	Tanjung Puting	3518	308	2.9	893
14	East Pembuang-Seruyan	7500	675	2.1	1418
15	W.Sampit floodplains	4327	162	2.7	437
16	Katingan floodplains	7402	740	1.8	1332
17	Sebangau catchment	5878	705	2.2	1551
18	W.Rungan floodplains	2217	199	2.0	398
19	Kahayan-Rungan catchment	2099	131	1.8	236
20	Sebangau-Kahayan floodplains	1932	135	2.0	270
21	Mangkutup (3575+307)	3882	194	1.8	349
22	Kapuas Murung-Barito plains	7745	290	1.8	522
23	Kahayan-Kapuas Murung hills	3115	140	1.2	168
24	Schwaner foothills east	1426	107	1.6	171
25	Schwaner foothills west	1460	109	1.8	196
26	Central Schwaner range	6241	468	0.6	281
27	Kahayan-Miri catchment	1478	55	0.8	44

Table XXVIII (b) (continued)

Re	C. Kalimantan polygons	Forest – km²	Cc unit	Estim. density	Total number
28	Bundang east	1285	80	0.8	64
29	Upper Dusun	1899	119	0.6	71
30	Busang Hulu (119+373)	492	61	0.6	37

Total central Kalimantan – 10,158

Table XXVIII (c)

Re	E. Kalimantan polygons	Forest – km²	Cc unit	Estim. density	Total number
31	Liangpran (633+765)	1398	126	1.6	202
32	Boh catchment	1755	184	0.6	110
33	Pari-Sentekan	1870	140	1.8	252
34	Belayan-Kedangkepala	1565	98	2.3	225
35	Central plains	617	31	2.5	77
36	Coastal Kutai	3237	121	2.3	278
37	Telen-Wahau	888	80	2.3	184
38	Tinda-Hantung hills	5729	344	1.8	619
39	Segah catchment	6619	331	2.5	828
40	Berau	2105	158	2.6	411
41	South Sembakung swamp	577	58	2.9	168
42	North Sembakung swamp	2202	165	2.8	462
43	Sebuku	2415	151	2.4	362
44	Sembakung Ulu	499	37	0.8	30

Total East Kalimantan – 4,208

Table XXVIII (d)

Re	Sabah polygons	Forest – km²	Cc unit	Estim. density	Total number
45	Ulu Kalumpang	408	36	1.2	43
46	Tabin (120+132)	252	22	1.9	40
47	Melangking	298	27	2.0	54
48	Southern Forest east	3710	278	2.0	556
49	Kinabatagan (610+977)	1587	143	2.3	329
50	Southern Forest north	982	44	1.8	79

Table XXVIII (d) (continued)

Re	Sabah polygons	Forest – km²	Cc unit	Estim. density	Total number
51	Southern forest west	1481	155	2.0	310
52	Trus Madi	394	35	1.6	56
53	Ulu Tunggud	163	15	1.0	15
54	Bidu-bidu hills	74	7	1.0	7
55	Sugut South	25	5	1.8	9
56	Sugut North	57	9	2.0	18
57	Lingkabau	459	55	2.0	110
58	Kinabalu	690	86	0.6	52
59	Crocker range	97	15	0.6	9

In Sabah total – 1,687

Table XXVIII (e)

Re	Sarawak-Kal. polygons	Forest – km²	Cc unit	Estim. density	Total number
60	Batang Lupar coastal swamp	106	11	2.0	22
61	Lanj. Entimau/Batang Aie	812	158	2.3	363

In Sarawak total – 385

Table XXVIII (f)

Re	Sumatra polygons	Forest – km²	Cc unit	Estim. density	Total number
1	North Aceh	1176	144	3.8	547
2	Woyla	1793	157	3.6	565
3	Gayo Mountains	1301	182	1.5	273
4	Bahbahrot swamp	192	14	4.5	63
5	Leuser	4445	933	2.8	2612
6	Singkil swamp	981	392	6.7	2626
7	Peusangan catchment	1086	98	3.2	314
8	Jambu Aie catchment	2014	126	2.6	328
9	Tamiang	1826	137	3.2	438
10	Sikundur-Langkat	1548	139	1.8	259
11	Alas valley	444	8	3.9	31
12	East Alas	174	16	3.6	58
13	Dolok Sembelin	148	22	3.4	75

TABLE XXVIII (F) (CONTINUED)

Re	Sumatra polygons	Forest – km²	Cc unit	Estim. density	Total number
14	Batu Ardan	206	31	2.2	68
15	Tapanuli Tengah	1564	391	6.1	2385
16	Anggolia	188	20	2.6	52
17	Simonangmonang	350	31	1.8	56
18	Sipirok	190	23	2.0	46
19	Angkola-Siondop	2327	332	3.9	1295
20	West Pasaman	405	28	1.8	50
21	Baruman	2167	260	1.4	364
22	Rimbo Panti	147	9	2.4	22
23	Habinsaran	1283	115	2.2	253

Total Sumatra – 12,770

Appendix 5

Table XXIX

List of research functions at the Ketambe Station, Aceh Tenggara.

Name researcher/ student assistants	Research topic	Name researcher/ student assistants	Research topic
H.D. Rijksen* 1971-74	Orang-utan	**R. Palombit*** 1988	Siamang
W.J.J.O. de Wilde 1972	Plant taxon.	P. Assink	L.t. macaque
B.J. de Wilde-Duyfjes	Plant taxon.	A. Gibson	Ecology
J. Krikken 1972	Insects	G. Grether	Ecology
E. Brotoisworo 1973	Siamang	**Barita O. Manullang**	Lianas
Y. Ruchiat 1973	Lar gibbon	R. Marlon	Figs/primates
A. Fernhout	Orang-utan	**L. Sterck*** 1988-92	Langur/mac.
A. Fernhout 1974-75	Orang-utan	A. Yahya	T. langur
C.L. Schürmann '75-79	Orang-utan	Yudha T.	L.t. macaque
Undang S. Halim	Orang-utan	K. Mandagi	L.t. macaque
C.P. van Schaik	Orang-utan	Ratna	L.t. macaque
M. A. van Noordwijk	L.t. macaque	**J. Cant** 1990	Orang-utan
C.P. van Schaik* '79-84	L.t. macaque	**P. Ungar** 1990	Orang-utan
M.v.Schaik-v.Noordwijk*	L.t. macaque	**F. Aurellli** 1990-91	L.t. macaque
T. Mitra Setia	Orang-utan	**T. Mitra Setia** '91-93	Orang-utan
Bambang Warsono	L.t. macaque	D. Priatna	Orang-utan
J. Sugardjito* 1979-83	Orang-utan	Y. Istiadi	Orang-utan
Nurhuda	Orang-utan	S.S.Utami Atmoko	Orang-utan
I.J.A. te Boekhorst	Orang-utan	Haryo Tabah W.	Orang-utan
R. Abdulhadi 1985	Forest ecol.	Rizal	Orang-utan
K. Kartawinata 1985	Forest ecol.	**R. Steenbeek** 1991-96	T. langur
S. Orbons 1986	Forest ecol.	**S.U. Atmoko*** 93-96	Orang-utan
Y. Robertson 1986	P.t. macaque	A. Hanifah Lubis	Orang-utan
J. de Ruiter* 1985-87	L.t. macaque	I. Hartadi	Orang-utan
L. Sterck	T. langur	Basrul	Orang-utan
Djojosudharmo '87-89	Orang-utan	C.N. Simanjuntak	Orang-utan
A. Senjaya Suhandi	Orang-utan	Syahwalludin Umar	Orang-utan
E. Perbatakusuma	Orang-utan	A. Sitompul	Orang-utan
Chairul Saleh	Orang-utan	J. Lawrence	Fig ecology
Kelly Harahap	Orang-utan	R. Blijdorp	Orang-utan
B.J. de Jong	L.t. macaque	S. Wich	Orang-utan
D. Vos	L.t. macaque	J. Oonk	Orang-utan
L. Hermanns	L.t. macaque	**A. Hanifah Lubis** 1996	Orang-utan
I. van Dijk	L.t. macaque	**S. Wich**	T. langur

* Persons indicated by an asterisk earned a PhD through their fieldwork at Ketambe.

REFERENCES

AARD-LAWOO, 1992. Acid sulphate soils in the humid tropics: water management and soil fertility. (SEVENHUYZEN, R.J., DAMANIK, M. and SUWARDJO, H. eds.) Jakarta. pp. 283.

AARD-LAWOO, 1992. Acid sulphate soils in the humid tropics: ecological aspects of their development. (KLEPPER, O., CHAIRUDDIN, G. and HATTA, M. eds.) Jakarta. pp. 100.

ABDULHADI, R. and KARTAWINATA, K. 1982. The pattern of forest vegetation at the Ketambe research station, Gn. Leuser National park. Unpubl. LBN–LIPI Report. Bogor.

ALLEN, G.M. and COOLIDGE, H.J. 1940. Asiatic primate expeditions collections: mammals. Bull. Mus. comp. Zool. 87: 131-166.

ALVIM, P.T. and ALGER, K. 1993. Forest in the life of the people. Paper presented at Beyond UNCED, Global Forest Conference, Bandung Febr. 1993. 14 pp.

ANON. 1969. Male orang-utan stabilizes group. Yerkes Reg. Prim. Res. centre Bull. June 1969: 22-24.

ASHTON, P. 1964. Ecological studies in the mixed dipterocarp forests of Brunei State. Oxf. For. Mem. 25.

ASHTON, P. 1976. An approach to the study of breeding systems, population structure and taxonomy of tropical trees. In: Tropical trees; variation, breeding and conservation (BURLEY, J. and STYLES, B.T. eds.) London.

AVELING, C. 1978. Expedition in South Langkat reserve, east of sungei Wampu. Progress report 15, IUCN/WWF proj. 1589. (unpubl.)

AVELING, C. and AVELING, R.J., 1979. The effects of logging on the fauna of lowland tropical rainforest. Joint PPA/WWF surveys in Sikundur Reserve, unpublished report.

AVELING, R.J. 1982. Orang-utan conservation in Sumatra, by habitat preservation and conservation education. In: The Orang-utan, its biology and conservation (DE BOER, L.E.M. ed.) The Hague. pp. 299-316.

AVELING, R.J. 1992. Tourism and Great Apes. In: Proceedings of the Great Apes Conference (Ministry of Forestry, Indonesia): 75-76.

AVELING, R.J. and MITCHELL, A.H. (1982). Is rehabilitating orang utans worth while? Oryx, 16 (3): 263-271

BALEN, J.H. VAN 1898. De orang-oetan. In: Album der Natuur Amsterdam. pp. 38-56.

BANKS, 1931. A popular account of the mammals of Borneo. J. Malay. Brch R. Asiatic Soc. 9 (2): 1-139.

BARD, K.A. 1993. Cognitive competence underlying tool-use in free-ranging orangutans. In: The use of tools by human and non-human primates (BERTHELET, A. and CHAVAILLON, J. eds.) Oxford.

BARBER, C.V., AFIFF, S. and PURNOMO, A. 1995. Tiger by the tail: Reorienting Biodiversity Conservation and Development in Indonesia. World Resources Inst.

BEAMAN, R.S. BEAMAN, J.H, MARSH, C.W. and WOODS, P.V. 1985. Drought and fires in Sabah in 1983. Sabah Soc. J. VIII (1): 10-30.

BECCARI, O. 1904. Wanderings in the great forests of Borneo. London: Constable.
BEECKMAN, D. 1718. A voyage to and from the island of Borneo, in the East Indies, pp. 37-38. London.

BELLWOOD, P. 1985. Prehistory of the Indo-Malaysian Archipelago. Sydney.

BEMMEL, A.C.V. VAN 1968. Contribution to the knowledge of the geographical races of *Pongo pygmaeus* (Hoppius). Bijdragen tot de Dierkunde 38: 13-15.

BEMMELEN, R.W. 1949. The Geology of Indonesia (2 Vols.) The Hague.

BENNETT, J.M. 1993. Orang-utans in Sarawak: Past, present and future. In: Orangutan Population and habitat viability analysis report of the captive breeding specialist group species survival commission of the IUCN. 18-20 January 1993.

BENNETT, E.L. and DAHABAN, Z. 1995. Wildlife responses to disturbances in Sarawak and their implications for forest management. In: Ecology, conservation and management of Southeast Asian rainforests (PRIMACK, R.B. and LOVEJOY, T.E. eds.). Yale London. pp. 66-86.

BEZEMER, K.W.L. 1929. Bedreigde diersoorten. In: Mededeeling No 7; Ned. Cie. Int. Natuurbescherming.

BLOUCH, R.A. 1997. Distribution and abundance of orangutans (Pongo Pygmaeus) and other primates in the Lanjak Entimau wildlife sanctuary, Sarawak, Malaysia. Tropical Biodiversity 4 (3): 259-274.

BOCK, C. 1882. The head-hunters of Borneo: a narrative of travel up the Mahakkamand down the Barito. London.

BODMER, R.E., SIDIK, L. and ISKANDAR, S., 1992 . Mammalian biomass shows that uneven densities of orang-utans are caused by variation in fruit availability. (Unpublished report).

BOER, L.E.M. DE, 1982. The Orang-utan. Its biology and conservation. Junk Publishers, The Hague.

BOER, L.E.M. DE and SEUANEZ, H.N. 1982. The chromosomes of the orang utan and their relevance to the conservation of the species. In DE BOER, L.E.M. (ed) The Orang Utan, its biology and conservation. Junk Publishers, The Hague.

BONNER, R. (1993). At the Hand of Man; peril and hope for Africa's wildlife. Simon and Schuster, London.

BONTIUS, J. 1658. Historiae Naturalis et Medicae Indiae Orientalis, Libri Sex, Pars V: Historia Animalum (ed. GUILIEMO PISONE) pp. 84-85. Elsevier, Amsterdam.

BORNER, M., 1976. Sumatra's Orang-utans. Oryx 13 (3): 290-293.

BORNER, M. and GITTENS, P. 1979. Round table discussion on rehabilitation. In: Recent Advances in Primatology. Vol 11. (CHIVERS, D. and PETERS, L. eds.), New York. pp. 101-105.

BORNER, M. 1985. The rehabilitated chimpanzees of Rubondo island. Oryx 19: 151-154.

BOTTEMA, J.W.T. 1995. Market formation and agriculture in Indonesia from the mid 19th century to 1990. PhD Thesis Nijmegen Univ. Jakarta.

BOURNE, G.H. 1971. The Ape People. London.

BRAMBELL, M.R. 1977. Reintroduction. Int. Zoo Yearbook 17: 112-116.

BRANDES, G. 1939. Buschi, vom Orang-Säugling zum Backenwülster, Leipzig.

BRANDON, K. 1996. Ecotourism and conservation: a review of key issues. World Bank Env. Dept. Paper 033.

BRANDON, K. 1997. Policy and practical considerations in land-use strategies for biodiversity conservation. In: Last Stand; protected areas and the defence of tropical biodiversity (KRAMER, R., SCHAIK, C.P. VAN and JOHNSON, J. eds.) Oxford UP, New York. pp. 90-114.

BROOKE, J. 1841. Letter relating to the orang-utan of Borneo. Proc. Zool. Soc. Lond. 1841: 55.

BROOKE, C.H. 1866. Ten Years in Sarawak, 2 Vols. London.

BROSIUS, J.P. 1986. River, forest and mountain: the Penan gang landscape. Sarawak Museum J. XXXVI (57): 173-184.

BROUWER, L.C. 1996. Nutrient cycling in pristine and logged tropical rain forest; a study in Guyana. Tropenbos Guyana Series 1. PhD thesis, Utrecht.

BROWDER, J. 1992. The limits of extractivism. Bioscience 42: 174-181.

BROWN, L. 1991. State of the world 1991; A Worldwatch Institute report on progress towards a sustainable society. Norton, New York. pp. 79-80.

BRUEN, D. L. and HAILE, N. S. 1960. Reports of the maias protection commission. Sarawak Government Printer.

BRUENIG, E.F. 1996. Conservation and management of tropical rainforests; an integrated approach to sustainability. CAB Int. Wallingford.

BRUSSARD, P.F. 1982. The role of field stations in the preservation of biological diversity. Bioscience 32 (5): 327-330.

BUGO, H. 1995. The significance of the timber industry in the economic and social development of Sarawak. In: Ecology, Conservation, and Management of Southeast Asian Rainforests (PRIMACK, R.B. and LOVEJOY, T.E. eds.) Yale U.P. New Haven. pp: 221-240.

BUIS, J. 1993. Holland Houtland; een Geschiedenis van het Nederlandse bos. Prometheus.

BURGESS, P.F. 1971. The effect of logging on hill Diperocarp forest. Malay. Nat. J. 24: 231-237.

CALDECOTT, J. and KAVANAGH, M. 1983. Can translocation help wild primates? Oryx 17: 135-139.

CAMPBELL, S. 1980. Is reintroduction a realistic goal? In: Conservation Biology: an evolutionally-ecological perspective. (SOULE, M.E. and WILCOX, B.A. eds.) Sinauer, Sunderland. pp. 263-269.

CAMPER, P. 1782. Natuurkundige Verhandelingen ... over den Orang-outang en eenige andere aap-soorten. Elsevier Amsterdam.

CARPENTER, C. R. 1938. A survey of wildlife conditions in Atjeh, North Sumatra. Med. Ned. Comm. Int. Nat. Besch. Amsterdam. No. 12: 1-33.

CATER, B., 1991. The case of the bartered babies. In: BBC Wildlife Vol. 9, no. 4. April 1991: 254-260.

CATTON, W.R. 1980. Overshoot; the ecological basis of revolutionary change. Univ. of Illinois, Urbana.

CAVALIERI, P. and SINGER, P. 1995. The Great Ape Project. In: Ape, man, Apeman: changing views since 1600 (CORBEY, R. and THEUNISSEN, B. eds.), Leiden. pp. 367-377.

CBSG/SSC/IUCN 1993. Orangutan Population and habitat viability analysis report of the captive breeding specialist group species survival commission of the IUCN. 18-20 January 1993.

CHASEN, F.N. 1940. A Handlist of Malaysian Mammals. Bull. Raffles Mus. 15:60-62.

CHIN, S.C. 1985. Agriculture and resource utilization in a lowland rainforest Kenyah community. Sarawak Mus. J. Special Monograph 4.

CHIVERS, D.J. 1973. An introduction to the socio-ecology of Malayan forest primates. In: Comparative Ecology and Behaviour of Primates (MICHAEL, R.P. and CROOK, J.H. eds.), London. pp. 101-146.

CHIVERS, D.J. 1980. Malaysian Forest Primates; ten years study in tropical rainforest. Plenum. NY.

CHIVERS, D.J. and BURTON, K.M. 1986. Some observations on the primates of Kalimantan Tengah, Indonesia. Primate Conservation 9.

CHODEN, K. 1997. Bhutanese tales of the Yeti. White Lotus, Bangkok.

CLEARY, M. and EATON, E., 1992. Borneo. Change and development. Oxford University Press, Oxford.

COCKBURN, P.F. 1978. The Flora. In: Kinabalu; Summit of Borneo (LUPING, D.M., WEN, C.W. and DINGLEY, E.R. eds.), Sabah Soc. Kota Kinabalu. pp: 179-198.

COLLINS, N.M., SAYER, J.A. and WHITMORE, T.C. 1991. The conservation atlas of tropical forests; Asia and the Pacific. IUCN, Macmillan.

COOMANS DE RUITER, L. 1932. Uit Borneo's Wonderwereld. Ned. Ind. Natuurhist. Ver. Batavia.

CORNER, E.H.J. 1978. The Plant Life. In: Kinabalu; summit of Borneo (LUPING, D.M., WEN, C.W. and DINGLEY, E.R. eds.), Sabah Soc. Kota Kinabalu. pp: 112-178.

COURTENAY, J., GROVES, C. and ANDREWS, P. 1988. Inter- or intra-island variation? An assessment of the differences between Bornean and Sumatran Orang-utans. In (SCHWARTZ, J.H. ed.) Orang-utan Biology. pp. 19-29. Oxford U.P. Oxford.

CROOK, J.H. 1965. The adaptive significance of avian social organizations. Symp. Zool. Soc. London 14: 181-218.

DAMMERMAN, K.W. 1932. De nieuw ontdekte orang pendek. De Tropische Natuur XXI (8): 123-131.

DAMMERMAN, K.W.1938. The Orang Utan. In: Nature Protection in the Netherlands Indies, compiled by the section Nature Protection of the Botanic Gardens, Buitenzorg. p. 1-6.

DAVENPORT, R. K. 1967. The orang utan in Sabah. Folia Primatol. 5: 247-263.

DAVIES, G. 1986. The orang-utan in Sabah. Oryx 20 (1): 40-45.

DAVIES. G. and PAYNE, J. 1982. A faunal survey of Sabah. IUCN/WWF Project report 1692. Kuala Lumpur.

DAVIS, D.D. 1962. Mammals of the lowland rain-forest of North Borneo. Bull. Nat. Mus. Singapore 31: 1-130.

DAVIS, G. and ACKERMANN, R. 1988. Indonesia; Forest, Land and Water: Issues in Sustainable Development. WB Doc Asian Regional Office (unpubl.).

DAVIS, W. and HENLEY, T. 1990. Penan, Voice of the Rainforest. WCWC Vancouver.

DEVEREUX, G. 1976. A study of abortion in primitive societies. New York.

DE SILVA, G.S. 1965. The East coast experiment. IUCN Publ. NS. 10: Conservation in Tropical Southeast Asia, Bangkok: 299-302.

DE SILVA, G. S. 1971. Notes on the Orang utan rehabilitation project in Sabah. Malay. Nat. J., 24: 50-77.

DE SOTO, H. 1993. The missing ingredient. In: The Future Surveyed; 150 Economic Years. Economist 11 Sept. 1993, pp 8-30.

DINERSTEIN, E., WIKRAMANAYAKE, E.D. and FORNEY, M. 1995. Conserving the reservoirs and remnants of tropical moist forest in the Indo-Pacific region. In: Ecology, conservation and management of Southeast Asian Rainforests (PRIMACK, B. and LOVEJOY, T.E. eds.) Yale, New Haven. pp. 140-175.

DITTUS, W.P.J. 1980. Population regulation: the effects of severe environmental changes on the demography and behavior of wild toque macaques. Int. J. Primatol. 3: 276.

DJOJOSUDHARMO, S. and VAN SCHAICK, C. P. 1992. Why are Orang-utans so rare in the highlands? Altitudinal changes in a Sumatran forest. Tropical Biodiversity 1: 11-22.

DOBSON, J. 1953-54. John Hunter and the early knowledge of the Anthropoid apes. Proc. zoo. Soc. London 123: 1-16.

DOVE, M.R. 1993. A revisionist view of tropical deforestation and development. Environmental Conservation 20 (1): 17-24 (56).

DRANSFIELD, J. 1972. The genus Borassodendron (Palmaue) in Malesia. Reinwardtia 8: 351-363.

EAST, R. 1981. Species-area curves and populations of large mammals in African savanna reserves. Biol. Conservation 21: 111-126.

EAST, R. 1983. Application of species-area curves to African savannah reserves. Afr. J. Ecol. 21: 123-128.

EDWARDS, G. 1758-64. Gleanings of Natural History, exhibiting figures of quadrupeds London.

EHRENFELD, D. 1978. The Arrogance of Humanism. Oxford.

ELLIS, S. and SEAL, U.S. 1995. Tools of the trade to aid decision making for species survival. Biodiversity and conservation 4 (6): 553-572.

EUDEY, A. 1987. Action plan for Asian primate conservation: 1987-91. IUCN/SSC Primate Specialist Group.

FAO, 1987. Special study on forest management, afforestation and utilization of forest resources in the developing regions. Asia-Pacific region. Assessment of forest resources in six countries. FAO, Bangkok Field Document 17.

FERRIS, S.D., BROWN, W.M., DAVIDSON, W.S., and WILSON, A.C. 1981. Extensive polymorphism in the mitochondrial DNA of apes. Proc. Nat. Acad. Sci. USA. 78: 6319-6323.

FOGDEN, M. P. L. 1972. The seasonality and population dynamics of equatorial forest birds in Sarawak. Ibis 114: 307-343.

FOSSEY, D. 1970. Making friends with mountain gorillas. Nat. Geographic M. 137: 48-67.

FOSSEY, D. More years with mountain gorillas. Nat. Geographic M. 140: 574-585.

FOSSEY, D. 1983. Gorillas in the mist. London.

FREDERIKSSON, G. 1995. Reintroduction of orangutans: a new approach. A study on the ecology and behaviour of reintroduced orangutans in the Sungai Wain nature reserve, Kalimantan; unpubl. MSc thesis Univ. Amsterdam.

FREEMAN, D. 1960. The Iban of western Borneo. In: Social Structure in Southeast Asia (MURDOCK, G.P. ed.), Chicago: pp. 65-87.

FREY, R. 1975. Sumatra's red apes return to the wild. Wildlife 17 (8): 356-363.

FREY, R. 1976. Orangutan rehabilitation centre Bohorok. WWF Annual Report 1976. pp. 12-23.

FREY, R. 1978. Management of orang-utans. In: Wildlife Management in Southeast Asia. Biotrop Spec. Publ. 8: 199-215.

GALDIKAS, B.M.F. 1978. Orangutan adaptation at Tanjung Puting Reserve, Central Borneo. PhD thesis. Univ. of California, Los Angeles.

GALDIKAS, B.M.F. 1980.Living with the great orange apes. Nat. Geographic M. 153: 830-853.

GALDIKAS, B.M.F. 1982. Orang utans as seed dispersers at tanjung Puting, Central Kalimantan: implications for conservation. In: The Orang Utan (DE BOER, L.E.M. ed.), The Hague: 285-298.

GALDIKAS, B.M.F. 1984. Adult female sociality among wild orangutans at Tanjung Puting Reserve. In: Female Primate Studies by Women Primatologists (SMALL, M.F. ed.): 217-235. New York.

GALDIKAS, B.M.F. 1985. Adult male sociality and reproductive tactics among orangutans at Tanjung Puting. Folia Primatol 45: 9-24.

GALDIKAS, B.M.F. 1991. Protection of wild orangutans and habitat vis a vis rehabilitation. Proceedings of the Great Apes Conference, Jakarta and Pangkalanbun Dec. 1991 (Min. of Forestry): 23-24.

GALDIKAS, B.M.F. 1995. Behavior of wild adolescent female orangutans. In: The Neglected Ape (NADLER, R.D. *et al.* eds.) New York. pp. 163-182.

GALDIKAS-BRINDAMOUR, B.M.F. 1975. Orangutans, Indonesia's "people of the forest". Nat. Geographic M. 148 (4): 444-472.

GHILIERI, M.P. 1985. The Chimpanzees of Kibale forest; a field study of ecology and social structure. Columbia U.P. New York.

GIESEN, W., DESCHAMPS, V. and DENNIS, R., 1994. Recommendations for modification of the boundary of Danau Sentarum Wildlife Reserve, West Kalimantan. Asian Wetland Bureau, Bogor, Indonesia (unpubl. report).

GILLMAN, G.P., SINCLAIR, D.F., KNOWLTON, R. and KEYS, M.G. 1985. The effect on some soil chemical properties of the selective logging of a North Queensland rainforest. For. Ecol. & Mgmt. 12: 195-214.

GOLDER, B. 1996. The people dimension. In: WWF; changing worlds (DE MATTOS-SHIPLEY, H., LYONS, J., BELLSHAM, C., HARDING, D. and JOHNSTON, M. eds.) London. pp. 151-159.

GOODALL, J. 1963. My life among wild chimpanzees. National Geographic 124:272-308.

GORDON, H.S. 1954. The economic theory of a common property resource; the fishery. J. of political economy 62: 124-142.

GRIFFITHS, M. and VAN SCHAIK, C.P. 1993. The impact of human traffic on the abundance and activity periods of Sumatran rain forest wildlife. Conservation Biology 7 (3): 623-626.

GROOMBRIDGE, B. (ed.), 1993. 1994 IUCN Red Data List of Threatened Animals. IUCN Gland, Switzerland and Cambridge, UK. 1VI + 286pp.

GROVES, C.P. 1971. Pongo pygmaeus. Mammalian Species 4: 1-6. The American Society of Mammologists.

GROVES, C.P. and HOLTHUIS, L.B. 1983. The nomenclature of the orang-utan. Zool. Meded. Leiden 59: 411-417.

GROVES, C.P., WESTWOOD, C. and SHEA, B.T. 1992. Unfinished business: Mahalanobis and a clockwork orang. J. of Human Evolution 22: 327-340.

HAGEN, B. 1890. Die Pflanzen und Tierwelt von Deli auf der Ostküste Sumatras. In: Naturwissenschaftliche Skizzen und Beiträge. pp. 66-70.

HAILE, N.S. 1973. The geomorphology and geology of the northern part of the Sunda shelf. Pacific Geology 6: 73-90.

HAILS, C. 1996. Keeping pace with change. In: WWF; changing worlds (DE MATTOS-SHIPLEY, H., LYONS, J., BELLSHAM, C., HARDING, D. and JOHNSTON, M. eds.) London. pp. 160-165.

HANNAH, L. 1992. African People, African Parks; an evaluation of development initiatives as a means of improving protected area conservation in Africa. Biodiversity Support Programme/Conservation International.

HAPGOOD, C.H. 1970. The path of the pole. Chilton, New York.

HARDIN, G. 1968. The Tragedy of the Commons. Science 162: 94-99.

HARDIN, G. 1982. Bookreview in The Ecologist 12 (1): 43.

HARLOW, H.F. and HARLOW, M.K. 1965. The affectional system. In Behavior of non-human primates (SCHRIER, A.M., HARLOW, H.F. and STOLLNITZ, F. eds.). Acad. Press New York.

HARRISSON, B. 1960. A study of Orang-utan behaviour in semi-wild state 1956-1960. Sarawak Mus. J. 9: 422-447.

HARRISSON, B. 1961. Orang utan: what chances of survival? Sarawak Mus. J. 10: 238-261.

HARRISSON, B. 1962. The immediate problem of the Orang utan. Malay Nature J. 16: 4-5.

HARRISSON, B. 1963. Education to wild living of young Orang utans at Bako National Park. Sarawak Mus. J. 11: 222-258.

HARRISSON, B. 1963. Report on censuses in sample habitat areas of North Borneo. World Wildlife Fund Report.

HARRISSON, B. 1965. Conservation needs of the Orang-utan. IUCN Publ. NS.10: 294-295.

HARRISSON, B. 1971. Conservation of non-human primates in 1970. WWF Basel.

HARRISSON, T. 1960. A remarkably remote Orang-utan: 1958-1960. Sarawak Mus. J. NS. 9: 448-451.

HARRISSON, T. 1965. Some quantitative effects of vertebrates on the Borneo flora. in: UNESCO Symposium on Ecological Research in Humid Tropics Vegetation (Kuching, 1963). pp. 164-169.

HERRERA, R., JORDAN, C.F., MEDINA, E. and KLINGE, H. How human activities disturb the nutrient cycles of tropical rainforest in Amazonia. Ambio 10: 109-114.

HERWAARDEN, J. VAN 1924. Een ontmoeting met een aapmensch. De Tropische Natuur XIII (7): 103-106.

HEYNSIUS-VIRULI, A. and VAN HEURN, F.C. 1935. Overzicht van de uit Nederlandsch-Indië ontvangen gegevens, met biologische aantekeningen omtrent de betreffende diersoorten. Ned. Comm. Int. Nat. Besch. Supplement op Med. no 10: 36-40.

HEURN, F. C. VAN and HEYNSIUS-VIRULY, A. 1935. Biologische aantekeningen over eenige belangrijke diersoorten van Java, Sumatra en Borneo; de anthropoide apen, de rhinocerossen en het baardzwijn. Med. Ned. Comm. Int. Nat. Besch. Amsterdam. No. 10: 1-28.

HOFFMAN, C.L. 1986. The Punan; hunters and gatherers of Borneo. Studies in Cultural Anthropology 12. UMI Research Press, Ann Arbor.

HONG, E. 1987 Natives of Sarawak; survival in Borneo's vanishing forests. Inst. Masyarakat Malaysia, Pinang..

HOOFF, J.A.R.A.M. 1995. The orangutan; a social outsider. In: The Neglected Ape (NADLER, R.D. *et al.*) New York. pp. 153-162.

HOOYER, D. A. 1948. Prehistoric teeth of man and of the Orang-utan from Central Sumatra, with notes on the fssil OU from Java and S. China. Zool. Mededeelingen. Rijksmuseum Leiden. 1-175.

HOOYER, D.A. 1961. The Orang-utan in Niah Cave pre-history. Sarawak Mus. J. NS. 10: 408-421.

HOOYER, D.A. 1962. Pleistocene dating and man. Advancement of Science, Jan. 62: 485-489.

HORNADAY, W.T. 1885. Two years in the jungle; the experiences of a hunter and naturalist in India, Ceylon, the Malay Peninsula and Borneo. Doubleday, London.

HORR, D.A. 1972. The Borneo Orang Utan. Borneo Research Bull. 4: 46-50.

HORR, D. A. 1975. The Borneo Orang Utan: Population structure and Dynamics in relationship to Ecology and Reproductive Strategy. In: Primate Behaviour 4 (L. A. ROSENBLUM ed.) New York: 307-323.

HOSE, C. and MCDOUGALL, W. 1912. The pagan tribes of Borneo, 2 Vols. London.

HRDY, S. 1977. The langurs of Abu; female and male strategies of reproduction. Cambridge (Mass.).

HUBBELL, S. 1979. Tree dispersion, abundance, and diversity in a dry tropical forest. Science 203: 1299-1309.

HURST, P. 1990. Rainforest Politics; ecological destruction in Southeast Asia. Zed, London.

INSAN. 1989. Logging against the Natives of Sarawak, Inst. of Social Analysis, Petaling Jaya.

ISTIADI, Y., LEKSONO, S.M. and DJANUBUDIMAN,G. 1997. Indikasi keberadaan dan pengambilan sampel genetika orangutan di Sumatera. Pusat Studi Biodiversitas dan Konservasi. UI. Unpubl. Report.

JACOBS, M. 1988. The Tropical Rain Forest: A First Encounter. Berlin.

JACOBSON, E. 1917. Rimboeleven in Sumatra - vervolg no 4. De Tropische Natuur VI (5): 69-72.

JANCZEWSKI, D.N., GOLDMAN, D. and O'BRIEN, S.J. 1990. Molecular divergence of orang utan (*Pongo pygmaeus*) subspecies based on isozyme and two-dimensional gel electrophoresis. J. of Heredity 81: 375-387.

JARVIS, C. 1968. International Zoo Yearbook Zool. Soc. London 8: 189.

JEPMA, C.J. and BLOM, M. 1990. Global trends in tropical forest degradation: the Indonesian case. IDE discussion paper, Int. Tropenbos Programme - Symp. Oct. 1990.

JOHNS, A.D. 1983. Ecological effects of selective logging in a West Malaysian rain forest. PhD thesis Univ. Cambridge.

JOHNS, 1983. Selective logging and primates; an overview. In: Proc. Symp. Conserv. of Primates and their habitats (HARPER, D. ed.). Vaughan Paper 31. Univ, Leicester.pp. 86-100.

JOHNS, A.D. 1985. Selective logging and wildlife conservation in tropical rainfforest: Problems and recommendations. Biological Conservation 31: 355-375.

JOHNS, A.D. 1986. Effects of selective logging on the behaviour and ecology of West Malaysian primates. Ecology 67: 684-694.

JOHNS, A.D. 1988. Effects of 'selective' timber extraction on the rainforest structure and composition, and some consequences for frugivores and folivores. Biotropica 20: 31-42.

JOHNS, A.D. 1992. Vertebrate Responses to Selective logging: implications for the design of logging system Phil. Trans. Roy. Soc. B 335: 437-442.

JOHNS, A. and MARSHALL, A.G. 1992. Wildlife population parameters as indicators of the sustainability of timber logging operations; in: Forest Biology and Conservation in Borneo (ISMAEL, G., MOHAMED, M. and OMAR, S. eds) Sabah Foundation, Kota Kinabalu:366-373.

JOHNS, A.D. and SKORUPA, J.P. 1987. Responses of rain forest primates to habitat disturbance - A review. Int. J. Primatol. 8: 157-191.

JONES, M.L. 1969. The geographical races of orangutan. Proc. 2nd Int. Cong. Primat. Vol 2: 217-223. Karger, Basel.

JONES, M.L. 1980. Studbook of the orang utan (*Pongo pygmaeus*). Zool. Society of San Diego, San Diego.

KALFF, S. 1925. Indische Monstruositeiten. De Tropische Natuur 14 (8): 113-121.

KARESH, W.B., FRAZIER, H., SAHJUTI, D., ANDAU, M., GOMBEK, F., ZHI, L., JANCZEWSKI, D., and O'BRIEN, S. J., 1997. Orangutan genetic diversity. In: Orangutan Species Survival Plan, Husbandry Manual (SODARO, C., ed.) Chicago Zool. Park.

KARTAWINATA, K. 1978. Biological changes after logging in a lowland dipterocarp forest. Biotrop Special Publ. 3: 27-34.

KARTAWINATA, K. and VAYDA, A.P. 1984. Forest conversion in East Kalimantan, Indonesia: the activities and impact of timber companies, shifting cultivators, migrant pepper farmers and others. In: Ecology in practice: Part 1, Ecosystem Management (DI CASTRI, F, BAKER, F.W.G. and HADLEY, M., eds.).UNESCO Paris: 98-126.

KAVANAGH, M., HILLEGERS, P.J.M, RIJKSEN, H.D., SAW LENG GUAN, and YONG GHONG CHONG, D. 1982. Lanjak-Entimau orang-utan Sanctuary; a management plan. WWF Malaysia proj 1995, Kuala Lumpur..

KEDIT, P.M. 1982. An ecological survey of the Penan. Sarawak Mus. J. 30 (51): 225-279.

KEMPE, C.H. 1976. Child abuse and neglect. In: Raising children in modern America; problems and perspective solutions. Boston. pp. 173-188.

KHAN, M. 1978. Man's impact on the primates of Peninsular Malaysia. In: Recent advances in primatology (CHIVERS, D. and LANE PETTER, C. eds.) Academic press, UK.

KING, V.T. 1978. Essays on Borneo Societies. Oxford.

KINGSLEY, S. 1982. Causes of non-breeding and the development of secondary sexual characteristics in the male orangutan: a hormonal study. In: The Orang Utan, its biology and behaviour (DE BOER, L.E.M. ed.), Den Haag. pp: 215-229.

KINNAIRD, M.F. and O'BRIEN, T.G. 1996. Ecotourism in the Tangkoko Duasaudara nature reserve: Opening Pandora's box. Oryx 30 (1): 65-73.

KLEIMAN, D.G. 1989. Reintroduction of captive mammals for conservation. Bioscience 39 (3): 152-161.

KNOX, R. 1840. Inquiry into the present state of our knowledge respecting the Orang-outang and Chimpanzee. Lancet, 2: 289.

KOENIGSWALD, G. H. R. VON 1981. Gibt's noch Orang-utans im China? [Are there still orangutans in China?] Natur und Museum 111(8): 260-261

KOESNOE, M. 1969. Musjawarah; een wijze van volksbesluitvorming volgens adatrecht. Publ. over Adatrecht, KU Nijmegen.

KONSTANT, W.R. and MITTERMEIER, R.A. 1982. Introduction, reintroduction and translocation of Neotropical primates: past experiences and future possibilities. Int. Zoo Yearbook 22: 69-77.

KRAMER, R., VAN SCHAIK, C. and JOHNSON, J. 1997. Last Stand, protected areas and the defense of tropical biodiversity. Oxford U.P. New York.

KROPOTKIN, P. 1902. Mutual Aid; a factor in evolution. Porter Sargent, Boston.

KRUK, R. 1995. Traditional Islamic views of Apes and Monkeys. In: Ape, Man, Apeman: Changing views since 1600 (CORBEY, R. and THEUNISSEN, B. eds.), Leiden Univ. pp. 29-42.

KURT, F. 1970. Final report to IUCN/WWF Int. of Proj. 596, Leuser Reserve (Sumatra): 1-68.

LAL, R. 1986. Deforestation and soil erosion. In: Land Clearing and Development in the Tropics (LAL, R., SANCHEZ, P.A., and CUMMINGS, R.W., eds.) Rotterdam.

LAMBECK, R.J. 1997. Focal species: a multi-species umbrella for nature conservation. Conservation Biology 11 (4): 849-856.

LAMMERTS VAN BUEREN, E.M. and DUIVENVOORDEN, J.F. 1996. Towards priorities of biodiversity research in support of policy and management of tropical rainforests. Tropenbos Foundation, Wageningen.

LAMMERTS VAN BUEREN, E.M. and BLOM, E.M. 1997. Hierarchical framework for the formulation of sustainable forest management standards. Tropenbos, Wageningen. pp.82.

LANGUB, J. 1989. Some aspects of life of the Penan. Sarawak Museum J. XL (61):169-189.

LARDEU-GILLOUX, I. 1995. Rehabilitation Centres: Their Struggle, Their Future. In: The Neglected Ape, (NADLER, R.D., GALDIKAS, B.F.M., SHEENAN, L.K. AND ROSEN, N. eds.), Plenum, N.Y. pp. 61-68.

LARDEUX-GILLOUX, I. 1996. How socially emotional are Bornean orang-utans? (unpubl. report).

LATHAM, R. 1978. The Travels of Marco Polo. Penguin, Harmondsworth.

LAUMONIER, Y. 1989. Search for Phytogeographic provinces in Sumatra. In: The Plant diversity of Malesia. (BAAS, P., KALKMAN, K and GEESINK, R. eds.) Proc. Flora Malesiana Symp.: 193-211. London.

LAUNER, A.E. and MURPHEY, D.D., 1994. Umbrella species and the conservation of habitat fragments; a case of a threatened butterfly and a vanishing grassland ecosystem. Biological Conservation 69: 145-153.

LEE, L.L., PHIPPS, M. and CHEN, P.C. 1991. The orangutan in Taiwan.; in Proceedings of the Great Apes Conference, MoF Jakarta: 105-111.

LEIGHTON, M. 1990.The Gunung Palung Research Project: Studies of ecological processes maintaining rainforest ecological diversity. Unpubl. report to LIPI.

LEIGHTON, M. 1993. Modeling dietary selectivity by Borneon Orangutans: Evidence for integration of multiple criteria in fruit selection. Int. J. Prim. 14: 257-313.

LEIGHTON, M. and DARNAEDI, D. 1996. Gunung Palung National Park, West Kalimantan, Indonesia special issue. Tropical Biodiversity 3(3): 141-143.

LEIGHTON, M. and LEIGHTON, D.R. 1983. Vertebrate responses to fruiting seasonality within a Bornean rain forest. In: Tropical Rainforest; Ecology and Conservation (SUTTON, S.L., WHITMORE, T.C. and CHADWICK, A.C., eds.) Oxford. p. 181-196.

LEIGHTON, M., SEAL, U.S., SOEMARNA, K, ADJISASMITO, WIJAYA, M, SETIA, T.M., SHAPIRO, G, PERKINS, L, TRAYLOR-HOLZER, K., and TILSON, R. 1995. Orangutan life history and Vortex analysis. In The Neglected Ape (NADLER, R.D., GALDIKAS, B.F.M., SHEENAN, L.K. and ROSEN, N. eds.), New York: 97-107.

LETHMATE, J. and DUCKER, G. 1973. Untersuchungen zum Selbsterkennen im Spiegel bei Orang-utans und einigen anderen Affenarten. Z. Tierpsychol. 33: 248-269.

LETHMATE, J. 1977. Problemlöseverhalten von Orang-utans (*Pongo pygmaeus*). Advances in Ethology (Berlin) 19: 1-69.

LING ROTH, H. 1896. The Natives of Sarawak and British North Borneo, 2 Vols., London.

LIPPERT, W. 1974. Beobachtungen zum Schwangerschafts und Geburtsverhalten beim

Orang-utan im Thierpark Berlin. Folia Primatol. 21: 108-134.

LOEB, E.M. 1935 (1972). Sumatra, its history and people. Kuala Lumpur.

LOW, H. 1848. Sarawak; its inhabitants and productions. London.

LU ZHI, KARESH, W.B., JANCZEWSKI, D.N., FRAZIER-TAYLOR, H., SAHJUTI, D., GOMBEK, F., ANDAU, M., MARTENSON, J.S., and O'BRIEN, S.J. 1996. Genomic differentiation among natural populations of orang utan. In press Curr. Biol. (pers. comm.).

LYON, M. W. 1911. Mammals collected by Dr. W. L. Abbott on Borneo and some of the small adjacent islands. Proc. U. S. Nat. Mus. 40 : 53 -146.

MACANDREWS, C. 1986. Land Policy in Indonesia. WB Report..

MACFIE, E.J. 1992. Veterinary considerations in Endangered Great Ape conservation; Experiences with Mountain Gorilla. In: Proceedings on the Great Apes Conference Jakarta and Pangkalanbun Dec. 1991 (Min. of Forestry, Indonesia): 79-82.

MACKINNON, J.R. 1971. The Orang-utan in Sabah today. Oryx 11 (2-3): 141-191.

MACKINNON, J.R . 1973. Orang-utans in Sumatra. Oryx 12 (2): 234-242.

MACKINNON, J.R. 1974. The Behaviour and Ecology of Wild Orang-utans. Anim. Behav. 22: 3-74.

MACKINNON, J.R. 1974. In search of the red ape. London.

MACKINNON, J.R. 1975. Distinguishing characteristics of the insular forms of orang-utan. Int. Zoo Yearbook 15: 195-197.

MACKINNON, J.R. 1977. The future of orang utans. New Scientist 23: 697-699.

MACKINNON, J.R. 1977. Pet orangutans; should they return to the forest?. New Scientist 74: 697-699.

MACKINNON, J.R. 1987. The ape within us. London.

MACKINNON, J.R. 1992. Species survival plan for the orangutan. in: Forest Biology and Conservation in Borneo (ISMAEL, G., MOHAMED, M. and OMAR, S. eds) Sabah Foundation, Kota Kinabalu: 209-219.

MACKINNON, J.R. and ARTHA, M.B., 1981. A National Conservation Plan for Indonesia (Vol. II: Sumatra and Vol. V: Kalimantan). UNDP/FAO - INS/78/061, Bogor, Indonesia.

MACKINNON, J.R. and WARSITO, 1982. Gunung Palung Nature Reserve. Kalimantan Barat. Preliminary Management Plan. UNDP/FAO Field Report.

MACKINNON, J.R., MACKINNON, K., CHILD, G., and THORSELL, J. 1986. Managing protected areas in the tropics. Gland.

MACKINNON, K. 1986. The conservation status of nonhuman primates in Indonesia. In:

Primates, the road to self-sustaining populations (BERNIRSCHKE, K. ed.). New York. pp. 99-126.

MACKINNON, K. & MACKINNON, J.R., 1991. Habitat protection and re-introduction programmes. In: Beyond captive breeding: Re-introducing endangered mammals to the wild (J. H. W. GIPPS ed.). Zoological Society London Symposia (Oxford) 62 : 173-198.

MACKINNON, K. and RAMONO, W.S., 1993. Orang-utans as flagship species for conservation. In: Orangutan Population and habitat viability analysis report. CBSG/SSC - IUCN.

MAIER, R. 1923. De orang pandak of orang pendek. De Tropische Natuur XII (10): 154-156.

MALLINSON, J.J.C. 1978. `Cocktail' orang utans and the need to preserve pure-bred stock. Dodo 15: 69-77.

MANSER, B. 1992. Stimmen aus dem Regenwald; Zeugnisse eines bedrohten Volkes. Bern.

MAPLE, T. 1980. Orang-utan behavior. New York.

MARGULIS, L. and SAGAN, D. 1986. Microcosmos. Summit, N.Y.

MARKHAM, R.J. 1980. An investigation into the differences in social behaviour between Sumatran (*Pongo pygmaeus abelii*) and Bornean (*Pongo pygmaeus pygmaeus*) sub-species of orang-utan in captivity. Thesis Univ. East Anglia.

MARKHAM, R.J. 1991. The husbandry and management of captive orangutans; problems and opportunities. In: Proceedings on the Great Apes Conference, Jakarta and Pangkalanbun, Dec. 1991: 118-131.

MARSH, C.W. and WILSON, W.W. 1981. A survey of Primates in Peninsular Malasian Forests. Univ. Kebangsaan Malaysia and Univ. of Cambridge.

MARTIN, E.B. and VIGNE, L. 1995. Nepal's rhinos - one of the greatest conservation success stories. Pachyderm 20: 10-26.

MASLOW, A.H. 1970. Motivation and Personality. 2nd ed. Harper and Row, New York.

MATHER, R. 1992. Distribution and abundance of primates in northern Kalimantan Tengah. In: Forest Biology and Conservation in Borneo (ISMAEL, G., MOHAMED, M. and OMAR, S. eds) Sabah Foundation, Kota Kinabalu: 175-189.

MATIKAINEN, M., HERIKA, D., and MUNTOKO, E. 1998. Logging trials in compartment 17 (RKT 1997/98). Report Berau Forest Management Project. Unpubl..

MCBETH, J. 1995. Swamp for sale; Indonesia woos private sector for huge rice scheme. Far Eastern Econ. Review Sept. 7: 58-59.

MCNEELY, J.A. 1989. Protected areas and human ecology: how national parks can contribute to sustaining societies into the twenty-first century. In Conservation for the twenty-first century (WESTERN, D. and PEARL, M. eds.) Oxford UP, New York. pp. 150-157.

MEDWAY, Lord, 1976. Hunting pressure on Orang-utans in Sarawak. Oryx, 13: 332-333.

MEDWAY, Lord, 1977. Mammals of Borneo. Monograph of the Malaysian Branch of the Royal Asiatic Society No. 7.

MEDWAY, Lord, 1979. The Niah excavations and an assessment of the impact of early man on mammals in Borneo. Asian Perspectives 20 (1): 51-69.

MEIJAARD, E. and DENNIS, R. A., 1995. The status and distribution of the Bornean Orang-utan (*Pongo pygmaeus pygmaeus*) in and around the Danau Sentarum Wildlife Reserve, West-Kalimantan. MOF-Tropenbos Kalimantan Project and ODA-MOF Indonesia-UK Tropical Forest Management Programme. Bogor, unpubl.

MEIJAARD, E. 1997. The importance of swamp-forest for the conservation of orang-utans (Pongo pygmaeus pygmaeus) in Kalimantan. Indonesia. In: Proceedings of the International Symposium on the Biodiversity, Environmental Importance and Sustainability of Tropical Peat and Peatlands (PAGE, S.E. and RIELEY, J.O., eds.):

MEREDITH, M. 1993. A faunal survey of Batang Ai National Park, Sarawak, Malaysia. Wildlife Conservation International. Kuching, Sarawak, Malaysia.

MILES, H.L.W. 1993. Language and the orang-utan; the old 'person' of the forest. In: The Great Ape Project (CAVALIERI, P. and SINGER, P., eds.), New York. pp. 42-57.

MILES, H.L.W. and HARPER, S.E. 1994. Chantek; the language ability of an acculturated orangutan. In: Proceedings of the Int. Conf. on Orangutans: the neglected ape. Fullerton 1994. pp. 209-219.

MILTON, O. 1964. The orangutan and rhinoceros in North Sumatra. Oryx 7: 177-184.

MITANI, J.C. 1985. Mating behavior of male orangutans in the Kutai reserve, East Kalimantan, Indonesia. Anim. Behav., 33: 392-402.

MITANI, J.C. 1989. Orangutan activity budgets: monthly variation and the effects of body size, parturition and sociality. Am.. J. Primatol. 18: 87-100.

MoF, 1990. Act of the Republic of Indonesia No. 5 of 1990 concerning conservation of living resources and their ecosystems. Ministry of Forestry of the Repunlic of Indonesia.

MoF/FAO, 1991. Indonesian Tropical Forestry Action Plan. 3 Vols, Ministry of Forestry. Jakarta.

MoF, 1994. The strategy for Orang-utan conservation; an overview of activities and plans. Ministry of Forestry of the Republic of Indonesia. Unpublished Report pp. 47.

MOHNIKE, O. 1883. Blicke auf das Pflanzen- und Tierleben in den Niederländische Malaienländern. Münster, Aschendorfschen Buchhandlung, pp 338-374.

MURPHEY, D.D. and WILCOX, B.A., 1986. Butterfly diversity in natural habitat fragments; a test of the validity of vertebrate-based management. In Wildlife 2000; modelling habitat relationships of terrestrial vertebrates (VERNER, J., MORRISON, M.J. and RALPH, C.J., eds.). pp: 287-292.

MURPHREE, M. 1994. The role of institutions in community based conservation. In: Natural connections: perspectives in community based conservation (WESTERN, D. and WRIGHT, M. eds.), Island Press, Washington.

MYERS, N. 1980. Conversion of tropical moist forests. Washington.

NADLER, R.D. 1995. Sexual behavior of orangutans; basic and applied implications. In: The Neglected Ape (NADLER, R.D., GALDIKAS, B.F.M., SHEENAN, L.K. AND ROSEN, N. eds.), New York. pp. 223-237.

NAPIER, J.R. and NAPIER, P.H. 1967. A handbook of Living Primates. London: 269-273.

NEWBERRY, D.M., RENSHAW, W. and BRUENIG, E.F. 1986. Spatial patterns of trees in Kerangas forest, Sarawak. Vegetatio 65: 77-89.

NEWBERRY, D.M., CAMPBELL, E.J.F., LEE, C.E., RISDALE, C.E. and STILL, M,J. 1992. Primary lowland dipterocarp forest at Danum valley, Sabah, Malaysia: structure, relative abundance and family composition. Phil. Trans. Roy. Soc. London. B 335: 341-356.

N.F.I. 1996. National Forest Inventory of Indonesia; Final Forest Resources Statistics Report Annex 2: Kalimantan. DG Forest Inventory and Land use planning - MoF, GOI and FAO. Jakarta, June 1996.

NIEUWENHUIS, A.W. 1900. In Centraal Borneo; Reis van Pontianak naar Samarinda 2 Vols. Leiden.

NOSS, R.F., 1990. Indicators for monitoring biodiversity; a hierarchical approach. Conservation Biology 4: 55-364.

NUSBAUM, R., ANDERSON, J and SPENCER, T.1995. Effects of selective logging on soil characteristics and growth of planted Dipterocarp seedlings in Sabah. In: Ecology, Conservation and Management of Southeast Asian Rainforests (PRIMACK, R.B. and LOVEJOY, T.E. eds.) Yale U.P. New Haven. pp. 105-115.

OHLSSON, B. 1990. Socio-economic aspects of forestry development. Indonesia Forestry Studies VIII-3, Min. of Forestry, FAO.

OSTROM, E., WALKER, J and GARDNER, R. 1992. Covenants without a sword: self-governance is possible. Am. Pol. Sci. Review 86: 404-417.

OWEN, R. 1835. On the osteology of the Chimpanzee and Orang Utan. Trans. zool. Soc. London 1: 343.

PAGE, S.E, RIELEY, J.O., HUSSON, S. and MORROUGH-BERNARD, H. 1995. The density of orang-utan in a peat-swamp forest in central Kalimantan, Indonesia (draft for presentation to Oryx).

PAGE, S.E. and RIELEY, J.O. (eds) 1997. Biodiversity and sustainability of tropical peatlands; Proceedings of the Intern. Symp. on the biodiversity, environmental importance and sustainability of tropical peat and peatlands. Samara, Cardigan.

PALOMBIT, R.A. 1992. Pair bonds and monogamy in wild Siamangs (*Hylobates syndactylus*) and White handed gibbons (*Hylobates lar*) in northern Sumatra. PhD Thesis Univ. California Davis.

PAYNE, J. 1987. Surveying orang-utan populations by counting nests from a helicopter; a pilot survey in Sabah. Primate Conserv. 8: 92-103.

PAYNE, J. 1988. Orang-utan Conservation in Sabah. WWF Malaysia, Kuala Lumpur.

PAYNE, J. and ANDAU, M. 1989. Orang-Utan. Malaysia's Mascot. Kuala Lumpur, Malaysia.

PEART, D.R. 1996. The sustainable management of rain forest lands: an overview of research at a major tropical rainforest national park. Tropical Biodiversity 3(3): 145-155.

PERKINS, L. 1993. Orangutan GASP Report. IUCN-CBSG News 4 (3): 16-17.

PERRY, J. 1976. Orang-utans in captivity. Oryx 13 (3): 262-264.

PERSOON, G. 1994. Vluchten of Veranderen; Processen van Verandering en Ontwikkeling bij Tribale Groepen in Indonesië. Dissertatie Leiden Univ. 446 pp.

PETERS, H. 1995. Orangutan reintroduction? Development, use and evaluation of a new method; unpubl. MSc. thesis University Groningen.

PHPA, 1994. Tanjung Puting National Park Management Plan. 1994-2019. Volume II. Ministry of Forestry. Directorate General of Forest Protection and Nature Conservation. Jakarta, Indonesia.

PIAZZINI, G. 1957. Expeditie Apo-Kayan, Amsterdam.

PIERCE COLFER, C.J. 1981. Women, men and time in the forests of East Kailmantan. Borneo Research Bull. 13 (2): 75-85.

PLEUMAROM, A. 1994. The political economy of tourism. The Ecologist 24 (4): 142-148.

POVILITIS, T. 1990. Is captive breeding an appropriate strategy for endangered species conservation? Endangered Species Update 8 (1): 20-23.

PPA, 1977. Laporan Survey Orientasi Areal Cadangan. Suaka Alam / Hutan Wisata. Hutan Lindung Gunung Tunggal di Kabupaten Sanggau, 1977. Direktorat Jenderal Kehutatan.

PURA, R. 1994. Suharto lawyers ask court to reject suit over Decree. Asian Wall Street Journal, Nov. 1:1.

PURI, R. K. 1992. Mammals and hunting on the Lur River: recommendations for management of faunal resources in the Cagar Alam Kayan-Mentarang. WWF, Indonesia.Unpublished Report.

RABINOWITZ, A. 1994. Helping a species go extinct: the Sumatran rhino in Borneo. Conservation Biology 9 (3): 482-488.

RADERMACHER, M.J.C. 1780. Beschrijving van het eiland Borneo. Verhandelingen van het Bataviaans Genootschap II.

RAO, M. and VAN SCHAIK, C.P. in press. The behavioural ecology of Sumatran orangutans in logged and unlogged forest. Tropical Biodiv.

RATHERT, G. 1997. Dynamics of regenerating tropical lowland rainforest in East Kalimantan. Berau Forest Management project, paper presented at CIFOR workshop: The management of secondary forest in Indonesia, Nov. 1997.

REDFORD, K.H. 1990. The ecologically noble savage. Orion 9: 24-29.

REPETTO, R. 1988. The forest for the trees? Government policies and misuse of forest resources. World Resources Institute, Washington.

REPETTO, R. 1992. Accounting for environmental assets. Sci. Am. Jun. 1992: 64-70.

RePPProT, 1987a. Review of phase I results. West-Kalimantan. Vol. 1, main report. Land Resources Department, Foreign and Commonwealth Office, England & Dir. BINA Program, Dep. Transmigrasi, Indonesia.

RePPProT, 1987b. Review of phase IB results. Central-Kalimantan. Vol. 1, main report. Land Resources Department, Foreign and Commonwealth Office, England & Dir. BINA Program, Dep. Transmigrasi, Indonesia.

RePPProT, 1990. The land resources of Indonesia. ODA/Min. of Transmigration, Jakarta.

REYNOLDS, V. 1967. The Apes. Cassell, London..

RICHARDS, P.W. 1952. The Tropical Rainforest. Cambridge U.P. London.

RIDLEY, M. 1996. The origins of virtue. Viking London.

RIJKSEN, H.D. 1974. Orang-utan conservation and rehabilitation in Sumatra. Biol. Cons. 6 (1): 20-25.

RIJKSEN, H.D. 1975. Gunung Leuser Reserve, Sumatra - overall programme. Project 733. In World Wildlife Fund Yearbook 1974/75: 167-170. Morges.

RIJKSEN, H.D. 1978. A field study on Sumatran Orang-utans (*Pongo pygmaeus abelli* Lesson, 1827): Ecology, behaviour and conservation. H. Veenman & Zonen, Wageningen.

RIJKSEN, H.D. 1978. Hunting behaviour in Hominids; some ethological aspects. In: Recent Advances in Primatology, vol. 3. (CHIVERS, D.J. and JOYSEY, K.A. eds.) Karger, Basel.

RIJKSEN, H.D. 1981. Infant killing; a possible consequence of a disputed leader role. Behaviour 78: 138-167.

RIJKSEN, H.D. 1982. Introduction to Conservation. SECM course I; 2nd ed.. BLK publication, Bogor.

RIJKSEN, H.D. 1982. How to save the mysterious 'man of the rainforest'? In: The Orang Utan; its biology and conservation (ed. L.E.M. DE BOER), Den Haag. pp. 317-341.

RIJKSEN, H.D. 1984. Conservation: not by skill alone. The Environmentalist 4 (7): 52-60.

RIJKSEN, H.D. 1986. Conservation of orang-utans; a status report 1985. In: Primates; the road to self-sustaining populations (BERNIRSCHKE, K. ed.) New York. pp. 153-159. RIJKSEN, H.D. 1987. Bosmens zonder Toekomst, Veenman, Wageningen.

RIJKSEN, H.D. 1991. Sustainable development and the extinction of a relative; witnessing the eradication of the people of the forest. In: Proceedings of the Great Apes Conference, Jakarta and Pangkalanbun Dec. 1991: 30-40.

RIJKSEN, H.D. 1993. The neglected ape?: Nato and the imminent extinction of our close relative. In: The Neglected Ape (NADLER, R.D.*et al.*), New York. pp. 13-21.

RIJKSEN, H.D. in prep. Beyond natural religion; the value of the nature conservation concept (originally SECM course manual, Bogor 1988).

RIJKSEN, H.D. and GRIFFITHS, M. 1995. Masterplan for the Leuser Development Programme. Wageningen.

RIJKSEN, H.D. and RIJKSEN-GRAATSMA, A.G. 1975. Orang-utan rescue work in North Sumatra. Oryx 13: 63-73.

RIJKSEN, H.D. and RIJKSEN-GRAATSMA, A.G. 1978. Rehabilitation; a new approach is needed. Tigerpaper 6 (1): 16-18.

RISWAN, S. and KARTAWINATA, K. 1987. Natural regeneration in primary and secondary mixed Dipterocarp forest in East Kalimantan, Indonesia. Paper presented to the international workshop on reproductive ecology of tropical forest plants, Bangi, Malaysia.

ROBERTSON, J.M.Y. and SOETRISNO, B.R. (1982). Logging on slopes kills. Oryx, 16 (3): 229-230.

RODMAN, P.S. 1973. Population composition and adaptive organization among orangutans of the Kutai reserve. In: Comparative Ecology and Behaviour of Primates (J.H. CROOK and R.P. MICHAEL, eds.), London: 171-209.

RODMAN, P.S. 1977. Feeding behavior of orangutans in the Kutai nature reserve, East Kalimantan. In: Primate Ecology (CLUTTON-BROCK, T.H. ed.). London. p. 383-413.

RODMAN, P.S. 1988. Diversity and consistency in ecology and behavior. In: Orang Utan Biology (SCHWARTZ, J.H. ed.) Oxford. p. 31-51.

RÖHRER-ERTL, O. 1983. Erforschungsgeschichte und Namengebung beim Orang-Utan, Pongo satyrus (Linnaeus, 1758). Spixiana 6 (3): 301-322.

RUMBAUGH, D.M. and GILL, T.V. 1970. The Learning Skills of Pongo. Proc. 3rd. Congr. Int. Prim. Soc., 3, Basel: 158-163.

RUSILA, Y. N and WIDJANARTI, E. H., 1995. A preliminary survey on the ecological potential of the Cagar Alam Muara Kendawangan. West Kalimantan. Unpubl. Report to PHPA/AWB.

RUSSELL, C.L.1995. The social construction of orangutans: an ecotourist experience. Society and Animals 3 (2): 151-169.

RUSSON, A.E., ERMAN, A. and SINAGA, P. 1996. Nest count survey of orangutans in and around the Danau Sentarum wetlands reserve, West Kalimantan, Indonesia. Unpubl. report to PHPA/ODA/WI.

SAJUTHI, D., KARESH, W., MCMANAMON, R., MARTIN, H., AMSEL, S. and KUSBA, J. 1991.

Recommendations to the Department of Forestry of the Republic of Indonesia on the medical quarantine of orang-utans intended for reintroduction. In Proceedings of the Great Apes Conference Jakarta and Pangkalanbun, Dec. 1991: 132-136.

SARICH, V.M. and WILSON, A.C. 1967. Immunological time scale for hominid evolution. Science 158: 1200-1203.

SARIJANTO, T. 1996. Policy governing the issuance of concession rights. Ministry of Forestry, Jakarta.

SAVAGE-RUMBAUGH, S. and LEWIN, R. 1994. Kanzi; the ape at the brink of the human mind. Wiley, New York.

SCHAERER, H. 1963. Ngadju religion; the conception of god among a south Borneo people. The Hague.

SCHAIK, C.P. VAN 1986. Phenological changes in a Sumatran rain forest. J. Trop. Ecol. 2: 327-347.

SCHAIK, C.P. VAN and AZWAR. 1991. Orangutan densities in different forest types in the Gunung Leuser National Park (Sumatra), as determined by nest counts. Unpubl. report to PHPA/LDP.

SCHAIK, C.P. VAN, AZWAR, and PRIATNA, D. 1995. Population estimates and habitat preferences of orangutans based on line transects of nests. In: The neglected ape (NADLER, R.D., GALDIKAS, B.F.M., SHEERAN, L.K. and ROSEN, N. eds.) New York. p. 129-147.

SCHAICK, C.P. VAN and MIRMANTO, E. 1985. Spatial variation in the structure and litterfall of a Sumatran rainforest. Biotropica 17: 196-205.

SCHAICK, C.P. VAN, PONIRAN, S., UTAMI, S., GRIFFITHS, M., DJOJOSUDHARMO, S., SETIA, T. M., SUGARDJITO, J., RIJKSEN, H.D., SEAL, U.S., FAUST., TRAYLOR-HOLZER, K and TILSON, R. 1995. Estimates of orangutan distribution and status in Sumatra. In: The neglected ape (NADLER, R.D., GALDIKAS, B.F.M., SHEERAN, L.K. and ROSEN, N. eds.) New York. p. 109-116.

SCHAICK, C.P VAN, FOX, E.A., and SITOMPUL, A. 1996. Manufacture and use of tools in wild Sumatran orangutans; implications for human evolution. Naturwissenschaften 83: 186-188.

SCHAIK, C.P. VAN, TERBORGH, J. and DUGELBY, B. 1997. The silent crisis: the state of rain forest nature preserves. In Last Stand; Protected areas and the defence of tropical biodiversity (KRAMER, R., SCHAIK, C.P. VAN, and JOHNSON, J.) Oxford UP, New York.

SCHALLER, G.B. 1961. The Orang-utan in Sarawak. Zoologica, 46: 73-82.

SCHEBESTA, P. 1928. Among the forest dwarfs of Malaya. London.

SCHENKEL, R. and SCHENKEL-HUELLIGER, L. 1969. Report on a survey trip to Riau area and the Mount Leuser reserve to check the situation of the Sumatran rhino and the orang-utan. Unpubl. Report PPA/WWF. Jakarta.

SCHLEGEL, H. and MUELLER, S. 1839-1844. Bijdragen tot de Natuurlijke Historie van de Orang-oetan (*Simia satyrus*). In: Verhandelingen over de Natuurlijke geschiedenis der Nederlandsche Overzeesche Bezittingen, door de Leden der Natuurkundige Commissie in Indië en andere Schrijvers. C.J. TEMMINCK. Zoologie, 2: 1-28. Leiden.

SCHOLZ, U. 1983. The natural regions of Sumatra, and their agricultural production pattern. CRIFC Bogor.

SCHÜRMANN, C.L. 1982. Mating behaviour of wild orang-utans. In: The Orang Utan (DE BOER, L.E.M. ed.). The Hague. p. 269-284.

SCHWARTZ, J.H. 1987. The Red Ape; Orang-utans and Human Origins. Houghton Mifflin, Boston.

SCHWARTZ, J.H., NGUYEN LAN CUONG, VU THE LONG, LE TRUNG KHA and TATTERSALL, I., 1994. A diverse hominoid fauna from the late middle pleistocene breccia cave of Tham Khuyen, Socialist Republic of Vietnam. Anthr.P. Am. Mus. of Nat. Hist. No. 73: 2 -11.

SELENKA, E. 1896. Die Rassen und der Zahnwechsel des Orang-utan. Math. Nat. Mitt. Akad. Wiss. Berlin 1896: 131-142.

SEUANEZ, H., EVANS, H.J., MARTIN,D.E., and FLETCHER, J. 1979. An inversion in chromosome 2 that distinguishes between Bornean and Sumatran orangutans. Cynogenet. Cell Genet. 23: 137-140.

SHAFFER, M. L., 1981. Minimum population sizes for species conservation. BioScience 31 (2): 131-134.

SHOLLEY, C.R. 1992. Conservation status of the Mountain Gorilla. In: Proceedings of the Great Apes Conference (Ministry of Forestry, Indonesia): 83-85.

SILVA, G.S. DE 1966. The east-coast experiment; an attempt to rehabilitate the orang-utan. Sabah Soc. J. 3 (2): 85-89.

SILVIUS, M. J., A. P. J. M. STEEMAN, E. T. BERCZY, E. DJUHARSA and A. W. TAUFIK. 1987. The Indonesian Wetland Inventory. A preliminary compilation of existing information on wetlands of Indonesia. PHPA - AWB/INTERWADER, EDWIN, Bogor, Indonesia.

SIMONS, H. 1987. Gunung Niut Nature Reserve. Proposed Management Plan. WWF Report.

SIMON, N. 1966. Red Data Book Vol 1: Mammalia. IUCN Morges.

SINT JOHN, S. 1862. Life in the forests of the Far-East. Vol 2. London.

SKORUPA, J.P. 1986. Responses of rainforest primates to selective logging in Kibale forest, Uganda: a summary report. In: Primates: the road to self-sustaining populations (BERNIRSCHKE, K. ed.) New York. pp. 57-70.

SKORUPA, J.P. 1987. Do line transect surveys systematically underestimate primate densities in logged forests? Am. J. Primatol. 13: 1-9.

SKORUPA, J.P. and KASENENE, J.M. 1983. Tropical forest management: can rates of natural treefall help guide us? Oryx 18: 96-101.

SMITH, R.J. and PILBEAM, D.R. 1980. Evolution of the orang-utan. Nature 284: 447-448.

SMITS, W.T.M. 1994. Dipterocarpaceae: Mycorrhizae and Regeneration. Tropenbos Series 9, Wageningen.

SMITS, W.T.M., LEPPE, D. YASMAN, I, and NOOR, M., 1992. Ecological approaches to commercial Dipterocarp forestry. In: Forest biology and conservation in Borneo (ISMAEL, G., MOHAMED, M. and OMAR, S. eds.). Yayasan Sabah, Kota Kinabalu. pp. 432-435.

SMITS, W.T.M., SUSILO, A. and HERIYANTO. 1993. Year report 1992; Orang-utan reintroduction at the Wanariset 1 station, Samboja, East Kalimantan.

SMITS, W.T.M., HERIANTO, and RAMONO, W.S. 1995. A new method for rehabilitation of Orangutans in Indonesia: A first overview. In: The Neglected Ape (NADLER, R.D., GALDIKAS, B.F.M., SHEERAN, L.K. and ROSEN, N. eds.) New York. p. 69-77.

SMIT SIBINGA, G.L. 1953. On the origin of the drainage system of Borneo. Geologie en Mijnbouw 15: 121-136.

SOULÉ, M.E. 1980. Thresholds for survival: maintaining fitness and evolutionary potential. In: Conservation Biology; an evolutionary-ecological perspective (SOULÉ, M.E. and WILCOX, B.E. eds.) Sinauer, Mass. pp. 151-169.

SOULÉ, M.E. 1986. The effects of fragmentation; the dimensions of the external threat. In: Conservation Biology; the science of scarcity and diversity (SOULÉ, M.E. ed.) Sinauer, Mass. pp. 233-237.

SOUTHGATE, D., COLES-RITCHIE, M. & SALAZAR-CANELOS, P. 1996. Can Tropical Forests be Saved by Harvesting Non-Timber Products?: a case study for Ecuador. In: Forestry, Economics and the Environment (eds. ADAMOWICZ, W.L. et al.) CAB Int. Wallingford UK.

SOEMARNA, K., RAMONO, W.S. and TILSON, R. 1995. Introduction to the orangutan population and habitat viability analysis (PHVA) workshop. In: The Neglected Ape (NADLER, R.D. et al. eds.) New York. pp. 81-83.

SOERIANEGARA, I. and LEMMENS, R.H.M.J. 1993. Plant resources of South-East Asia: Timber Trees; major commercial timbers. Pudoc, Wageningen. p: 454-457.

SPENCER, F. 1995. Pithekos to Pithecanthropus: an Abbreviated Review of Changing Scientific Views on the Relationship of the Anthropoid Apes to Homo. In: Ape, Man, Apeman; Changing Views since 1600 (CORBEY, R. and THEUNISSEN, B. eds.) Leiden Univ. pp. 13-27.

STEARMAN, A.M. 1994. Only slaves climb trees; revisiting the myth of the cologically noble savage in Amazonia. Human Nature 5: 339-357.

STEVENS, W. 1969. Report to the Government of Malaysia on Game Conservation. pp. 67. (Unpubl.)

STEVENSON, M, FOOSE, T.J. and BAKER, A. 1991. Global captive action for primates (Discussion edition) 2 vols. CBSG/SSC.

STORM, P. 1996. The evolutionary significance of the Wajak skulls. PhD thesis VU Amsterdam.

STORK, N.E. 1995. Inventorying and monitoring of biodiversity. In: Global Biodiversity Assessment (HEYWOOD, V.H., ed.) UNEP, Cambridge U.P. pp. 453-544.

STOTT, K. and SELSOR, C. J., 1961. The Orang-utan in north Borneo. Oryx 6: 39-42.

STRUHSAKER, T.T. 1976. A further decline in numbers of Amboseli vervet monkeys. Biotropica 8: 211-214.

SUGARDJITO, J. and NURHADA, N. 1981. Meat-eating behaviour in wild orang-utans. Primates 22 (3): 414-416.

SUGARDJITO, J. 1986. Ecological Constraints on the Behaviour of Sumatran Orang-utans in the Gunung Leuser National park, Indonesia. Thesis Utrecht.

SUGARDJITO, J. and VAN SCHAICK, C. P., 1992. Orangutans: Current population status, threats and conservation measures. In: Proceedings of the Great Apes Conference (Jakarta, Pangkalanbun) 1991. Jakarta.pp. 142-152.

SUGIYAMA, Y. 1968. Social organization of chimpanzees in the Budongo forest, Uganda. Primates 9: 197-225.

SUMARDJA, E.A. 1992. Conservation management in Kalimantan. In: Forest Biology and Conservation in Borneo (ISMAEL, G., MOHAMED, M. and OMAR, S. eds) Sabah Foundation, Kota Kinabalu: pp. 404-407.

SUMARDJA, E.A.1997. Biodiversity as development capital in Indonesia. Paper presented at panel discussion on biodiversity. Jakarta, Indonesia March 1997. 17pp.

SURYOHADIKUSUMO, D. 1992. Opportunities and constraints of sustainable forest management in Indonesia. In: One century of sustainable forest management in with special reference to teak in Java (SIMON, H., FATTAH, A., SUMARDI, DIPODININGRAT, S., and ISWANTORO, H.). Proc. Int. Symp. on Sustainable Forest Management, Yogyakarta. pp: 55-66.

SUTLIVE, V.H. 1978. The Iban of Sarawak. Illinois.

SUZUKI, A. 1989. Socio-ecological studies of orang-utans and primates in Kutai National park, East Kalimantan in 1988-89. Overseas Res. Rep. of Studies on Asian Non-Human Primates 7: 1-42.

SUZUKI, A. 1992. The population of orangutans and other non-human primates and the forest conditions after the 1982-83's fires and droughts in Kutai National park, East Kalimantan, Indonesia. In: Forest Biology and Conservation in Borneo (ISMAIL, G., MOHAMED, M. and OMAR. S., eds.) Kota Kinabalu: 190-205.

TE BOEKHORST, I.J.A., SCHÜRMANN, C.L. and SUGARDJITO, J. 1990. Residential status and seasonal movements of wild orang-utans in the Gunung Leuser Reserve (Sumatera, Indonesia). Anim. Behav., 39: 1098-1109.

TERBORGH, J. 1975. Faunal equilibria and the design of wildlife preserves. In: Tropical Ecological Systems: Trends in terrestrial and aquatic research (GOLLEY, F. and MEDINA, E., eds.) Springer, NY. Pp. 369-380.

TERBORGH, J. 1986. Keystone plant resources in the tropical forest. In: Conservation Biology; the science of scarcity and diversity (SOULE, M. ed.) Sinauer, Sunderland. pp. 330-344.

THOMAS, K. 1983. Man and the Natural World; a history of the modern sensibility. London.

TILLEMA, H.F. 1990. A journey among the Peoples if Central Borneo in Word and Picture. Oxford UP. Singapore.

TULP, N. 1641. Een Indiaansche Satyr. In: Geneeskundige Waarnemingen 3e boek. Amsterdam pp: 370-379.

TURNER, J.M. 1996. Species loss in fragments of tropical rain forest: a review of the evidence. J. of Applied Ecol. 33: 200-209.

UHL, C., JORDAN, C., CLARK, K., and HERRERA, R. 1982. Ecosystem recovery in Amazon caatinga forest after cutting, cutting and burning, and bulldozer clearing treatments. Oikos 38: 313-320.

UTAMI, S. and MITRA SETIA, T. 1995. Behavioral changes in wild male and female orangutans (*Pongo pygmaues abelii*) during and following a resident male take-over. In: The Neglected Ape, (NADLER, R.D., GALDIKAS, B.F.M., SHEENAN, L.K. AND ROSEN, N. eds.), Plenum, N.Y. pp. 183-190.

VALENTIJN, F. 1726, Beschryvinge van het eyland Borneo, en onze handel aldaar. Oud en Nieuw Oost-Indiën, vervattende een naaukeurige en uitvoerige verhandelinge van de Nederlandsche mogendheyd in die gewesten 3 (2): 236-252.

VALKENBURG, J.L.C.H. VAN, 1996. Non-timber forest products of East Kalimantan; potentials for sustainable forest use. Tropenbos Series 16. Wageningen.

VAYDA, A.P., PIERCE COLFER, C.J. and BROTOKUSUMA, M. 1980. Interactions between people and forest in East Kalimantan. Impact of Science on Society 30 (3): 179-190. UNESCO.

VEDDER, A. and WEBER, W. 1990. Mountain Gorilla Project, Rwanda. In: Living with Wildlife; wildlife resource management with local participation in Africa. (KISS, A. ed.). World Bank, pp. 83-90.

VERSTAPPEN, H.T. (1975) On paleo climates and landform development in Malesia. In: Modern Quarternary Research in Southeast Asia. (G.-J. BARTSTRA AND W.A. CASPARIE eds.) Rotterdam, pp. 3-35.

VINES, G. 1993. Planet of the free apes? New Scientist June 5th 138 (1876): 39-42.

VOOGD DE, C.N.A and RENGERS HORA SICCAMA, G.F.H.W. 1936-38. Onderwerpen van lokalen aard, residentiegewijs gerangschikt. In: Drie Jaren Indisch Natuur Leven (Ned. Ind. Vereeniging tot Natuurbescherming), Batavia. pp. 113-115.

VRIES, O. DE, VAN WATERSCHOOT VAN DER GRACHT, W.A.J.M., KLEIN, W.C., VAN HEURN, F.C., BIJLMER, H.J.T. and ESHUIS, W. 1939. Report concerning the possibilities of protecting the primitive natives, especially the mountain Papuan tribes in Dutch New Guinea. Special report Neth. Committee for International Nature Protection.

WAKKER, E. 1998. Introducing zero-burning techniques in Indonesia's oil palm plantations; report prepared for WWF-Indonesia programme. AIDEnvironment, Amsterdam.

WALLACE, A.R. 1856. On the habits of the Orang utan of Borneo. Ann. Mag. Nat. Hist., 18: 26-32.

WALLACE, A.R. 1869. The Malay archipelago, new. ed. London: Macmillan.

WAVELL, S. 1996. Tiger faces extinction as WWF aid misfires. The Sunday Times 20.10.1996.

WCMC, 1993. The WCMC biodiversity map library: Availability and distribution of GIS data sets. World Conservation Monitoring Centre, Cambridge.

WCMC, 1995. Threatened species data summary. World Conservation Monitoring Centre. Cambridge.

WELLS, M. 1997. Indonesia ICDP Study; interin report of findings. World Bank, Jakarta pp. 16 (unpubl.).

WESTENENK, L.C. 1918. Orang pandak (boschmenschen) op Soematra. De Tropische Natuur VIII (7): 108-110.

WESTENENK, L.C. 1962. Waar mens en tijger buren zijn. Leopold, Den Haag. pp. 254.

WESTERLUND, C.G. 1889. Wer ist der Verfasser der '*Dissertatio academica nova testaceorum genera sistens*' Lundae 1788? - Nachrichtsblatt der Deutschen Malakozoologischen Gesellschaft 1889 (1/2): 21-23

WESTERMANN, J.H . 1936-1938. Natuur in Zuid- en Oost Borneo. In: Drie Jaren Indisch Natuurleven, 11e jaarverslag (1936-1938). Ned. Ind. Ver. tot Natuurbescherming,

WHITEHEAD, J. 1893. Exploration of Mt. Kina Balu, North Borneo. London: Gurney & Jackson.

WHITESIDES, G.H., OATES, J.F., GREEN, S.M., and KLUBERDANZ, R.P. (1988). Estimating primate densities from transects in a West African rain forest. A comparison of techniques. J. of Anim. Ecol. 57: 345-367.

WHITMORE, T.C. 1982. Tropical rainforests of the Far East. Oxford.

WHITMORE, T.C. 1995. Comparing Southeast Asian and other tropical rainforests. In: Ecology, Conservation and Management of Southeast Asian Rainforests (PRIMACK, R.B., and LOVEJOY, T.E. eds.) New Haven. p. 5-15.

WHITTEN, A.J., DAMANIK, S.J., ANWAR, J. and HISYAM, N. 1984. The ecology of Sumatra. GAMA Bandung.

WHITTEN, A.J., HAERUMAN, H., ALIKODRA, H.S. and THOHARI, M. 1987. Transmigration and the Environment in Indonesia; the past, present and future. IUCN.

WIENS, J.A. 1977. On competition and variable environments. Amer. Sci. 65: 590-593.

WILSON, C.C.. and WILSON, W.L., 1973. Census of Sumatran primates. Final report to LIPI, Jakarta (unpubl.).

WILSON, C. C. and WILSON, W. L.., 1975. The influence of selective logging on primates and some other animals in East-Kalimantan. Folia Primat. 23: 245-275.

WILSON, A.C., CANN, R.L., CARR, S.M., GEORGE, M., GYLENSTEN, U., HELM-BYCHOWSKI, K.M., HIGUCHI, R.G., PALUMBI, S.R., PRAGER, E.M., SAGE, R.D., and STONEKING, M. 1985. Mitochondrial DNA and two perspectives on evolutionary genetics. Biol. Journ. of the Linnean Society 26: 375-400.

WITKAMP, H. 1932. Het voorkomen van enige diersoorten in het landschap Koetai. De Tropische Natuur 21: 169-175.

WITKAMP, H. 1932. Orang Utan. Korte Mededelingen, De Tropische Natuur, 21: 210.

WOODS, P.V. 1992. Effects of fire in logged and primary forests of Sabah - tree mortality and post fire succession. In: Forest Biology and Conservation in Borneo (ISMAIL, G., MOHAMED, M. and OMAR, S. eds.) Yayasan Sabah, Kota Kinabalu: 505-506.

WORKING GROUP Orangutan Life History and Vortex analysis. 1993. In CBSG Orangutan Population and habitat viability analysis report, Medan: 31-42. Unpubl.

WORLD BANK, 1988. Indonesia; adjustment, growth and sustainable development, WB document 7222-IND, Jakarta.

WORLD BANK, 1988. Forests, land and water; issues in sustainable development, WB - Indonesia. Jakarta (unpubl.).

WORLDBANK, 1997. Investing in Biodiversity; a review of Indonesia's Integrated Conservation and Development Projects, WB, Jakarta (draft, unpubl.).

WRANGHAM, R. W. 1975. The behavioral ecology of chimpanzees in Gombe National park, Tanzania. Ph.D. dissertation Univ. of Cambridge.

WRANGHAM, R.W. 1979. On the evolution of ape social systems. Soc. Sci. Inform. 18: 335-368.

YANUAR, A., SALEH, C , SUGARDJITO, J. and WEDANA, I. M., 1995. Density and abundance of primates with special focus on West, Central and East Kalimantan's rain forests. (in prep.)

YASUMA, S., 1994. An invitation to the mammals of East-Kalimantan. PUSREHUT Samarinda.

YEAGER, C., 1997. Orangutan rehabilitation in Tanjung Puting National Park, Indonesia. Conservation Biology 11 (3): 802-805.

YOSHIBA, K. 1964. Report of the preliminary survey on the Orang utan in North Borneo. Primates 5: 11-26.

ZONDAG, J.L.P. 1931. Het voorkomen van enige diersoorten in de Zuider en Ooster Afdeling van Borneo. De Tropische Natuur, 20: 221-223.

ZUCKERMAN, S. 1932. The social life of monkeys and apes. London.

INDEX

adat (customary tribal laws) 113, 115, 223, 318, 321, 323, 326, 341, 385
alluvial forest 38, 69, 76, 126, 256, 348, 372
altitude 40, 43-48, 54, 68, 76, 136, 179, 183, 187-189, 198, 199, 214, 219, 221, 225, 226, 229, 230, 239-242, 247-251, 257-261, 267, 268, 284, 293, 300-302, 355, 356
altitudinal limit 68, 186, 241
AMDAL (environmental impact assessment report) 181, 285, 324
Angkola 23, 35, 55, 58, 59, 110, 249, 258-262, 272, 278, 303, 312, 399, 425, 426, 440
Antiaris 112
Apo Kayan 45, 50, 225, 226, 311, 313, 419
arena 88-90, 132
arson 26, 101-105, 195, 197, 200, 202, 205-229, 272-277, 283, 284, 299-301, 326-328, 341, 353, 355, 403-405, 437
awareness 79, 134, 138, 142, 149, 159-165, 171, 251, 253, 276, 332, 334, 344, 363, 374, 375, 379, 380, 385, 394, 396, 398, 401, 403, 411, 412

Bangka-Belitung-Karimata 32, 34, 38-40
Bappedal 273
bearded pig 73, 112, 193
Bendang 66
Bentuang Karimun 198, 300, 301, 304, 405
biodiversity 78, 79, 99, 108, 132, 144, 198, 276, 292, 296, 315, 320, 346, 359, 368, 372, 385, 391, 392
birds 27, 88, 89, 137, 201, 252, 330, 344, 354, 387
Bohorok 93, 146, 155, 156, 161, 164, 166-168, 171, 255, 412-415
Bombax 66, 70
Borassodendron 66, 184
Bukit Baka 136, 199, 200, 213, 219, 221, 269, 274, 277, 278, 300, 301, 304, 305, 357, 406, 437
Bukit Lawang 166, 168, 256, 413, 415
Bukit Raya 133, 134, 136, 199, 200, 212, 219, 220, 221, 277, 278, 300, 301, 304, 305, 406

calling 144, 149
Captive Breeding Specialist Group. See CBSG
captivity 28, 61, 86, 91, 121, 122, 152, 283, 425
carrying capacity units 268, 269, 272, 273, 276, 283, 285, 300, 301
Castanopsis 54, 68
CBSG 125, 146, 201, 229
certification 360, 364, 365, 381, 408-410
chimpanzee 29, 30, 32, 33, 67, 71, 82-84, 86, 97, 124, 158, 403, 421, 424, 425
chromosome 30, 33
CITES 119, 121, 122, 143
coal mining 182, 355
colonial government 27, 133, 134, 137
commuters 80, 82, 84, 90, 91, 106, 107, 297
competition 70, 73, 76, 77, 82, 84-91, 109, 156, 173, 297, 413
confiscation 152, 156, 161, 172, 174, 191, 230, 324, 342, 343, 352, 401, 403
conservation concession 136, 370, 391
constitution 321, 368, 379
conversion forest 215, 227, 284, 291, 292, 323, 324, 326, 393

Danau Sentarum 135, 136, 196, 197, 277, 300-304, 347, 355, 399, 405
Danum Valley 93, 136, 179, 239, 277, 301, 356, 406
Dayak 112-114, 120, 126, 192, 193, 219, 220, 223, 232, 309, 311
deme 34, 72-80, 83, 86, 88-92, 106, 107, 126, 157, 187, 243, 258, 282, 297, 413
desertification 299
development 140-144, 193-198, 204, 205, 207, 211, 215-218, 220-224, 240-248, 253-255, 284-289, 317-319, 322-329, 332-338, 341-349, 372-379, 384, 386, 389, 390, 393, 399, 400, 404-407, 410-413
diet 60, 65, 66, 67, 71, 72, 76, 78, 100, 106, 126
Dipterocarp forest 93, 96, 198, 238, 239, 241-243
DNA 30, 32, 33, 46
Durio 60, 71, 101, 184

eco-tourism 166, 176
Eusideroxylon 361, 386

FAO 131, 132, 141, 182, 183, 233, 234, 237, 287, 289, 295, 323, 336
Ficus 54, 66, 68, 71, 72, 74, 184
fire 27, 77, 102-105, 150, 173, 189-191, 206, 224-229, 272-277, 283-286, 290, 297, 299, 303, 310, 313, 338, 355, 356, 362, 377
fire-arms 112, 312
flooding 74
flowering 71, 74
flying foxes 73, 77
Food and Agricultural Organisation See FAO
forest types 43, 97, 182, 224, 239, 269, 283, 303
fossil finds 41, 42
fragmentation 26, 96, 101, 182, 185, 258, 262, 265, 268, 274, 296, 297-299, 303, 315, 329, 353, 381, 403
freshwater swamp forest 201, 217, 242
fruit 53, 60, 65-79, 83, 84, 87-90, 93, 95, 98-101, 109, 112, 120, 123, 126, 147, 175, 184, 194, 201, 227, 297, 322, 330, 347, 365, 386, 397
fruit productivity 69, 73, 74, 101

gaharu/geharu 188, 198, 315, 330
genetic diversity 34
gibbon 27, 47, 66, 73, 79, 100, 246, 382, 441
GIS 182, 189, 221, 225, 267, 268, 276, 383, 384
Gonystylus 193
gorilla 29, 30, 32, 82, 86, 124, 158, 412, 413, 427
group formation 83
Gunung Niut 136, 194, 195, 196, 269, 274, 300-302, 305, 311, 355, 437
Gunung Palung 93, 133, 136, 191, 201, 202, 277, 300, 301, 347

habitat 25-28, 34, 39, 42-48, 65-81, 86-102, 129-132, 135-141, 146-149, 152, 153, 156, 157, 167-170, 173-175, 183-188, 194-196, 199-201, 211-220, 225-234, 242-244, 249, 250
habitat patchiness 72, 95, 186, 297
hair 33, 34, 60, 61, 84, 110, 113, 179, 403, 424, 425
head-hunting 31, 109, 113, 115, 192, 309, 314
Heritiera 70, 71
hill forest 51, 56, 96, 100, 195, 196, 199, 220, 259, 262, 277
Holocene 32, 42, 109, 285, 286
home range 73, 78, 80, 86, 87, 89, 151, 173
hornbill 73, 75, 77, 79, 188
human population growth 316, 319, 320, 377
hunting 29, 40, 45-47, 52, 66, 78, 101, 106, 109, 111-117, 122-125, 130, 134, 143, 144, 146, 154, 170, 175, 185, 193, 201, 229, 231, 232, 244, 297, 309, 311, 312, 329, 333, 334, 342, 353, 396
Hutan Negara 137, 322, 385
hybrids 122, 403

Iban 48, 70, 110, 113, 114, 115, 192, 196, 198, 223, 225, 234, 235, 301, 311, 312, 419, 420
ICDP 378
IFAW 149, 150, 339
Imperata 103
inbreeding 122, 297, 299
indigenous people 30, 101, 116, 325, 348
institutional deficiencies 347, 348, 353, 374, 375
International Studbook 121
ITTO 198, 200, 220, 324, 346, 399
IUCN 118, 125, 130-132, 139, 144, 146, 147, 149, 172, 201, 229, 233, 268, 285, 320, 335, 337, 395, 401, 409, 433
IWF 131, 144, 156, 181, 392

jaga wana 324, 385

karyotype 30, 33
kawasan plasma nutfah 364
Kayan 43-45, 48, 51, 109, 135, 146, 175, 188, 223-225, 229-232, 300, 311, 347, 354, 419
Kenyah 45, 48, 112, 126, 188, 223, 225, 231, 234, 311, 419

Ketambe 7, 14, 65, 66, 71, 74, 75, 80, 81, 88, 93, 140, 155-157, 161, 171, 179, 256, 257, 372, 413, 441
Kubu 109, 110, 264
Kutai national park 142, 156, 191, 224, 228, 282, 300, 362

ladang 104, 140, 189, 225, 226, 244, 257, 273, 293, 325, 329, 365
laissez-faire 132, 289, 292, 296, 305, 320, 325, 332, 335, 337, 345, 358
Lake Toba 34, 56, 57, 59, 109, 140, 247, 248, 262, 285, 407, 426
Lanjak-Entimau 277, 303, 357
law-enforcement 169, 172, 342
lek 88, 89, 90
Leuser Ecosystem 67, 68, 81, 136-138, 250, 251, 252, 253, 254, 255, 256, 258, 262, 265, 277, 278, 299, 300, 303, 304, 309, 354, 389, 407, 408
Licuala 184
Limited Production Forest 99, 206, 290, 323, 359
Lithocarpus 54, 68, 72
logging 78, 96-108, 120, 167, 180, 181, 184-187, 193-196, 199-202, 205, 211-214, 217, 218, 224-234, 238-241, 255-261, 265, 267, 272, 294-298, 311, 324, 326, 328, 341, 348, 355, 362-366, 372, 374, 380, 381, 398
Lun Dayeh 223, 419

Macaranga 103
Madhuca 66
Mamaq 109, 110, 264
Massenerhebung 68, 95, 251
masyarakat 381, 385, 408
mineral licks 67
Ministry of Environment (MoE) 273
Ministry of Forestry (MoF) 144, 149, 168, 171, 172, 183, 276, 295, 345
montane forest 93, 189, 199, 224, 239, 242, 284
mycorrhizae 69, 471

National Forest Inventory see NFI
Neesia 67, 72, 184
nests 58, 92, 184, 185, 188, 199, 214, 215, 217, 226, 235, 238, 249, 261, 264, 265
NFI 182, 183, 189-293

nutrient 43, 48, 69, 73, 74, 100, 102, 108, 204, 251

OFI 166, 167, 171, 412
one-million-hectare-project see PLG
orang-pendek 59, 60, 61, 62, 63, 64, 264, 278
orang-utan numbers 25, 54, 186
Orang Utan Recovery Service see OURS
Orang-utan Survival Programme see OUSP
Orangutan Foundation International see OFI
Ot Danum 134, 208, 219, 311, 419
OURS 130
OUSP 25, 26, 149, 170, 367, 395, 398, 400, 404

participation 132, 175, 197, 201, 228, 296, 333-337, 341, 379, 386, 390, 397, 408, 474
participatory management 146, 316, 332, 333, 340
peat 25, 26, 50, 52, 54, 68, 72, 90-94, 97, 99, 103, 109, 177, 187, 189, 191, 194, 196-198, 201-210, 213-218, 224, 226, 229, 232, 233, 241, 252, 262, 276, 277, 283, 284, 301, 377, 395, 399, 405
Penan 48, 54, 112, 126, 130, 188, 234, 246, 311, 341
PHPA 29, 131, 135, 143, 144, 155, 156, 164, 167, 168, 171, 172, 181, 183, 193-198, 200-202, 211, 217, 218, 224-227, 230, 251, 255, 257-259, 299, 343, 345, 346, 364, 367, 389, 393, 400, 401, 404
physiognomy of orang-utans 33, 35, 37, 61, 86, 90
pig-tailed macaques 126
Pleistocene 32, 34, 37-42, 47, 188
PLG 98, 191, 205, 207, 216-218, 284, 405
poaching 54, 101, 109, 114, 118, 119, 122-125, 132, 135, 143, 148, 152, 164-169, 172, 175, 181, 194-196, 199-202, 205, 210, 211, 214-217, 219, 226-231, 235, 250, 252, 253, 257, 258, 267, 272, 277, 292, 296, 299, 309-315, 319, 324, 327-331, 335, 340-343, 346, 352-355, 357, 359, 364, 366, 372, 375, 379, 384, 385, 388, 401-403, 410, 412

479

Pothos 184
Production Forest 99, 135, 173, 206, 220, 276, 277, 284, 290, 292, 295, 323, 324, 336, 355, 359-361, 380-383, 389, 397, 406, 409
Protection Forest 48, 56, 58, 59, 133-135, 172, 198, 200, 206, 221, 239, 240, 249, 257-261, 265, 272, 277, 284, 289, 290, 292, 323, 324, 344, 355, 383, 405, 406
Punan 45, 54, 112, 126, 188, 198, 223, 225, 231, 232, 311, 354, 419

Ramin 193
rattan/rotan 66, 76, 184, 188, 243, 315, 319, 330, 344, 411
re-introduction 343, 354, 398, 402
reforestation fund 344, 392, 393
regeneration 97, 99, 102, 103, 108, 187, 201, 206, 220, 222, 228, 229, 234, 276, 287, 290, 296, 345, 346, 356, 360, 362, 363, 365, 366, 381, 393, 399, 408, 409
rehabilitation 91, 121, 122, 140, 146-157, 160-176
research 28, 32, 34, 48, 67, 69, 77, 81, 90, 91, 97, 107, 112, 121, 131, 137, 139, 144, 149, 155, 161, 166, 172, 179-181, 200, 201, 211, 214, 220, 222, 256, 288, 326, 363, 371-374, 379, 382, 383, 396, 398, 400, 411, 413, 441
reserves 55, 112, 130, 132, 133, 135, 136, 147, 148, 198, 200, 205, 207, 237-243, 276, 300-305, 321, 335, 354-357, 363, 364, 371, 382, 388, 389, 395, 397, 399, 404, 405, 407, 409
residents 73, 75-77, 80, 82, 84, 87-89, 106, 107, 173, 219, 322
Rhaphidophora 184
rhinoceros 112, 140, 453, 462
rubber 60, 140, 262, 273, 287, 319, 344, 393

satellite imagery 45, 180, 182, 183, 189, 195, 198, 214, 217, 218, 220, 221, 225, 233, 238-242, 254, 262, 268, 273, 287, 291, 294, 300-302, 361
Scindapsus 184
Sebuku-Sembakung 277, 406
selective logging 96-102, 108, 214, 259, 293, 353, 365, 366, 404, 405, 410
Semengok 155, 313
Sepilok 155, 164, 166, 167, 170, 239, 243, 313
shifting cultivators 48, 112, 125, 126, 192, 325, 327
skull 112, 113, 119, 120, 143, 144, 195, 199, 311, 313, 314, 331, 342, 426
slash-and-burn agriculture 48, 205, 210, 234, 240
smuggling 119, 165, 313
social arena 88-91, 201, 253-257, 299
social status 76, 77, 80, 81, 83, 86, 87, 153
soil fertility 74, 76, 94, 207
Southern Forest Reserves 239, 277, 406
spatial patchiness 91
squirrel 66, 67, 184
SSC Primate Specialist Group 144
State Forest land 103, 136, 137, 182, 189, 223, 231, 248, 251, 276, 284, 289, 290, 291, 296, 316, 317, 322-328, 331, 335, 341, 342, 345, 347, 354, 356, 358, 379, 380, 384-386, 389, 391, 393, 396, 406, 408, 411
Strychnos 112, 184
sub-species 461
Sundaland 32-35, 37, 39, 41
sustainability 175, 292, 296, 318, 320, 328, 336, 346, 359, 360, 362, 363, 368, 377, 378, 381, 384, 385, 388, 409, 410
sustainable development 25, 132, 183, 254, 316, 331, 333, 336, 350, 359, 376, 377, 390
sustainable use 197, 335, 340, 345, 368, 408

tanah kosong 322, 325, 326, 363, 379
Tanjung Puting National Park 191, 211, 212, 412
Tanjung Puting rehabilitation centre 162
Tata guna hutan kesepakatan see TGHK

taxonomy 32, 443
terrapins 330
Tetramerista 71, 72, 74, 97, 365
TGHK 183, 290-293, 323, 324, 359, 383
timber concessions 62, 99, 123, 180, 199, 200, 216, 220, 249, 253, 254, 259, 268, 273, 293, 294, 303, 309, 312, 327, 330, 396, 406
trade 29, 47, 60, 114, 118, 119, 122, 124-126, 129, 130, 137, 152, 154, 156, 165, 169, 180, 185, 195, 198, 230, 246, 247, 259, 311-315, 320, 327, 334, 343, 377, 383, 388, 396, 403, 411, 425, 426
transmigration 60, 120, 123, 141, 143, 181, 182, 194-196, 207, 210, 215, 217, 230, 232, 248, 250-254, 257, 264, 284, 290, 291, 296, 298, 303, 304, 316, 328, 341, 383-387, 406
Trigonobalanus 54, 68

umbrella species 78, 108, 132, 134, 151, 367, 375, 383, 391, 408
uning 67, 251

volcano 55, 158, 247

Walhi 29, 392
Wanariset Samboja 149, 156, 170, 172, 174, 230, 402
wanderers 80, 82, 84, 91, 106, 107
WCMC 183, 185, 189, 198, 288
WNF 131, 144, 145, 149, 150, 155, 391
World Bank 143, 148, 234, 287, 290, 316, 328, 345, 369, 370, 389
World Conservation Strategy 132, 146, 332, 333, 337
WSPA 149, 150
WWF 131, 132, 142-144, 146, 149, 150, 155, 167, 184, 198, 231, 233, 238, 240, 335, 338, 339, 345, 399, 405-407

zoo 61, 91, 117, 118, 121, 122, 129, 130, 137, 139, 157, 163, 402, 403

Printed in Great Britain
by Amazon.co.uk, Ltd.,
Marston Gate.